# Mortuary Ritual and Society in Bronze Age Cyprus

## Monographs in Mediterranean Archaeology

Series Editor: A. Bernard Knapp

Aimed at the international archaeological community, *Monographs in Mediterranean Archaeology (MMA)* seeks significant new contributions from the multicultural world of Mediterranean archaeology. In general, we publish problem-oriented studies that present a solid, extensive corpus of archaeological data within a sound theoretical and/or methodological framework. *MMA* volumes will deal with major archaeological issues related to the islands and lands or regions that border (or have had a demonstrable impact on) the Mediterranean Sea. No constraints are placed on the period of focus, from Palaeolithic through early Modern. We encourage contributions that treat the social, politico-economic and ideological aspects of local or regional production and development; issues related to social interaction and change or exchange; or more specific and contemporary issues such as gender, agency, identity, representation, phenomenology and landscape.

# Mortuary Ritual and Society in Bronze Age Cyprus

**Priscilla Keswani**

Equinox Publishing Ltd

LONDON • OAKVILLE

Published by

Equinox Publishing Ltd.
UK:   Equinox Publishing Ltd., Unit 6, The Village, 101 Amies St., London SW11 2JW
USA:  DBBC, 28 Main Street, Oakville, CT 06779

www.equinoxpub.com

First published 2004

British Library Cataloguing-in-Publication Data

A catalogue record for this book is available from the British Library.

ISBN   1 904768 03 2   (hardback)

Library of Congress Cataloging-in-Publication Data

Keswani, Priscilla.
  Mortuary ritual and society in Bronze Age Cyprus / Priscilla Keswani.-- 1st ed.
      p. cm. --  (Monographs in Mediterranean archaeology)
  Includes bibliographical references and index.
  ISBN 1-904768-03-2 (hardback)
 1.  Funeral rites and ceremonies--Cyprus--History. 2.  Cyprus--Social life and customs.
3.  Cyprus--Antiquities.  I. Title. II. Monographs in Mediterranean archaeology (Equinox Pub.)
  GT3271.5.K47  2004
  393'.095693--dc22

Typeset by ISB Typesetting, Sheffield
www.sheffieldtypesetting.com
Printed and bound in Great Britain by Antony Rowe Ltd, Chippenham, Wiltshire

# Contents

# List of Figures

# List of Tables

# Preface and Acknowledgments

This book is the culmination of a project that began 20 years ago with my doctoral research on mortuary practices and social hierarchy in Late Bronze Age Cyprus. Although much of the material from my dissertation is included here, particularly in Chapter 5, this is a very different work in format and analytical approach. Moreover, the study of mortuary practices has been expanded to include a detailed consideration of the evidence from the Early and Middle as well as the Late Cypriot periods.

Thanks are due to many people for their help along the way. I would especially like to thank Dr Bernard Knapp, the editor of this series, for his encouragement to undertake the present work, his invaluable critical commentary, and his unfailing patience while the manuscript was in progress. I am also deeply indebted to the anonymous reviewers of this work for their constructive criticism, and to Janet Joyce, Valerie Hall, Heidi Robbins and the rest of the Equinox team for their efforts during the production process. Going further back in time, Ms Alison South and Dr Ian Todd made it possible for me to launch my studies in Cypriot archaeology and have extended many forms of assistance, along with friendship and hospitality, over the years. Dr Stuart Swiny and Dr Vassos Karageorghis both offered enthusiastic support for my research and eased its progress considerably. Dr Veronica Tatton-Brown of the British Museum, Dr Marie-Louise Winbladh of the Medelhavsmuseet in Stockholm, along with Dr Paulos Flourentzos, Mr Grigoris Christou, and the entire staff of the Cyprus Museum, graciously facilitated my studies of archaeological collections. Members of my doctoral committee at the University of Michigan, chaired by Dr Henry T. Wright, offered insightful comments on preliminary drafts of my dissertation. I am also grateful to the Archaeological Institute of America and the Fulbright-Hays Commission for the doctoral and post-doctoral fellowships which supported various phases of my research.

More than anyone else, however, I must thank my husband Sushiel, who made it all possible and patiently endured the entire process.

# Introduction

## Objectives and arguments

Mortuary rituals have much to tell us about the social life and ideology of the communities in which they are staged. Beyond merely 'giving the dead their due', funerals are often focal contexts for the negotiation of social hierarchies, the affirmation of social alliances, and the articulation of sanctified propositions about relationships between individuals, groups, their ancestors, and the cosmos. As such, they help to restore the order of daily life that is disrupted by death, reasserting the crucial values upon which social bonds are premised. Yet inasmuch as social relations, economy, and politics are always in some degree of flux, the order that is restored in the context of mortuary ritual may differ subtly or substantially from the one that prevailed before. Death rites can thus afford an occasion on which traditional social relationships are reproduced, or a forum in which the social order of the past is actively reinterpreted and sometimes subverted.

For archaeologists, the reconstruction of the social structures and cultural meanings associated with ancient mortuary practices is hardly a simple task, constrained as we are by the incompleteness of our information about the past and by the limitations of our own frame of reference. However, the systematic description of funerary rituals as archaeologically materialized—the disposition of human remains in geographical and social space, the physical treatment of corpses, the characteristics of the burial facilities in which bodies or bones were arranged, and the types and quantities of goods, if any, bestowed upon the dead—is a first step towards understanding the social practices represented in those rituals. The next, more problematic step of eliciting the cultural meanings with which past mortuary rites were imbued requires some knowledge of context, both the context of a particular set of mortuary practices as situated within a long-term local sequence, and the 'macro-processual' or regional historical context of demographic, technological, economic, and political change in which particular funeral ceremonies were enacted. Given such knowledge, we may begin to propose provisional interpretations of what some have called 'the lived experience', or the ways in which human actors dynamically constructed and altered their social relations and ideology in the midst of ongoing historical change.

The archaeological record of Bronze Age Cyprus is an attractive subject for mortuary analysis, not only because of the sheer volume of its funerary data, but also because of the important cultural transitions that it encompasses. In the second half of the third millennium BC, the traditional lifeways and economy of the island's Chalcolithic communities were transformed by the introduction of cattle, the onset of plow agriculture, the expansion of copper metallurgy, and many other technological changes (Knapp 1990a; Webb and Frankel 1999; Frankel *et al.* 1996). These developments may have been partly associated with new episodes of colonization from the Anatolian mainland. Contemporaneous changes in kin and community structure are reflected in new forms of domestic architecture and settlement organization at places such as Sotira *Kaminoudhia*, Marki *Alonia* and Alambra *Mouttes*, where agglomerative rectilinear houses that contrast markedly with the 'roundhouses' of Chalcolithic sites have been found (Swiny 1989; Frankel and

Webb 1996; Coleman 1996; Swiny *et al.* 2003). Over the course of the Early and Middle Cypriot periods (see Table 1.1 for chronology), local metallurgical production expanded, and exchange contacts with the Aegean and Near East seem to have increased. By the transition to the Late Cypriot period, the first towns had been founded at Enkomi on the east coast and Morphou *Toumba tou Skourou* in the northwest, presumably to take advantage of long distance trading opportunities. Thereafter a series of urban centers with varying types of public buildings, storage, and industrial facilities were established all along the south coast of the island. It is probable that Cyprus or one of its major polities is to be identified with the so-called kingdom of *Alashiya*, known from Egyptian and Near Eastern textual references as a major supplier of copper to the Mediterranean world (Knapp 1985; 1986a; 1994b; 1996b; 1997; Muhly 1972; 1982; 1985; 1986; Peltenburg 1996; Keswani 1996; Goren *et al.* 2003). Thus, between the beginning and the end of the Bronze Age, Cypriot communities underwent major shifts in economy and organizational complexity, moving from hoe cultivation to plow agriculture and specialized industrial production, from village to urban life, and from relative insularity to considerable prominence in international trade and politics.

The rough outline of these macro-processes alone suggests that major changes in social structure must have taken place concurrently, changes that might be greatly illuminated through the study of mortuary data. But in addition to elucidating the particulars of cultural change in Bronze Age Cyprus, the study of this mortuary complex may also contribute to a more general understanding of how ritual practices may be deployed and transformed in situations of economic intensification and increasing political complexity, a topic of broader anthropological and historical interest. The Cypriot case is of particular importance to this discussion because, as I will argue, it affords an example of how mortuary rituals, as local arenas of status competition and display, generated a funerary economy that influenced the larger trajectory of cultural development and urbanization.

The general objectives of this work are to examine the details and long-term trends of mortuary practices in Bronze Age Cyprus, to explore the aspects of social structure and ideology expressed in mortuary rituals, and to elucidate the ways in which social structure and social change were actively produced in the ritual context. I propose, based on the character and complexity of ritual practices and the increasing levels of mortuary expenditure evident in Early–Middle Cypriot tombs, that funerary celebrations served as the premier venues in which Cypriot kin groups created and enacted their ancestral ideologies and asserted their prestige. I further hypothesize that the growing elaboration of mortuary festivities and their crucial importance in the negotiation of status hierarchies directly stimulated the intensification of copper production and the expansion of interregional exchange relations, setting the stage for other political and economic transformations at the beginning of the Late Cypriot period. I attempt to demonstrate that fundamental changes in social structure took place early in the Late Cypriot period when, for the first time, the presence of hereditary elites making use of a distinctive complement of status symbolism becomes apparent in the archaeological record. I also propose, based on changing patterns of tomb location, ritual treatment of the dead, and mortuary expenditure, that the centrality or communal focus of mortuary ritual gradually receded over the course of the Late Cypriot period, as urban institutions multiplied and the affirmation of kin group identity became less important in the construction of social prestige and status hierarchies.

## Overview of the work

The analysis of the Bronze Age mortuary sample from Cyprus cannot be undertaken without some preliminary consideration of the problems and potential of mortuary studies. In Chapter 2, therefore, I present a synopsis of theoretical discussions of funerary evidence and its relationship to the social structure of living communities, beginning with early attempts by processual archaeologists to define regular relationships between mortuary variability and social organization, and more recent ethnological and post-processualist arguments against such an enterprise. At the crux of this controversy is the processualist assertion that differences in mortuary treatment are direct reflections of status differentials in the living society, countered by the post-processualist contention that in the course of ritual practice, societies may variably deny, reflect, or exaggerate certain attributes of their internal structure and systems of authority (Morris 1987:39; see also Parker Pearson 1982; Shanks and Tilley 1982). Post-processualists also argue that even within a particular cultural context, ritual practices and their meanings are subject to continuous revision over time. Consequently, the significance of mortuary variability (or the lack thereof) and the meanings of mortuary practices can only be interpreted contextually, and preferably in a long-term historical framework (Parker Pearson 1984; 1993). While concurring with the post-processualist position, I would also observe that there are some important cross-cultural similarities (not to be equated with universal laws) in the ways that certain aspects of mortuary practice articulate with macro-processes of economic intensification and increasing sociopolitical complexity. I discuss some of these patterns, and exceptions to them, in order to make explicit the body of background knowledge that informs, but does not exclusively determine, my interpretations of the Cypriot archaeological data. I focus specifically on fluctuations in the spatial positioning of burial grounds, practices involving secondary treatment and collective burial, and trends in mortuary expenditure, variables that figure prominently in my analysis of the Cypriot case.

In Chapter 3, I address the problems of the Cypriot mortuary record as an archaeological sample. The overwhelming majority of burial features from the Early through the Late Bronze periods in Cyprus were rock-cut chamber tombs that were generally used for multiple burials, often over extended periods of time. The natural and cultural processes of post-depositional disturbance affecting these burial groups must therefore be considered when interpreting the preserved remains of both humans and grave goods. Moreover, although many hundreds of tombs have been investigated at sites throughout the island, the quality of information available from these investigations varies considerably, having frequently been compromised by the limitations of previous research agendas, excavation techniques, and standards of reporting. The scarcity of detailed analyses of human skeletal remains is an especially serious problem arising from the nature of past research. In addition, mortuary samples from different regions and time periods are often uneven in terms of size and the representation of ritual treatments, demographics (i.e. age and gender variation), and other dimensions of local mortuary variability. All of these issues have necessitated the development of a specialized methodology for interpreting programs of mortuary treatment, the relationship between collective tombs and living social groups, and differences in mortuary expenditure when tomb groups, rather than individual burials, are the primary subjects of analysis. An outline of this methodology is presented in the latter part of the chapter, followed by a brief note on chronological issues.

Chapter 4 is concerned with the principal characteristics of Early–Middle Cypriot mortuary practice—the widespread appearance of extramural cemeteries and rock-cut chamber tombs, the development of rituals involving secondary treatment and collective burial, the increasing elaboration of tomb construction, and the rising consumption of ceramics, copper, and other goods in funerary celebrations. Throughout the discussion of these changes in mortuary ritual, I attempt to address a number of currently controversial questions, such as the extent to which new forms of burial practice represent customs introduced by settlers from the mainland versus locally initiated modes of symbolic expression, and the problem of whether the unusually elaborate tomb complexes and metal-rich burial assemblages found at Vasilia *Kafkallia* and Lapithos *Vrysi tou Barba* in the north should be interpreted as evidence for the presence of a highly stratified social order dominated by an hereditary elite. In evaluating these issues, I tentatively conclude that EC–MC mortuary rituals, while undoubtedly influenced by traditions experienced or encountered in neighboring regions, nevertheless grew into a unique and locally rooted complex of social practices and ancestral ideologies. I also argue that the wide distribution of certain prestige goods and the broadly based, diachronic increases in mortuary expenditure apparent in tomb architecture and grave goods are more likely to represent a highly competitive social system in which rank and prestige were dynamically created and revised, rather than a fixed hierarchical order with institutionalized inequality. Mortuary ritual appears to have been the most prominent forum for the assertion of social status at this time, and the associated intensification of competitive prestige displays probably stimulated both the growth of the local copper industry and of long distance trade through which foreign prestige goods might be acquired. Thus developments in the realm of mortuary practice had far-reaching consequences for the trajectory of sociopolitical and economic change in the following era.

In Chapter 5, I turn to developments in Late Cypriot funerary rites, focusing, as in Chapter 4, on the spatial dimensions of mortuary practice, the ritual treatment of corpses, the grouping of the dead, and changing patterns of mortuary expenditure and material symbolism. Foremost among the contrasts with earlier practices was the location of tombs in close proximity to residential, administrative, and workshop areas within the emerging town centers, rather than in formal, spatially reserved burial grounds. Periods of tomb reuse lengthened considerably, sometimes continuing over hundreds of years. Meanwhile, traditional rituals involving secondary treatment and collective burial persisted for a long time, and even seem to have intensified at rural sites in LCI, but may have declined in the second half of the Late Cypriot period. Also evident is the emergence of elite groups making use of a very distinctive complement of prestige goods, with certain types of ornate and iconographically rich valuables displaying a highly restricted distribution within the mortuary sample overall. Yet even as large quantities of gold jewelry and other exotic valuables were disposed of during elite funerals, there were reductions in other areas of mortuary expenditure such as tomb construction and ceramic consumption, and by the end of the Late Cypriot period, the disposal of valuables had also decreased considerably. Unimpressive pit or shaft graves containing single or small numbers of high and low status burials came into use during the LCIII period at some localities, and the cycle of mortuary obsequies may have been shortened overall. I propose that these developments represent the altered significance of mortuary ritual in the midst of socially heterogeneous and stratified urban polities, where status differentials were no longer created mainly through periodic, ritualized exhibitions staged by competitive kin groups, but were instead based on differential access to copper, trade networks, and roles or positions within various court and temple institutions.

In Chapter 6, I review the long-term changes in mortuary ritual from the beginning to the end of the Cypriot Bronze Age, attempting to forge an integrated construct of the ways in which local ideologies and funerary practices were affected by, and reflexively acted upon, supralocal or macro-processes of economic and sociopolitical change. This enterprise might be construed as an exploration of the process of 'structuration', in which human agents, operating on the basis of the prevailing system of cultural meanings and social interests, variably reproduce or transform their social structure through the intended and unintended consequences of their actions. The concept of such a project derives from the work of Giddens (1979, 1984) and related discussions in sociology and cultural anthropology (e.g. Marcus 1988; Marcus and Fischer 1986; Knorr-Cetina 1981; Sahlins 1981; Bourdieu 1977). Its objective is one that can never be fully realized in an ethnographic case, let alone in an archaeological context, but in pursuing it I hope to devise a provisional model that will help to elucidate the archaeological record and serve as a stimulus to future research. I begin by sketching the major developments in demography, settlement patterns, household and community organization, subsistence practices and other forms of economic production that can be discerned from non-mortuary, archaeological evidence for the EC–MC and LC periods respectively. I then consider the significance of mortuary practices and their transformations in the context of those developments, and the ways in which Cypriot communities experienced and altered the trajectory of social change at the local level through the enactment of death rituals. I conclude with a brief epilogue on directions for future research and the need for greater methodological rigor in mortuary excavations.

# 2 Mortuary Ritual and Society: Some Theoretical Considerations

## The ethnological foundations of mortuary analysis

Anthropologists have long been interested in the significance of death rites. In the late nineteenth century, Tylor (1979:14) interpreted them as evidence for beliefs about the afterlife and Frazer (1886:74–75) explained them as attempts to appease or control the souls of the deceased. Some years later, Hertz (1960) construed them as a series of transformations in the relationships between the living and the dead, ending in the unification of the soul with the society of the ancestors and the reintegration of the bereaved in the society of the living. Similar ideas were expressed by Van Gennep (1960:146–65) in a more general work on rites of passage. Malinowski (1925:53) regarded mortuary rituals as society's means of consoling the survivors, and Radcliffe-Brown (1922:285–6) emphasized their role in restoring a community's solidarity following the loss of one of its members (see also Bartel 1982).

Subsequently, many ethnographers have emphasized the centrality of mortuary ritual in the social life of traditional societies, where it often constitutes 'the supreme event, providing the major occasions and themes for cultural elaboration and emphasis' (de Laguna 1972:531, quoted in Kan 1989:4; see also Weiner 1976:61; Douglas 1969:219; Palgi and Abramovitch 1984:395). In such communities, funerary rites may serve as the principal arena for the negotiation of status hierarchies among the living (Metcalf 1982; Kan 1989) and for the reaffirmation of social alliances among kin groups whose bonds are temporarily sundered by death (Forman 1980; Traube 1980; Weiner 1980).

The importance of mortuary ritual in contemporary 'post-industrial' societies is by contrast much attenuated, the result of complex historical transformations in Western attitudes towards death and long-term changes in sociopolitical structure, economy, and demography (Ariès 1974; 1977; Parker Pearson 1982; Palgi and Abramovitch 1984:408; Metcalf and Huntington 1991: 206–10; Walter 1994). Because status is mediated less by the hereditary transmission of rank and wealth than by education, occupation, and achievement, the amount of symbolic capital to be derived from celebrating an individual's death is often small. The predominance of nuclear families over extended kin groups further limits the scale of social readjustment associated with the death of an individual. Moreover, the increasing ages to which people now live, along with the attendant infirmities, tend to relegate the elderly to the periphery of social and economic life long before death occurs. But in societies where people are more likely to die in the prime of life, death usually involves major social readjustments, and mortuary ritual is likely to be an important arena for the reallocation of authority and assets (Woodburn 1982:206; see also Douglas 1969:202, 217).

## The processual approach to mortuary variability

'New Archaeologists' were quick to recognize the potential of mortuary data in reconstructing the social organization of past societies. In the late 1960s and early 1970s, North American processualists such as Saxe (1970) and Binford (1972) combed the ethnographic literature and amassed dozens of cases affirming the relationship between variations in mortuary treatment and status differentials among the living (e.g. Hertz 1960; Van Gennep 1932; Radcliffe-Brown 1922; Rattray 1927; Bendann 1930; Goody 1962), and a surge of archaeological analyses followed (e.g. Brown 1971; Chapman *et al.* 1981; O'Shea 1984). According to Binford, variations in dress and accompanying implements or tool kits were likely to reflect horizontal status distinctions relating to age and sex, while kin or sodality affiliations might be indicated by the location of burial or the presence of distinctive totemic items. Cause of death or social deviance might be symbolized in the locus of burial and the treatment of bodily remains (Binford 1972:233–5). Evidence for ranking could be derived from an analysis of the material and energetic costs entailed in executing the burial program and constructing the burial facility (Tainter 1973; 1975), from the quantity and quality of goods interred with the deceased, and from the presence of status insignia or socio-technic items serving as badges of rank (Peebles 1971; Binford 1972; Shennan 1975; Franken-stein and Rowlands 1978). Social hierarchy might also be inferred from locational hierarchy, that is, the differential distributions of burials within and between cemeteries, or with respect to public or ceremonial areas of the settlement (Goldstein 1976). Although aspects of horizontal status vari-ation were sometimes found to have a low aptitude for archaeological preservation, it was argued that vertical or rank-related status distinctions were more readily discernible, partly because of the levels of energy expenditure by which they were characterized (O'Shea 1981:50), and partly because they tended to accrue redundantly to certain individuals, heightening their archaeological visibility (Saxe 1970:75,195; O'Shea 1984:30).

To explain the linkage between mortuary variability and social status, Saxe and Binford drew heavily on Goodenough's (1965) discussion of role theory, in which it was proposed that every individual has a social persona, a composite of his or her various social identities. Each social identity entails a specific relationship with others: parent-child, friend-friend, master-servant, supplicant-benefactor etc. The expression of status in mortuary ritual, it was held, is realized through these identity relationships or reciprocal links between the deceased and survivors (cf. Saxe 1970:5; Binford 1972:225–6). Persons of high social status would participate in a larger number of duty-status relationships than persons of low status, and so it was predicted that the size and composition of the social group acknowledging status responsibilities to the deceased would have an important influence upon the level of corporate disruption engendered by mortu-ary rites and upon their form.

In the processual construction of mortuary theory, the relationship between status differenti-ation and mortuary treatment was regarded as being direct, with every component of mortuary variation corresponding to some aspect of the social persona of the deceased. Formal analysis of these components of variation would permit the definition of distinct social personae, the orga-nizing principles of the mortuary system, and the organizing principles of the social system at large. Furthermore, the overall complexity or variability of the mortuary ritual within a given society would be a direct reflection of social complexity in terms of both vertical or hierarchical and horizontal or non-hierarchical groupings (Binford 1972:232–5).

The initial acclaim for mortuary analysis as a key to unlocking social organization was soon moderated by critiques from within and without the domain of processual archaeology. Insiders voiced cautionary notes concerning the archaeological invisibility of many high status burial programs (Braun 1981; Brown 1981), as well as the multiplicity of biases that post-depositional processes, sampling problems, and recovery techniques impose upon the available archaeological record (O'Shea 1984). Outsiders, mainly members of the symbolic-structuralist and post-processualist movement, raised still more fundamental arguments about the alleged isomorphism between social organization and mortuary ceremonialism.

## The post-processual critique and ethnographic examples

Perhaps the most important objection raised by post-processualists against processualist mortuary theory concerns the strong emphasis placed on the representation of the social personae of the deceased in mortuary ritual and the tendency to overlook the importance of death rites as occasions for display, social self-promotion, and the transfer or reallocation of rights and positions among the living (but see Saxe 1970:10). Many post-processualists have argued that mortuary variation is not invariably an exact, direct or systematic representation of social roles and social hierarchy within the living society. Rather, the mortuary record is the product of ritual practices which may be manipulated according to sectional interests, and at times it may actually mask or overstate the quotidian system of power relations within the living society (Pader 1982; Parker Pearson 1982; 1984; 1993; Shanks and Tilley 1982; Morris 1987; 1992; for an alternative approach emphasizing the construction of individual identities, see Meskell 2000 and references therein).

As an example of this, Parker Pearson (1982) discusses the case of contemporary British society, in which mortuary practices indeed vary according to social class, but with a paradoxical twist. The upper and middle echelons of British society tend to dispose of the dead swiftly and invisibly via cremation—death being viewed as a failure of modern medicine—whereas in some segments of the lower classes, death is still an important occasion for prestige competition and display, with expenditures on monuments, coffins etc often exceeding the means of the deceased or their kin. Thus the patterning of mortuary expenditures is not a direct manifestation of the hierarchical structure of British society in the late twentieth century.

Practitioners of processual archaeology may be inclined to shrug this off by arguing that such extreme inversions—whether associated with differing class-based attitudes towards death or masking egalitarian ideologies—are rarely characteristic of the non-state, pre-state, and early state societies with which most archaeologists are concerned (e.g. Trigger 1990:126), and therefore the point is moot. But there are other, even more compelling cautionary tales from the ethnographic literature of the very middle-range societies which command so much archaeological attention.

Metcalf's account of mortuary ritual and mausoleum construction among the Berawan of Borneo (Metcalf and Huntington 1991; Metcalf 1982) offers a striking example of how the social status of the deceased may be manipulated and even misrepresented in mortuary ritual. Berawan society is not rigidly stratified, but it evinces a considerable degree of internal status rivalry. Political leaders and would-be leaders attempt to demonstrate their prestige through their capacity to marshall community labor for the construction of funerary monuments for their followers. This practice has some startling ramifications in light of processualist mortuary theory:

> All mausoleums are known by the person for whom they were prepared, and paradoxically these people often turn out to be nobodies. Further investigation reveals their special merit: they were relatives of emergent leaders (affines or consanguines, but residents of the same apartment) who happened to die at the right moment. In honoring them with a mausoleum the leader ennobles himself (Metcalf and Huntington 1991:150).

Metcalf also offers this observation:

> If the longhouse life cycle runs its course, the great man may himself occupy a humble grave, because the community is left disunited by his passing. The exceptions are cases where the important man shares the tomb he originally built for another, or where he succeeded in begetting, adopting, or recruiting a young man capable of commanding respect in a similar fashion. In this regard, the building of mausoleums is also important because it ennobles the heir as it establishes the legatee (Metcalf and Huntington 1991:150).

In the Berawan case mortuary ritual would seem to express the political interests of the living to a much greater extent than the social personae of the deceased.

Another sobering story, at least for those concerned with inferring status hierarchies from variations in energy expenditure, comes from Dillehay's (1990) research on the Mapuche chiefdoms of south-central Chile. When a Mapuche chief dies, his body is first interred in a shallow earth grave. After a year his grave is taken over by the lineage and is built up by periodic soil capping or *cueltun* rites, held every four to eight years. These rites are supervised by a shaman and entail the participation of all of the dead chief's affinal and consanguineal relatives. When the next chief dies, the focus of ritual shifts to his own mound. Consequently, Dillehay notes:

> ...there seems to be little, if any, correlation between the size of a mound and the political power of its interred chief. Instead, mound size and growth seem to be related to the duration of office of a dead chief's successor and to the number of relatives participating in capping episodes during this period (Dillehay 1990:233–4).

One might argue that the size of a chief's kin base (i.e. the size of the group acknowledging duty-status relationships to her/him) was probably an important attribute indeed of 'political power' (Sjögren 1986), but the issue of the successor's longevity is one which is seldom considered in the interpretation of mortuary variability. The Mapuche case again illustrates the material consequences of mortuary ritual used as a means of legitimizing the status of the living.

## Towards an interpretive rapprochement (1): balancing multiple lines of evidence

The preceding cautionary tales hardly warrant abandoning attempts at social reconstruction through established methods of mortuary analysis. However, they should redirect our attention to the ritual context of mortuary practice and the attendant opportunities for social maneuvering and manipulation. Moreover, the importance of examining multiple lines of evidence in interpreting the mortuary record ought to be readily apparent.

A consideration of the degree of structure or redundancy that a mortuary complex evinces may help in distinguishing entrenched hierarchical differentiation from competitive elaboration of the Berawan variety. Over time, in societies with relatively stable social hierarchies and highly structured mortuary practice, the archaeological record should afford a distinctively patterned or tiered representation of status hierarchy in the living society, with strong correlations between the different elements of mortuary variability (locational, architectural, ritual, artifactual). But in societies characterized by a high degree of social competition, these correlations may be less apparent, reflecting the fluctuating hierarchical relations between groups. Here the distribution of wealth and status symbols may also be characterized by continuous variation rather than by stepped or discrete status positions (Rathje 1973; Shennan 1986).

In addition, a comparison of settlement and mortuary data is essential. The evaluation of cost and energy differentials in mortuary ritual versus expenditures in constructing the houses of the living may yield a more balanced picture of the actual order of stratification within the living society (cf. Bloch 1971:112–3; Metcalf and Huntington 1991:110; Parker Pearson 1984:71). Differential distributions of prestige goods (and subsistence commodities, e.g., faunal remains) in settlement contexts may also illustrate variations in 'real wealth' within the living society.

Finally, where circumstances permit, chemical and paleopathological studies on human bone may reveal important correlations between mortuary treatment, nutrition, and the physical condition of the deceased, reflecting differences in diet and lifestyle associated with social hierarchy (Powell 1988; Stuart-Macadam 1989; Lukacs 1989; Price 1989; Schoeninger and Moore 1992;

Schurr 1992; Gamble *et al.* 2001). Paleopathological studies of bone may further elucidate variation in occupational stress (Kennedy 1989) and trauma resulting from wounds, accidents, and deliberate deformations of the skeleton (Merbs 1989). The relationship between these physical parameters and variations in mortuary treatment could be an indication of the extent to which social hierarchy was associated with differential access to food resources, different patterns of physical activity, and differential participation in rites of passage and other social ceremonies.

## Towards an interpretive rapprochement (2): understanding mortuary ritual as a dynamic cultural system

The post-processual critique and related ethnographic accounts of mortuary ritual pose a new problem for practitioners of mortuary analysis. Before we attempt to interpret the significance of mortuary variability with respect to status differentiation, we need to develop an understanding of the way in which past systems of mortuary ritual were articulated with other aspects of social structure, ideology, and economic life. Some theorists argue that what is represented in mortuary ritual is not social organization *per se*, but a ritualized expression of social structure, in which empirical relationships of authority may be denied, reflected, or exaggerated (e.g. Morris 1987: 39). This in turn prompts the question: Under what circumstances do societies deny, reflect, or exaggerate their social structure, and are there any patterns to such occurrences? Or is it possible to explain, as Kan puts it, 'how a particular type of sociopolitical order tends to generate, through the logic of its workings, a certain type of cultural perception of death and the ancestors, and a certain type of mortuary complex' (Kan 1989:15)?

Many post-processualists would probably object to such an enterprise, inasmuch as mortuary practices are held to be the products of culturally specific meanings and historical contingencies, and the proposal of cross-cultural generalizations thus becomes an intellectually sterile foray into behaviorism (Morris 1991). Certainly every culture is unique, and there is no invariant relationship between particular social concerns (e.g. the masking or affirmation of status differentials) and the ritual or other symbolic forms through which they are expressed. Nevertheless, there are recurrent (if by no means universal) patterns of mortuary ritual associated with processes of sociopolitical and economic competition, and the exegesis of these patterns contributes to an undestanding of how systems of inequality are created and naturalized through ritual performances and material symbolism. My aim in pursuing a comparative analysis in this context is not to establish any 'universal laws' but rather to synthesize the body of background knowledge that informs my interpretation of the Cypriot archaeological data. Such knowledge is also susceptible to reflexive modification in light of the peculiarities of many different individual cases.

Woodburn's (1982) analysis of mortuary practices among socially unstratified hunter-gatherers offers a useful point of departure for this line of inquiry. Woodburn observed interesting differences between the mortuary complexes of groups that he characterizes as 'immediate' and 'delayed-return' societies. Immediate return societies include groups such as the !Kung, the Baka and Mbuti Pygmies, and the Hadza, whose procedures for the treatment of the dead go only a little beyond the practical requirements for getting rid of a corpse. The body may be disposed of simply by pulling down a hut over the remains or by simple interment in a shallow earth grave with little, if any, prior physical preparation. The tasks associated with burial do not seem to be allocated to specific kinsmen or affines, and there are no death ritual specialists. There are few, if any, rules concerning the behavior of survivors, and the possessions of the dead are either disposed of with the body or are 'quickly and casually' distributed. Postmortem obsequies, which may be irregularly performed, bear little connection to beliefs about the afterlife or the fertility of humans and the natural world (Woodburn 1982:189–204).

Woodburn explained the limited development of mortuary ritual in these groups to the 'immediate-return' focus of their socioeconomic life, in which subsistence and social activities are strongly oriented to the present, food and other resources obtained are rapidly consumed, and individuals are not dependent on the intergenerational transfer of property or status. By contrast, in the more common 'delayed-return' systems characteristic of sedentary or intensive food collectors, mortuary rites are more formalized and elaborated, and death is often linked with conceptions of fertility, the ancestors, and the reproduction of social relations and alliances within the living society (Woodburn 1982:205–6). Delayed-return societies are concomitantly distinguished from immediate-return groups by greater investments in labor, long-lasting artifacts and facilities, and by long-term debts and obligations to kinsmen, exchange partners, and other corporate groups. Woodburn argues that 'other things being equal, where death involves major social readjustments and the risk of conflict and disorder, death beliefs will be more elaborate and more ritualised than where such adjustments involve no reallocation of authority or of assets but are largely matters of personal feelings' (1982:206).

Woodburn's analysis tends to the conclusion that the intensification of subsistence activities and the formalization of kinship and other social relations are likely to be associated with increasing formalization and investment in mortuary ritual. Contrary to Morris (1991:152), I would argue that such an hypothesis need not imply an extreme form of economic determinism, inasmuch as the causes of socioeconomic change may be deeply rooted in social and ideological concerns, and mortuary ritual may feature as the principal arena in which social actors create and transform both social relations and ancestral ideologies.

Although Woodburn does not specify the lines along which the 'formalization' of mortuary ritual will occur, current archaeological and ethnographic research suggests that the appearance of formally bounded cemeteries, the elaboration of ritual practices involving secondary treatment and collective burial, and overall increases in the level of mortuary investment and consumption (e.g. labor involved in monument construction, material expenditure in grave goods) may often be involved. The relationships between these dimensions of mortuary practice and broader transformations in sociopolitical and economic structure require further investigation, not merely in terms of the conditions associated with mortuary elaboration or intensification, but also conversely in terms of the conditions associated with the de-emphasis, simplification or restraint of ritual displays.

## Formal cemeteries, descent groups, property, resources, and identity

A number of archaeologists, beginning with Saxe (1970) in his controversial Eighth Hypothesis, have proposed that the appearance of extramural or formally bounded ('spatially reserved' in the terminology of Morris 1987) cemeteries is linked to the emergence of descent groups and the lineal transmission of property and rights of access to critical resources (Goldstein 1976; 1981; Renfrew 1976; Chapman 1981a; 1995; Charles 1995). Renfrew (1976) interpreted the megaliths of Neolithic Europe as the territorial markers of expanding agricultural populations. Chapman (1981b) made explicit use of Woodburn's contrast between immediate and delayed-return societies in relating the inception of cemeteries and monuments in Neolithic Europe to the shift from hunting and gathering to sedentary agriculture. Saxe's original hypothesis and its permutations have been discussed at length in two recent reviews (Morris 1991; Chapman 1995) and will not be revisited at length here. Most significant are the critiques, both processualist and post-processualist, and related empirical evidence brought to bear on the arguments linking cemeteries, corporate descent groups, and various kinds of resources.

From within the processualist camp, Goldstein (1981) observed that although the presence of formally bounded cemeteries usually implies the existence of corporate kin groups legitimizing control over resources through assertions of lineal ties to the dead, the *absence* of formal cemeteries need not imply the absence of corporate descent groups or concerns with property, territory, etc. Different societies may ritualize particular aspects of their social organization in different fashions (Goldstein 1981:61). To put this another way, as in language where the relationship between signifier and signified is usually arbitrary, there is no invariant relationship between a given concept or social concern and its material symbolism in different cultures. The fact that certain symbols such as formal cemeteries recur cross-culturally may be partially attributable to the limited potential for diversity in material symbolism relative to language.

Hodder (1982) attacked the alleged linkage between monumental cemeteries and sedentary agriculture, observing that the collective burial mounds and megaliths of the TRB phases of the Dutch Neolithic were associated with a relative impermanence of settlement. He argued that the construction and protracted reuse of monumental tombs reflected contradictions between 'the social processes and productive capacities of society', and the need to legitimize social group structure and territorial rights through mortuary symbolism (Hodder 1982:169–76). In the TRB, he noted, 'Control over dispersed and changing settlement was achieved by the dominant lineages through links with tombs and the ancestors' (1982:170; see Sherratt 1990 for a similar argument; cf. Whitley 2002 on the alleged misuse of 'ancestors' in European prehistory). Hodder contrasted the TRB mortuary system with the cemeteries of individual, primary inhumations in the LBK–Rossen sequence of the loess areas to the east and southeast, where there was comparatively greater nucleation and stability of settlement (1982:174).

Rather than contradicting the basic argument of Saxe and others, Hodder's discussion serves to refocus our attention on the importance of mortuary practices in symbolizing the identities of corporate groups (whose critical resources may vary from place to place). Nor does the association between megaliths and groups with mobile or shifting subsistence-settlement complexes obviate the significance of the formal, albeit non-monumental cemeteries maintained by more stable agricultural communities as statements of group identity and rights to critical resources. It does, however, raise interesting questions about the circumstances associated with 'monumentality' in mortuary practice (cf. Sjögren 1986; Trigger 1990; Chapman 1995; Parker Pearson 1999:39–40). It also presages some of the more recent discussions concerning how human communities actively create 'social memory' and expressions of group identity through the construction and maintenance of mortuary landscapes (e.g. Silverman and Small 2002).

Morris (1991), while acknowledging that, 'In any group where social and economic power is transferred through claims of lineal descent from the dead, the placing of their physical remains can hardly fail to be significant', nevertheless contends that, 'The functions of placing the dead vary, and cannot be 'retrodicted' from cross-cultural generalizations' (1991:150). He argues that the functions of cemeteries do not exist independently of the beliefs and ideas of the prehistoric actors/social agents, and cannot be explained without reference to localized systems of meaning (1991:147–8). Chapman rebuts the post-processual arguments of Hodder, Morris and others by denying the possibility of recovering meaning from the archaeological record and, more judiciously, by noting that ethnographic analogies suggest 'a (not *the*) function for cemeteries and monumental tombs' (1995:37). Ironically, both Chapman and Morris arrive at the same conclusion—that cross-cultural generalizations are useful guides to research but their applications must be customized, adjusted, reinterpreted for every specific case, making reference to contextual, non-mortuary data.

As a starting point for analysis, not only of particular archaeological cases, but for developing a theory of the relationship between mortuary practices and their socioeconomic context, we might propose that the appearance of formal cemeteries is one change that may be associated with economic intensification, the formation or solidification of descent group identities, and/or competition between descent groups. Over time, the meaning of corporate burial grounds from the standpoint of the actors may change, as in the Merina case where the importance attached to the location of a person's tomb has recently shifted from the expression of property rights to the adherence to a traditional social order in a period of rapid socioeconomic change (Bloch 1971:128–9). These changes in meaning need not necessarily be inaccessible to archaeologists equipped with long sequences of data from both mortuary and non-mortuary contexts; indeed, as Parker Pearson writes, 'The meaning of any practice can *only* be recovered by situating it within a changing sequence of traditions' (1993:204, emphasis mine).

The circumstances associated with the disappearance of formal cemeteries have received relatively less attention than those associated with their appearance. It seems legitimate to ask, in light of the supposed connection between cemeteries, descent groups, and concerns over territorial or resource control, whether the reversion to the use of spatially unreserved burial grounds may not reflect a relaxation of those concerns and/or changes in social structure. Ramsden (1990) touches upon this problem in his discussion of Middle and Late Woodland mortuary practices in Ontario. He proposes that Middle Woodland cemeteries, frequently associated with the major summer macroband campsites of mobile foragers, may have served to mark and identify band territories. With the transition to shifting agriculture and higher population densities in the early Late Woodland period, there was a shift to scattered burial areas within palisaded villages, a change which he ascribes to the increasing importance of group identity and community, as opposed to territorial boundaries:

> In a situation of resource stability and population density, and the consequent need to evolve mechanisms for stabilizing local populations and distancing them from other such local populations, a community may come to place more emphasis on who the group is rather than where it is. With the changing nature of land use and the changing nature of social identity, it may become more important to place boundaries around groups of people rather than around territories, and to maintain space between groups (Ramsden 1990:176).

Morris (1987:181–2) suggests that the shift from extramural cemeteries to intra-settlement burial in tenth-century BC Greece was related to increasing intra-group competition and, following Fustel de Coulanges (1980), the tendency for descent groups to define themselves by adopting closer relationships to their ancestors. These studies suggest that changes in the placement of the dead are integrally related to definitions of group membership and levels of social competition, but the significance of such changes must be evaluated, once again, on a case by case basis.

## Mortuary elaboration and secondary treatment

Another important development in mortuary ritual that may be associated with increasing socio-economic complexity is the emergence and elaboration of what Kan (1989) refers to as systems of 'dual obsequies'. These belief systems, first defined by Hertz (1960), entail an initial phase in which the individuality of the corpse, the pollution and deterioration of the flesh, and the separation or removal of the deceased from the society of the living are emphasized. This is followed by a liminal or transitional period of variable duration, and finally there is a terminal celebration in which the purification of the deceased, his or her reunion with the immortal ancestors, the reproduction of the living society, and the end of mourning are predominant themes. The final celebration often, but not invariably involves secondary treatment of the corpse, for instance, exhumation after a certain period of time and subsequent reburial. Often too, but again not invariably, when individuals are reburied they are interred in a common tomb maintained by their kin group or some larger corporate body to commemorate the ancestors as a collective entity.

Kan (1989) is concerned with explaining the development of mortuary rituals involving 'dual obsequies' and secondary treatment of the dead among the Tlingit, a matrilineal ranked society of the Northwest Coast of North America whose economy was based largely on fishing and semi-sedentary hunting and gathering. The first phase of Tlingit mortuary ritual traditionally involved a primary funeral of four to eight days duration, ending with the cremation of the deceased and the placement of the remains inside the lineage gravehouse. This was followed by a second set of mortuary observances held a year or more later, depending on the length of time required to accumulate the necessary resources. On this occasion the cremated remains of the deceased were placed in new boxes inside new or rebuilt gravehouses, an act referred to as 'finishing the body', after which a potlatch was held to memorialize the dead and to publicly feast and remunerate members of the opposite moiety who had performed the death services. It was believed that at this time the ghost of the deceased settled forever in the cemetery and the spirit moved to the village of the dead. The name and regalia of the deceased were bestowed upon members of the matrikin, and the rank, title, wealth, and widow of the deceased might be given to his/her successor (Kan 1989:41–43).

Kan notes some important differences between Tlingit mortuary rituals and those of the unranked, semi-nomadic Athabaskan hunting 'bands' of the adjacent interior western sub-Arctic. Like the Tlingit, with whom they seem to have shared a common cultural origin, the Athabaskan groups practiced cremation of the dead followed some time later by a memorial potlatch which effected the final dispatch of the ghost of the deceased (Kan 1989:260). However, among the Athabaskans, the corpse and the ghost of the deceased were perceived as presenting a much greater danger to the well-being of the living, and secondary treatment of the physical remains was rarely practiced. Also, prior to the increased circulation of wealth associated with the fur trade, the Athabaskan potlatches were more modest in scale and less competitive than those of the Tlingit, with most of the food and gifts being provided by one person. The Athabaskan mortuary rites were less formalized, and ritual expertise was broadly distributed within the population, rather than being concentrated within an aristocracy (Kan 1989:265–9).

Contrasting the Tlingit and Athabaskan mortuary rites, Kan posits a relationship between the plenitude of subsistence resources, the emergence of hierarchical ranking, and the elaboration of mortuary ritual. He argues, along the lines of Rosman and Rubel (1983, 1986), that the more complex social organization of the coastal groups may have been linked to the greater abundance and variety of subsistence resources which they enjoyed. This in turn promoted increases in population, greater sedentariness, the development of crests and other group-owned prerogatives symbolizing claims to subsistence areas, the formation of lineages and sub-lineages within clans, and the emergence of hereditary rank (Kan 1989:269–71). In conjunction with these changes, the role of the dead and the centrality of mortuary ritual within the sociocultural order greatly increased. Kan writes:

> Clan ancestors became not only the owners of but the symbols of the sacred heritage that made the living clan members what they were. The dead became less threatening, the importance of their idiosyncratic characteristics diminishing, and that of their immortal attributes (names, regalia, and similar items) increasing. The scale of the mortuary rites also increased dramatically, including the number of participants, the amount of wealth involved, and the duration and formalization of the funeral and the potlatch. The mortuary rites, and especially the potlatch, became more prestige-oriented. The aristocracy began to use them, as well as other life cycle rites, to strengthen its dominant role in society and increase its status (Kan 1989:271).

As in the Berawan society described by Metcalf, mortuary ritual became the primary arena for the negotiation and reproduction of status hierarchies.

Although the ideological complexes, ritual practices, and material correlates of dual obsequies vary considerably from one culture to another, the origins and elaboration of such protracted mortuary systems may derive from similar sociopolitical processes. Both Bloch (1982) and Kan (1989) have argued that there is a logical interrelationship between the development of dual obsequies and the rise of political systems legitimized by traditional and ancestral authority. In a discussion which closely parallels the work of Bloch, Kan writes:

> To make the deceased into a valuable cultural resource, the ritual must separate his perishable and polluting attributes from the immortal and pure ones. The funeral begins this process, but usually some time is needed for all the elements constituting his total social persona to be separated from each other, for the perishable and impure ones to be discarded, and for the immortal ones to be channelled back into the social order of the living. As Bloch (1982:223–224) suggests, whenever the sociopolitical order and its authority structure are grounded in a larger ideal and unchanging ancestral order, the funeral rituals have to overcome in one way or another the individuality of a particular corpse. Consequently, in such societies, there will always be a double aspect to funerals (Kan 1989:289).

The transitional period between obsequies permits the transformation of the corpse from perishable flesh to immutable bone (or spirit), and effects a distancing between the actual social persona of the deceased and her/his role as an ancestral spirit.

The development of systems of 'dual obsequies' has important practical as well as ideological ramifications. The time lag between death and the final ceremonies helps to shift the focus of secondary rituals from the mourning of the deceased to the political interests of the living, and to render the dead 'susceptible to manipulation' (cf. Parker Pearson 1982:101). It also facilitates higher levels of mortuary consumption than might otherwise be possible, since it allows the heirs more time to amass the resources (manpower and material for the construction of monuments, food for feasting, gifts for distribution to the living or for disposal with the dead) necessary for a celebration consistent with either their social status or their aspirations. In societies characterized by dynamic status competition, the potential for 'exaggerating' as opposed to 'reflecting' social status is consequently increased, although the extent of 'exaggeration' which is socially permissible may vary (cf. Metcalf and Huntington 1991:150; Kan 1989:168).

However frequently the elaboration of mortuary processing may be associated with the intensification of prestige competition and ranking, it is important not to equate the presence or absence of secondary treatment with the presence or absence of 'Hertzian' mortuary ideologies or the complex of dual obsequies discussed by Bloch and Kan. Goldstein (1989; 1995:116), for example, notes that secondary treatment of individuals is sometimes related to their circumstances of death. The mortuary ideology of the LoDagaa of northwestern Ghana is distinctly Hertzian in its structure and associated beliefs about the soul and its transition to the afterlife, and yet the LoDagaa do not practice secondary treatment; rather, the final passage of the deceased from the society of the living to the world of the ancestors is accomplished through the establishment of an ancestor shrine—a wooden statue placed inside the compound byre (Goody 1962). Even in societies with both Hertzian ideologies and secondary treatment, secondary treatment is not invariably practiced; the Berawan frequently substitute grand primary funerals for *nulang* or secondary funerals, and in other societies ethnographers have noted that secondary treatment, albeit the cultural ideal, is most often accomplished for elite members of the community (Harrison 1962; Freedman 1966; Watson 1988:208; *cf.* Morris 1991:153). Such caveats emphasize the necessity of examining the entire mortuary context of the society under study before attempting to interpret the significance of particular elements.

The diminution of complex ritual programs is yet another phenomenon requiring explanation. Extended sequences of mortuary rites may be left incomplete in circumstances of socioeconomic stress (e.g. Miles 1965), or supplanted by elaborate primary rituals in periods of culture change, as when an older order of ranking and prestige is being challenged (Hudson 1966) or new 'strategies of representation' focusing on the individual ego and particular named ancestors come into play (Thomas 1991:116, 129, 138). In other cases ruling elites may impose restrictions on popular mortuary practices so as to emphasize the status differences between themselves and subordinate populations. Law (1989) discusses an example of the latter in the context of King Wegbaja's ban on the decapitation of corpses in eighteenth-century Dahomey:

> The decapitation of corpses in earlier times was probably related to the practice of separate burial and subsequent veneration of the deceased's head as part of the *ancestor cult* of his own lineage. The suppression of this practice by the kings of Dahomey can be understood in terms of their desire (for which there is other evidence) to downgrade the ancestor cults of the component lineages of Dahomey, in order to emphasize the special status of the public cult of the royal ancestors, and more generally to concentrate or monopolize ritual as well as political and judicial power in the hands of the monarchy (Law 1989:415).

Rawski (1988:29–30) alludes to similar restrictions on ancestral rites imposed by the ruling classes on the commoners prior to the Sung period in China. Clearly, the causes of mortuary 'simplification' may be case specific, but there are cross-cultural parallels nonetheless.

## Secondary treatment and collective burial

Hertz was probably the first scholar to recognize the close interrelationship between the practices of secondary treatment and collective burial. He regarded the transfer of the bones of the dead from the isolated primary grave to the familial or communal tomb as the logical finale of the deceased's passage from the society of the living to the society of the ancestors:

> The gradual destruction of the earthly body, which prolongs and completes the initial assault, expresses concretely the state of bewilderment and anguish of the community for so long as the exclusion of the deceased has not been completed. On the other hand, the reduction of the corpse to bones, which are more or less unchangeable and upon which death will have no further hold, seems to be the condition and sign of the final deliverance. Now that the body is similar to those of its ancestors, there seems to be no longer any obstacle to the soul's entering their community (Hertz 1960:82–83).

He argued that this could only be achieved by the physical transfer of the bones:

> The material on which the collective activity will act after death, and which will be the object of the rites, is naturally the very body of the deceased. The integration of the deceased into the invisible society will not be effected unless his material remains are reunited with those of his forefathers. It is the action of society upon the body that gives full reality to the imagined drama of the soul (Hertz 1960:83).

Secondary treatment, collective burial, and beliefs about the soul and the afterlife are thus inextricably linked in Hertz's formulation.

Chapman (1981a) has argued that the appearance of communal tombs, as well as other types of corporate burial grounds, may be linked to the development of corporate descent groups which legitimize their access to critical resources by asserting their lineal descent from the dead. A survey of the ethnographic literature (Keswani 1989b) indeed suggests that collective tombs express significant social groupings and interests within the society at large. Moreover, the periodic recurrence of collective burial rites may serve not only to express but actively to create a sense of lineage or descent group identity and membership (Rawski 1988; Watson 1988) or 'social memory' (Hendon 2000; Chesson 2001).

In some cases primary units of social organization such as residence or descent groups are translated more or less directly into mortuary groupings. Among the LoDagaa, for example, mortuary units tend to reflect the close ties among brothers born of the same mother and father (as opposed to the same father only); these brothers reside together, work together, and share common property rights. Just as these groups endure for a single generation only, so too the LoDagaa tombs contain only a single generation of individuals (Goody 1962). Among the Bara, ranked patrilineal descent groups of considerable time-depth are prominent sub-units of the living society, and patrilineal descent determines the assignment of individuals to particular tombs over a period of several generations (Huntington 1973; Metcalf and Huntington 1991:121). In the Massim societies of Kiriwana (Weiner 1976; 1980) and Tubetube (Macintyre 1989), matrilineal groups appear to have been similarly represented. Rukuba tombs are associated with particular sub-clans of the community (Muller 1976), and Kwaio skullhouses are maintained by cognatic descent groups (Keesing 1982). For the most part, these tomb groups express not only aspects of kinship structure but also related concerns with the inheritance of territorial and resource rights. One of the most overt examples of this relationship can be seen in the Greek islands, where collective tombs are often directly associated with the family lands (Kenna 1976).

However, in other societies the relationship between tomb groups, kin groups, and 'property' is more complicated. Among the Ma'anyan and the Merina, tomb group membership is not specifically prescribed, and factors such as residence, personal affections, and status considerations may influence the individual's choice in association. Among the Tlingit of the Northwest Coast of America, members of the same matrilineal group may be buried together in the same mausoleum or gravehouse, but individuals accorded high status treatment are often buried alone in very elaborate burial structures (Kan 1989:40–1). Similarly, among the Berawan, mortuary units may be used for highly variable numbers of interments, with members of the same residence group often reusing older tombs as a matter of convenience, while other individuals are 'ennobled' by the construction of costly new mausolea (Metcalf and Huntington 1991; Metcalf 1982).

The Choctaw of the southeastern US (Bushnell 1920:95–99), the Huron of the upper Great Lakes region (Biggar 1929:160–3; Wrong 1939:211–4; Thwaites 1896–1901:279–311; Hickerson 1960), and the northern Greek community of Potamia (Danforth 1982) offer examples of mortuary groupings which express the interests of larger, village-level or pan-communal groups, masking the identities and hierarchical relationships among smaller constituent kin groups. Occasionally such burial groups may give an impression of group cohesion or solidarity which is not entirely consistent with quotidian social realities; the Basque community of Murelaga, for example, makes use of a village ossuary, and yet, according to Douglas, the village is actually 'a rather insignificant social context for the rural population' which is inwardly focused around individual homesteads and hamlets (1969:207).

Although collective tombs tend to symbolize contemporary social concerns such as group identity, social status, territorial or other economic rights, and broad political alliances, it is important to remember that the ideological significance of collective burial groups may be altered over time. In both the Ma'anyan and the Merina societies, tombs groups portray an ancient order of social hierarchy and descent group structure which is increasingly at variance with contemporary social realities. The reuse of ancestral tombs marks an adherence to tradition in a period of major socioeconomic change. In other societies the 'meaningfulness' of collective tombs has been almost altogether lost. While in rural northern Greece reinterment in the village ossuary is said to symbolize the complete incorporation of the deceased into the world of the dead and the 'ultimate unity of the village dead', elsewhere the use of communal ossuaries seems to be more of an adaptation to limited village or urban mortuary space (Danforth 1982:56–7; Douglas 1969:73). The interpretation of collective burial units thus requires an evaluation of mortuary patterns over an extended timespan, taking into account independent evidence for household and settlement organization along with spatial and economic factors.

If the appearance of collective burial units is linked to expressions of group interests and identity, the question of whether their disappearance represents a dissolution of those concerns necessarily arises. Kristiansen (1984), for example, attributes the shift from the 'Megalithic Culture' (*c*. 4000–2800 BC) in Neolithic Denmark to the 'Single Grave Culture' (*c*. 2800–2400 BC) to the transition from a hierarchical sociopolitical system supported by intensive shifting cultivation to a more mobile pastoral economy dominated by relatively autonomous, competitive local lineages (see also Thomas 1987). The shift to primary inhumations in earthen barrows, according to Kristiansen, reflects an inversion of the megalithic mortuary complex based on communal ritual and ancestor worship:

> Now the individual and his property are the focus of burial rites. Equipped with his personal items of rank, with liquor for feasting, he is once and for all separated from the world of the living. Leaving for another world he is socially and ritually equipped to re-occupy his place there and even brings land with him, symbolized by the half to one ha of turf covering him (Kristiansen 1984:84, see also Thomas 1991).

O'Shea (1988) describes a similar shift from multi-stage programs of secondary interment in collective ossuaries to primary individual inhumations accompanied by items of personal wealth during the Late Woodland period in the Great Lakes region of North America. He relates these changes to factors such as depopulation and the destabilization of traditional communities and authority patterns resulting from European contact, and the development of new avenues for individual wealth acquisition in conjunction with the fur trade. Morris, following the work of Strathern on the Melpa of New Guinea, notes that the adoption of individualized cemetery burial at the prompting of Christian missions 'weakened rituals of agnation and ties to the land, since they were set up at the expense of lineage skull cults' (which underwent a renaissance in the 1970s; Morris 1991:151; Strathern 1981:207–8; 1984: 47; Valentine 1965:171). These cases of 'dissolution' consistently support the notion that collective burial rites not only express but may actively strengthen or create bonds of group identity and membership.

## Long-term variations in mortuary elaboration

In societies where, as Woodburn puts it, death involves 'major social readjustments and the risk of conflict and disorder' (1982:206), mortuary ritual is likely to be an important arena for the reallocation of authority and assets. In general, but not invariably, the greater the social importance of the deceased, the larger the size of the group affected by his or her death will be, and the greater the number of persons who will have interests in the order which is established or re-established in the aftermath of the death. This in turn may have material consequences in terms of the relative elaboration and levels of labor and material expenditure associated with mortuary observances. We would thus expect to see greater levels of mortuary expenditure in 'delayed' as opposed to 'immediate' return societies, but it does not necessarily follow that levels of expenditure will continuously escalate as 'delayed' return societies become more complex, or that decreasing expenditure is associated with decreasing social hierarchy.

Many researchers have argued that the elaborate expenditure of wealth in mortuary ritual is most characteristic of societies in transition to increasing social hierarchy or new orders thereof (e.g. Metcalf and Huntington1991:159–61; Parker Pearson 1982:112; Trigger 1990). Childe (1945) was perhaps the first to observe this phenomenon, noting that in ancient civilizations such as Egypt and Mycenaean Greece, the highest levels of expenditure in mortuary ritual were associated with the earlier phases of state formation and consolidation (e.g. the age of pyramids in the Egyptian Old Kingdom, the Shaft Grave era in Greece), when it was essential for the ruling elite to provide a constant demonstration of power and magnificence. As Trigger puts it:

> In human societies, the control of energy constitutes the most fundamental and universally recognized measure of political power. The most basic way in which power can be symbolically reinforced is through the conspicuous consumption of energy. Monumental architecture, as a highly visible and enduring form of such consumption, plays an important role in shaping the political and economic behaviour of human beings (Trigger 1990:128).

In subsequent periods characterized by greater political complexity and stability, the need for ostentatious display may have declined, or else, as Trigger suggests, 'as political relations of domination changed, the type of buildings by means of which that power was expressed also altered' (1990:128) such that mortuary monuments were no longer the most striking features of the ceremonial landscape.

Recent discussions of 'cycles' in mortuary consumption (e.g. Pollock 1983; Parker Pearson 1984; Cannon 1989; Morris 1992) have suggested that phases of competitive elaboration (exacerbated by lower class emulation of elite prestige symbolism) will lead to such a level of 'expressive redundancy' that greater conservatism and/or changes in mortuary symbolism must ensue. The trajectory of 'expressive redundancy' or the shift to 'mortuary restraint' is often accompanied by important social and ideological changes. As noted earlier in the context of rituals of secondary treatment, ruling elites sometimes impose restrictions on mortuary expenditure, establishing sumptuary laws to limit consumption and display by commoners or lesser members of the elite, thus emphasizing their own exalted prerogatives. Alternatively, elites may adopt an altogether different complex of ritual sanctions or ideologically based stylistic symbolism to demarcate themselves from the commoners, as Parker Pearson (1984:68) argues was the case in first-century AD Jutland. In other instances such as Archaic Greece, declining consumption in mortuary ritual seems to have been accompanied by an increasing emphasis on the construction of civic or religious monuments dedicated to the gods who sustained and legitimized the polity, and by votive offerings to those deities (Morris 1987:186; see also Parker Pearson 1984:88), exemplifying the changing basis in the relations of political domination to which Trigger alludes.

An issue of current dispute is the extent to which egalitarian or levelling ideologies may serve to mask overt expressions of rank or the actual order of stratification and power relations within societies. Morris suggests that this factor was operative during the Archaic period in Argos and Corinth (1987:184–6), and it is further attested during the Classical period in Athens and other Greek city-states, where individuated elite mortuary displays were subdued in favor of civic memorials to the war dead and the glorification of the polis (Morris 1992). Such muting of private in favor of public 'monumentalism' was of course consistent with the egalitarian ideology of the polis. Trigger, following Crone (1989), argues that egalitarian ideals were otherwise of little importance in early civilizations, in contrast to 'some later pre-industrial societies and all modern ones' (Trigger 1990:126).

Others (e.g. Hodder 1982; 1984; 1990; Shanks and Tilley 1982; Tilley 1984) have nonetheless suggested that masking egalitarian ideologies also restricted individual displays of wealth in the putatively ranked societies of Middle Neolithic Europe. In this argument, as summarized by Trinkaus, 'communal burials are an inverted indicator of growing status differentiation, used to mask already well-established asymmetries of power' based on differential access to land and centralized control over the production of utilitarian goods (Trinkaus 1995:59). Yet Trinkaus' analysis of lithic exchange and production revealed very little evidence for centralized control over production and distribution that might have provided an 'enforceable basis' for social inequality (Trinkaus 1995:70–71). To the extent that rank had developed amongst these Middle Neolithic societies, it is possible that what was at stake was not so much 'false consciousness' or a masking ideology, but the phenomenon which Renfrew (1974) describes as 'group-oriented' versus 'individualizing' chiefdoms. In these societies, authority is intimately bound up with ritual leadership, veneration of the ancestors, and religious ideology—due in some measure perhaps to the absence or limited availability of exotic valuables essential to 'individualizing' displays.

In many if not all societies, the development of social rank is closely associated with prestige competition and material display. The creation of rank in effect depends upon the creation and continual assertion of social differences, which is inconsistent with the masking or denial of those differences. It is possible that 'masking' is more characteristic of societies which have already experienced some degree of social stratification but are undergoing a political shift which compels the dominant power groups to downplay their wealth and status prerogatives, and, as in the case of Archaic and Classical Greece, to embrace egalitarian ideals as a means of securing popular support.

## Mortuary ritual and society reconsidered

The initial goal of processual mortuary analysis was the reconstruction of social organization and complexity based on the evidence of variability within a given mortuary complex. Yet post-processualists have contended that the mortuary record presents not an isomorphic representation or reflection of status variation in the living society but rather a ritualized expression of social structure that is subject to manipulation by 'sectional interests' or power groups within the living society. The forms and meanings of mortuary practices are evolved and modified in conjunction with other contextual variables. It is only by defining the major dimensions of change taking place in mortuary ritual over long diachronic sequences, and by observing their articulation with independent evidence for developments in sociopolitical structure and economy, that we can begin to understand the structure and ideological significance of mortuary ritual within a particular society, and in turn to interpret the significance of mortuary variability and the spatial arrangement of mortuary complexes observed in the archaeological record.

The collision of the processual and post-processual perspectives creates the opportunity to forge new and more productive directions for research. Mortuary remains can no longer be viewed as direct portayals of social roles and statuses, thus serving—from the archaeologist's perspective—mainly as a convenient line of evidence for social reconstruction. In effect, mortuary practice is a dynamic cultural phenomenon which needs to be studied in and of itself. Rather than attempting to develop or sustain generalizations about the relationship between mortuary and social variation, we should be more concerned with the relationships, recurrent and otherwise, between changes in mortuary ritual and the broader sociopolitical and economic contexts in which they occur.

The remainder of this work is devoted to a detailed examination of a particular case, Bronze Age Cyprus, in which mortuary rituals underwent a number of major transitions along with macro-processes of economic intensification and increasing sociopolitical complexity. These macro-processes included the shift from hoe to plow agriculture, the development of a local copper industry, the emergence of urban centers and elites, and various types of cultural interaction, ranging from migration to more casual trade contacts, with neighboring parts of the Mediterranean world. A close analysis of the attendant changes in mortuary practice—the shift to and away from extramural cemeteries, the elaboration and de-elaboration of rituals of secondary treatment and collective burial, and long-term fluctuations in the quantity and quality of material expenditures—may contribute not only to an understanding of the 'lived experience' of social transformations within the Cypriot context, but also to a more general understanding of mortuary ritual as an arena for the production of social life.

# 3 The Archaeological Record of Mortuary Practice in Cyprus: Formation Processes, Sampling Issues, and a Methodology for Interpretation

## Introduction

The general objective of this work is to develop an understanding of long-term changes in social structure and mortuary ritual during the Cypriot Bronze Age, and of the ways in which structure and ritual were dynamically articulated, each producing and transforming the other in the course of cultural practice. Social structure and mortuary ritual, however, are only the starting points in the creation of the archaeological record. As discussed in the previous chapter, the material remains of mortuary rituals do not constitute a direct account of the social and ideological systems with which they were associated, for in the performance of those rituals prestige differentials, power relations, and other aspects of social structure may have been variably overstated or downplayed. Moreover, not all mortuary practices are necessarily 'materialized' in archaeologically visible fashion—feasts, processions, songs, dances, and other ceremonies honoring the dead may leave few, if any, archaeological traces. And even in the case of mortuary practices that *are* susceptible to archaeological preservation, the material evidence is likely to have been transformed prior to archaeological recovery by post-depositional processes of both natural and cultural origin, especially when tombs were reused over long timespans. Finally, the archaeological record as it is known to us has been shaped by historically situated research methodologies, excavation and recording practices, and the contingencies of sampling at the local and regional levels. A judicious interpretation of the mortuary record must therefore be prefaced by a consideration of 'what portion of the total funerary system is being observed' and the types of questions which can and cannot be addressed with the available information (O'Shea 1984:27; Clarke 1973).

This chapter begins with an examination of some of the intermediating screens or filters that have affected the representation of mortuary practices in Bronze Age Cyprus. Problems of preservation associated with collective burial units in general and with the chamber and pit tombs typically found on the island are considered first. The limitations of past and present approaches to tomb excavation and mortuary analysis are then reviewed. This is followed by a discussion of sampling problems in regional, chronological, and local perspective. Lastly, I outline the basic principles of an analytical methodology that may be helpful in eliciting aspects of ritual systems and social structure from the complex and sometimes problematic mortuary record of the Cypriot Bronze Age.

## Post-depositional processes affecting the preservation of collective tombs and other mortuary features in Cyprus

The most commonly preserved mortuary features of the Cypriot Bronze Age are rock-cut chamber tombs cut into the hard white calcareous subsoil known as *havara* or *kafkalla*. The chambers, extremely variable in size and plan, were usually accessed through the chamber wall by an entrance tunnel known as a *dromos*, which might take the form of a shallow, pit-like cutting, a deeper bathtub-shaped shaft, an elongated corridor, or, in the case of the so-called 'chimney' tombs, a shaft opening into the roof of the chamber. The dromos opened onto one or more chambers through an orifice known as a *stomion*, which was usually closed with large limestone slabs or smaller stones. It is unclear as to whether the *dromoi* were generally left open or deliberately filled after each interment in the tomb chambers (cf. Gjerstad *et al*. 1934:78; Stewart 1962a:216; Herscher 1978:783–4), yet the groups that buried their dead within seem to have had no difficulty in relocating them periodically, as many tombs were used for multiple burials, sometimes taking place over several generations or even centuries. From a structural perspective, rock-cut chamber tombs have a high propensity for long-term archaeological preservation, but the associated burial deposits were subject to many complex and frequently destructive post-depositional processes, as discussed below.

In some parts of the island, pit graves or pit tombs dug in havara or a matrix of earth and stony conglomerate were employed instead of, or in addition to, rock cut chamber tombs. The highest frequency of pit tombs seems to occur in a band of sites stretching from Akhera to Alambra in the central part of the island (e.g. Gjerstad 1926:48; Stewart 1962a:216; Kara-georghis 1965a; 1965b), where their use may in some instances have been dictated by the con-straints of the local geology, although in other cases ground suitable for cutting chamber tombs was located nearby (see Stewart 1962a:322 on Nicosia *Ayia Paraskevi*; Sneddon 2002:102, 114 on Marki *Davari*). Pit tombs tended to be relatively small and simple in plan, with entrances at the top that were presumably filled in with earth and/or capped with stones. Like the chamber tombs, they were often reused for multiple burials. However, the comparative instability of the matrix into which some of the pit tombs were cut and the exigencies of fill removal in the course of repeated usage would have rendered their burial deposits even more susceptible to disturbance than those associated with chamber tombs.

With the exception of the earthen and rubble tumuli surmounting the Middle Cypriot III tombs of Korovia *Paleoskoutella* (Gjerstad *et al*. 1934: Figs 163, 166) and a few of the Late Cypriot tombs at Enkomi (Gjerstad *et al*. 1934: Tombs 7, 10, 11, 13, and 21), monumental markers of tomb locations are nowhere in evidence today. It is possible, however, that markers and many other surface features associated with ritual activity, such as the so-called 'funeral pyres' observed above the Late Cypriot Tombs 1 and 2 at Morphou *Toumba tou Skourou* (Vermeule 1974; Vermeule and Wolsky 1990:169, 245) and the 'cult areas' found beneath Tumuli 1, 3 and 6 at *Paleoskoutella* (Gjerstad *et al*. 1934), have simply not been preserved, their remains having been eradicated by centuries of exposure, erosion, plowing, and other forms of cultural distur-bance. Similarly, any other above ground, or subterranean but relatively shallow, burial features such as single inhumation earth graves would have had a low aptitude for archaeological preser-vation. The fact that such features have indeed been encountered at some sites (for example, a series of possibly evacuated pit graves was found at Lapithos *Vrysi tou Barba*, see Herscher 1978:818–9) nevertheless suggests that they may also have been present elsewhere in the past.

As in other parts of the Mediterranean world where chamber tombs, tholos tombs, and other types of built tomb were used for collective burials, the mortuary features and practices charac-teristic of Bronze Age Cyprus pose a variety of problems not typically encountered in the study of single inhumations. With a few important exceptions, collective burial groups were usually formed as a series of depositional events rather than as a single episode of interment, and consequently the associated burial strata were subject to continuous alteration over the entire period of tomb use. Each time a tomb was opened to insert a new burial or burials, previously interred corpses would have been vulnerable to displacement, disarticulation, and other physical damage. In some cases these disturbances may have been the 'accidental' effects of making room for newcomers, but in other cases they may represent secondary phases of mortuary treatment, during which the remains of previously interred burials were deliberately rearranged, removed, or otherwise handled as part of an extended program of ritual celebrations. Under either set of circumstances, a loss of information regarding the condition and position of individual corpses at the time of interment, as well as a confusion of the associations between particular grave goods and particular individuals, would inevitably have ensued. Additionally, certain types of grave goods (e.g. metal objects) associated with earlier burials may sometimes have been removed in the course of tomb reuse. Provided that they can be distinguished from substantially later looting activity, such practices are in themselves quite interesting in terms of the ideologies (or counter-ideologies) that permitted them, and perhaps as indicators of prevailing economic circumstances. However, like the disturbance of skeletal remains, they tend to obscure the specific distinctions accorded to individuals within the tomb. In the study of collective tombs, therefore, the focus of analysis must be shifted from the comparison of individual burials to the study of differences between larger burial groups (see also Chapman 1977:26; 1981a:398, 406).

In addition to the complications associated with the repeated reuse of collective tombs, the preservation of burial remains will also have been affected by post-depositional disturbances occurring in later periods of antiquity and more recent times. Repeated episodes of flooding would have accelerated the deterioration of skeletal remains through physico-chemical processes such as leaching, carbonate encrustation, and the cementation of bones within the surrounding sedimentary matrix. Swirling flood waters are further alleged to have induced the physical displacement of skeletal remains and grave goods at sites such as Myrtou *Stephania* (Hennessy 1964) and Morphou *Toumba tou Skourou* (Vermeule and Wolsky 1990:161). Natural processes of erosion that hastened the structural collapse of burial features would similarly have been detrimental to the preservation of bone and artifactual material. Moreover, tombs such as those of the Late Cypriot period that were frequently located within urban habitation zones would have been particularly vulnerable to looting and destruction in the course of later building and tomb cutting activity. The disturbance or obliteration via over-building of older tombs may often have been inadvertent, but some instances may be interpreted as intentional expressions of sociopolitical hegemony, as Manning (1998a) has noted.

Tomb looting has probably contributed the most to the decimation of burial deposits from antiquity through the present time. The frequency of tomb destruction and plundering has risen considerably as the result of modern construction activity, and the number of tombs excavated in the course of rescue operations rather than in the context of systematically planned research is continually increasing. Much to the credit of the Department of Antiquities and the various foreign missions working in Cyprus, the results of these salvage operations have usually been published in a timely and detailed fashion, but the depredations of looting and the constraints of time and resources surrounding such excavations frequently restrict the conclusions which can be drawn from the data.

## Past and present approaches to the excavation and study of Bronze Age tombs in Cyprus

The reconstruction of the mortuary practices of Bronze Age Cyprus is, like most archaeological endeavors, compromised not only by factors of preservation but also by the limitations of past and current research agendas and technologies, the interpretive models brought to bear upon the data, and the vagaries of sampling. Although hundreds of tombs have been excavated at localities throughout the island over the last 125 years, and many have been published in voluminous reports, the quality of information that can be extracted from this extensive corpus is, nonetheless, extremely variable. Many important cemeteries are known largely or exclusively from the work of late nineteenth- and early twentieth-century excavators such as Luigi Palma di Cesnola (1877), Max Ohnefalsch-Richter (1893), and representatives of the British Museum (Murray *et al.* 1900), most of whom were preoccupied with the recovery of fine pottery, metal artifacts, and imported valuables for museums and private collections (Goring 1988). Working in haste and with limited objectives, they devoted little attention to the details of tomb architecture, the stratigraphy of burial deposits, the position and preservation of skeletal remains—which were almost never saved—or the associations between objects and burials. Their reports highlighted finds of exceptional artistic and chronological interest, but seldom included complete inventories of all of the goods recovered from any particular tomb, let alone descriptions of the burials. To the benefit of later scholarship, Einar Gjerstad, a leading member of the Swedish Cyprus Expedition, summarized much of the extant information concerning tomb architecture and skeletal remains from the fieldnotes of these and other early excavations in his doctoral dissertation (1926), but the descriptions which he obtained were for the most part extremely brief. Thus, a great deal of information concerning mortuary practices at many important sites has been lost forever.

The publication of the Swedish Cyprus Expedition's (SCE) research from 1927–31 (Gjerstad *et al.*1934; Gjerstad 1980) marked a major advance in the standards of archaeological practice in Cyprus. While the SCE's work may have fallen somewhat short of the ideals of contemporary scientific archaeology, it remains exemplary in terms of the attention that was devoted to tomb stratigraphy, burial postures and placement, and the spatial positioning of grave goods. Detailed plans of tomb architecture and illustrations of the arrangement of burials and goods within the tomb chambers were presented for nearly all intact and many disturbed mortuary features, along with largely comprehensive catalogues of intact, restored, and fragmentary tomb finds. The level of descriptive detail is sometimes sufficient to discern complex programs of mortuary ritual and to make comparisons of mortuary expenditure between groups while controlling for variations in tomb group size and chronology. Also, and quite remarkably for the time, some of the cranial material from the SCE excavations was saved for study by physical anthropologists, with notably large, albeit incomplete samples drawn from Late Cypriot tombs at Ayios Iakovos *Melia* and Enkomi. More skulls were added to this collection in the aftermath of Schaeffer's 1933 excavations at Enkomi (Schaeffer 1936). Although the resulting analyses (Fürst 1933; Hjörtsjö 1946–7) were concerned primarily with issues of race and ethnic origins, the demographic identifications of the specimens published are of considerable interest, as they suggest potentially significant biases in eligibility for chamber tomb burial based on age and sex at these sites, patterning that has recently been reconfirmed by Fischer in his (1986) restudy of material from the SCE and French excavations.

Subsequently, even though the majority of excavators have modeled their excavation and recording practices on the standards set by the SCE, the recovery and description of grave goods have continued to receive priority over the recovery and description of human skeletal remains. Meanwhile, many fieldworkers have implicitly assumed that all or most of the burials in any particular chamber occurred as sequential primary interments of initially intact corpses. Therefore, any indications of skeletal disarticulation or anomalies in the representation of body parts have been dismissed as the results of post-interment disturbances associated with tomb reuse, looting, or flooding, rather than being considered as possible consequences of ritual practices. The importance of detailed reporting of the skeletal remains and their associated depositional matrix has in turn been overlooked. While the presumption of primary burial may be valid in many cases, it has led to the loss of much extremely valuable information concerning complex programs of mortuary treatment in others. Furthermore, until very recently, most studies of human skeletal material have maintained a very narrow focus on questions of race or cultural origins and the related significance of practices of cranial deformation (e.g. Buxton 1920; 1931; Charles 1960; 1965a; 1965b; Schwartz 1974; Domurad 1986; 1989). Detailed and comprehensive analyses of osteological collections including both cranial *and* post-cranial skeletal remains, such as those undertaken by Angel (1972) on Late Cypriot skeletal remains from Kourion *Bamboula*, have only begun to be practiced since the 1980s, most notably in the context of Early–Late Cypriot tomb excavations in the Kalavasos area (e.g. Todd 1985; 1986; Schulte-Campbell 1986; South 1997; 2000; South *et al.* n.d.), at the Philia–EC cemetery of Sotira *Kamioudhia* (Swiny *et al.* 2003; Schulte-Campbell 2003), at the Early–Middle Cypriot sites of Alambra *Mouttes* (Coleman 1996; Domurad 1996) and Marki *Alonia* (Frankel and Webb 1997; 1999; 2000b; Moyer 1997), and in the case of one important Late Cypriot III shaft grave burial at Hala Sultan Tekke (Niklasson 1983; Schulte-Campbell 1983). Because so many excavators have failed to appreciate the insights to be derived from the study of human skeletal remains, there are enormous gaps in our knowledge of tomb demographics, age and gender-based burial treatments, details of human health and life expectancy, and ritual practices in general.

In light of the historical circumstances surrounding tomb excavations in Cyprus, along with the art historical preoccupations of most of the scholars who have worked there until recently, it is not surprising that the major synthetic publications dealing with Bronze Age mortuary data (Stewart 1962a; Åström 1972a; 1972b; Åström and Åström 1972) have been concerned mainly with particular types of grave goods rather than with the mortuary practices accompanying the disposal of those goods. Studies devoted to the ritual and social dimensions of mortuary practice are a relatively new phenomenon in the field of Cypriot archaeology. To be counted among the earliest forays into ritual practices are an essentially descriptive survey of Bronze Age burial customs by Cassimatis (1973) and the discussion of funerary symbolism in Early–Middle Cypriot ceramic models and tomb carvings by Frankel and Tamvaki (1973). The latter theme was also pursued by Åström (1988). A more extensive overview of the evidence for complex burial programs and social hierarchy in Middle and Late Cypriot mortuary contexts was undertaken in earlier works by this author (Keswani 1989a; 1989b), followed by a short work by Webb (1992), who came to a contrastingly negative series conclusions regarding the evidence for ritual practices in the Middle Cypriot mortuary data. In the last few years, a growing interest in mortuary symbolism and the representation of social prestige has become apparent in a number of works concerned with varying aspects of grave goods, tomb architecture, and locational symbolism (e.g. Goring 1989; Swiny 1989; Manning 1993;1998a; Peltenburg 1994; Keswani 1996; Baxevani 1997; Sneddon 2002). Davies (1995; 1997) and Bright (1995) have also recently undertaken quantitative analyses of Early–Middle and Late Bronze Age burial data with the aim of eliciting elements of mortuary variation and social structure, but their studies, while promising, are as yet largely programmatic and descriptive in character, and both authors tend to emphasize the shortcomings rather than the potential of the existing data. The realization of that potential depends, I would argue, on the development of a theoretically and ethnographically informed, systematic analytical methodology, one which focuses on the positive evidence for mortuary practices and ideology evinced even within a highly problematic dataset. The main points of this methodology are discussed at the conclusion of this chapter.

## Sampling issues: problems of regional, chronological, and local coverage

Although dozens of cemeteries and hundreds of tombs have been investigated in Cyprus over the last 125 years, the mortuary samples available for different regions and time periods vary considerably in size and detail of reporting. Prior to 1974, research efforts were most intensively concentrated in the northern and eastern parts of the island now under Turkish control. Subsequently, the focus of research shifted to the south, providing a much needed balance in regional coverage. However, many questions which might be raised about sites and areas previously investigated in the Turkish controlled part of the island must remain unanswered until this area becomes accessible to archaeologists once again.

The Early–Middle Cypriot mortuary sample (see Table 3.1 and Fig. 3.1) is dominated by evidence from the north coast of the island, and in particular by the extensively excavated and relatively well-published cemeteries of Bellapais *Vounous* (Schaeffer 1936; Dikaios 1940; Stewart and Stewart 1950; Dunn-Vaturi 2003), and Lapithos *Vrysi tou Barba* (Gjerstad *et al.* 1934; Myres 1940–45; Herscher 1978). These are the only EC–MC mortuary complexes within which it is possible to undertake detailed analyses of variations in mortuary practice, expenditure, and symbolism at the local level, both within and between specific chronological periods. It is, therefore, inevitable that interpretations of mortuary ritual and social structure in the Early–Middle Cypriot periods will be dominated by the trends observed within these larger site samples, although their characteristics may not be entirely representative of mortuary practices in other contemporaneous communities.

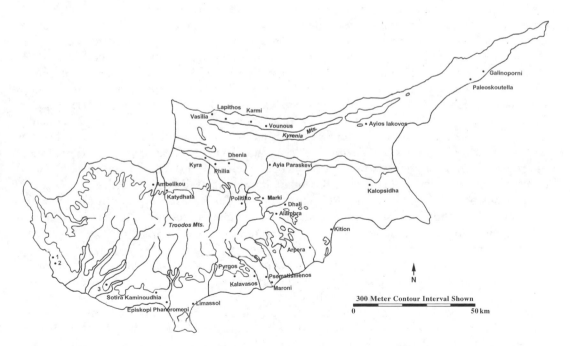

**Figure 3.1**    Map of Cyprus showing Philia and Early–Middle Cypriot sites referred to in the text and tables. Chalcolithic sites referred to in comparative discussions are shown as numbered localities: (1) Kissonerga *Mosphilia*; (2) Lemba *Lakkous*; (3) Souskiou *Vathyrkakas* and *Laonas*.

Smaller samples of mortuary data recovered from other areas nevertheless afford useful comparative material, permitting at least some preliminary conclusions concerning regional continuities and variations in mortuary traditions. For example, the Swedish Cyprus Expedition's excavations at Ayios Iakovos *Melia* and Korovia *Paleoskoutella* (Gjerstad *et al.* 1934) yielded much detailed information concerning Middle and early Late Cypriot mortuary practices in northeastern Cyprus, even though the small number of tombs investigated at these sites makes the assessment of mortuary variability within the associated communities problematic. A large sample of ECIII and Middle Cypriot tombs has been excavated, mainly in rescue operations, in the south coast village of Kalavasos (Karageorghis 1958; Todd 1986), and detailed analyses have been published for the human skeletal remains from some of these tombs (Schulte-Campbell 1986). The publication of additional tombs excavated in the areas of the old elementary school and the cinema in 1987 (Todd 1988:134) should contribute further to our knowledge of Middle Cypriot mortuary practices. Reports of EC–MC tombs excavated at Sotira *Kaminoudhia* (Swiny 1985; Manning and Swiny 1994; Swiny *et al.* 2003) and Episkopi *Phaneromeni* (Weinberg 1956; Carpenter 1981; Herscher 1981), along with surveys of a number of cemetery sites in adjacent areas (Swiny 1981), offer insights concerning mortuary practices in southwestern Cyprus.

Perhaps the most serious gap in coverage of EC–MC mortuary practice falls in the central part of the island, along the Mesaoria plain and the copper-rich foothills of the northern and eastern Troodos. Although large numbers of EC–MC tombs were opened here by late nineteenth- and early twentieth-century excavators such as Cesnola (1877), Ohnefalsch-Richter (1893), Myres (1897), and Markides (Åström 1989) at sites in the vicinities of Katydhata, Politiko, Nicosia *Ayia Paraskevi*, Kalopsidha, and Alambra, previous looting activity and the poor standards of excavation and reporting associated with these archaeological investigations place serious limitations on the conclusions that can be drawn from their finds. However, as minimal as the evidence may be, these and other cemeteries noted in Table 3.1 still offer important information concerning mortuary rituals and the complex of prestige symbolism that emerged as contacts with mainland regions intensified in the Middle Cypriot period and subsequently (Courtois 1986a; Merrillees 1986).

The Late Cypriot mortuary sample (see Table 3.2 and Fig. 3.2), like the earlier EC–MC corpus, is quite uneven in terms of the representation of various regions and chronological phases. The Late Cypriot I period is best documented in the northern part of the island, particularly at sites such as Morphou *Toumba tou Skourou* (Vermeule and Wolsky 1990), Ayia Irini *Palaeokastro* (Pecorella 1977; Quilici 1990), Myrtou *Stephania* (Hennessy 1964), Pendayia *Mandres* (Karageorghis 1965c), Akhera *Chiflik Paradisi* (Karageorghis 1965b), Ayios Iakovos *Melia*, and Korovia *Nitovikla* (Gjerstad *et al.* 1934). Tomb excavations at these sites offer important evidence for complex mortuary practices, including the so-called instances of 'mass burials' (probably episodes of collective secondary interment, as discussed in Chapters 4 and 5), and for transformations in mortuary prestige symbolism during this important transitional era. However, as is so often the case, the small number of contemporaneous tombs excavated at each site, ranging from a single instance to slightly over a dozen tombs, makes it difficult to evaluate the range of mortuary variation within any one community.

By contrast, mortuary practices of the Late Cypriot II–III periods are best known from the urbanized coastal settlements of eastern and southern Cyprus. The largest and most extensively reported mortuary sample comes from the east coast center of Enkomi *Ayios Iakovos*, where hundreds of tombs and other mortuary features spanning the entire LC period were uncovered over more than 70 years of excavations, beginning with the British Museum Expedition in 1897 and followed by the work of Swedish, French, and Cypriot archaeological teams (Murray *et al.* 1900; Gjerstad *et al.* 1934; Schaeffer 1936; 1952; 1971; Lagarce and Lagarce 1985; Courtois 1981; 1986b; Dikaios 1969). The British Museum also conducted large-scale tomb excavations at the south coast sites of Hala Sultan Tekke (Bailey 1972), Maroni (Johnson 1980), and Kourion

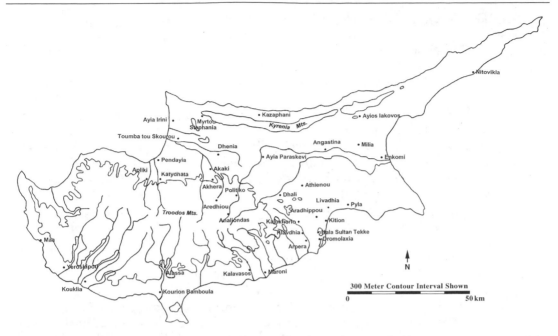

**Figure 3.2**     Map of Cyprus showing Late Cypriot sites referred to in the text and tables.

(Murray *et al.* 1900), recovering a broad range of exotic goods which illustrate the wealth and wide-ranging long distance exchange contacts of the elite groups which emerged in this period. However, as noted earlier, much valuable information concerning tomb architecture, the human skeletal remains, and their physical disposition was also destroyed in the course of these excavations. Fortunately, more recent work at Hala Sultan Tekke (Karageorghis 1968; 1972b; Åström 1983; Niklasson 1983), Kition (Karageorghis 1960b; 1974), Kalavasos (South 1997; 2000; South *et al.* 1989; n.d.; Pearlman 1985; McClellan *et al.* 1988; Goring 1989), Kourion (Benson 1972), and Kouklia (Catling 1968; 1979; Maier and Wartburg 1985; Karageorghis 1967; 1983; 1990a) has considerably broadened our knowledge of mortuary practices along the south coast. The samples from Kalavasos and Kourion in particular are sufficiently large to permit some assessment of local mortuary variability, and the south coast mortuary sample overall attests to both important similarities and contrasts between regions.

As was the case with the EC–MC periods, Late Cypriot mortuary practices in the central part of the island remain the least well-known. At sites such as Akhera *Chiflik Paradisi* (Karageorghis 1965b), Dhenia *Kafkalla* and *Mali* (Karageorghis 1977; Hadjisavvas 1985), Politiko *Ayios Iraklidhios* (Karageorghis 1965a), Nicosia *Ayia Paraskevi* (Kromholz 1982; Flourentzos 1988), Dhali *Kafkallia* (Overbeck and Swiny 1972), Milia *Vikla Trachonas* (Westholm 1939a), and Angastina *Vounos* (Karageorghis 1964b; Nicolaou 1972), very few well-documented tombs are available for consideration, and the conclusions which can be based on such a sample are necessarily limited. Yet when considered as a group, these sites suggest important contrasts in the availability and deployment of prestige goods between coastal and inland regions, a pattern which may be indicative of hierarchical, or at least inegalitarian, sociopolitical relationships between different parts of the island.

In general, while the regional and chronological representation of Bronze Age mortuary practices may be less than optimal, it is nonetheless adequate to allow the definition of similarities and differences between regions and to elucidate long-term transformations in mortuary practice. Other important questions that need to be addressed, however, concern the extent to which the sample from any single site or region illustrates: (1) the entire sequence of ritual practices associated with a particular mortuary complex (assuming the possibility of a multi-phase ritual system); (2) the range of variation in mortuary treatments associated with different age and gender groups; and (3) the range of variation in mortuary treatments associated with different sub-groups (e.g. kin segments) of potentially different social status or resources *within* a given community.

The first question is perhaps the easiest to answer. Burial programs involving secondary treatment of the dead are evident in chamber and pit tomb complexes of numerous sites dating to both the Early–Middle and Late Cypriot periods, as discussed at length in Chapters 4 and 5, but mortuary features associated with the preliminary phases of burial processing have been observed at relatively few localities. Clearly, therefore, we are most often presented with the latest or terminal stage of the ritual system. Yet from the diverse mortuary features and apparent redeposition of burials and/or grave goods at the best-documented sites, it is at least possible to discern the practice of multiple phases of ritual and to identify the types of archaeological evidence that may have been lost or misinterpreted in the past, but may yet be recovered in the course of well-informed future research.

With regard to the second question, it is possible to say with some confidence, even in the absence of detailed osteological analyses, that infants and young children are significantly under-represented in the mortuary populations of both the EC–MC and the LC periods, and most strikingly in the earlier era. The majority of twentieth-century excavators made note of the presence of child burials whenever they recognized them, even if they did not pursue further studies of the human skeletal remains recovered in their excavations. While the skeletal remains of juveniles are to some extent more fragile and less amenable to preservation than those of adults, this factor is not sufficient to account for the dearth of young individuals in EC–MC tombs, and their still infrequent, albeit more numerous presence in tombs of LC date. It is impossible to determine how most EC–MC children were disposed of at death; some may have been exposed outdoors or placed in abandoned settlement contexts (as some recent examples from Marki *Alonia* suggest; see Frankel and Webb 1997; 1999; 2000b; Moyer 1997). Others may have been buried or stored in pithoi, as was the case with some transitional Philia Culture burials at Kissonerga *Mosphilia* (Peltenburg *et al.* 1998:72) and Marki *Alonia* (Frankel and Webb 2000a:764; 2000b: 68–70), and as may also have been the case with a number of EC burials conceivably redeposited in chamber tombs at Lapithos *Vrysi tou Barba* (these and other pithos burials are discussed more extensively in Chapter 4). The latter site also affords some of the earliest examples of the use of dromos 'cupboards' (very small chambers or cuttings in the dromos walls) for child burials (Herscher 1978:703–4), a practice which became more common in the succeeding LC period. In the LC period, it is evident that children were occasionally buried in pit graves (e.g. Enkomi Cypriot Tomb 8, 8A, and 20, Dikaios 1969) and, as also attested at Enkomi, in jars or other pots placed in pits beneath building floors (Dikaios 1969:431–2; Courtois 1981:291–2; see also Keswani 1989a:386). Sometimes, too, there is evidence for their remains having subsequently been exhumed from primary burial facilities and redeposited in tomb chambers used for adult burials (most notably in Kalavasos *Ayios Dhimitrios* Tomb 11, South 2000:352). Older children may have been more routinely buried within the principal tomb chambers at the time of death.

The question of whether adult males and adult females were accorded chamber tomb burial with similar frequency is more problematic. Analyses of small samples of crania from excavations at the EC–MC sites of Bellapais *Vounous* (15 adults and 1 adolescent from the excavations of Stewart and Stewart 1950:374; 4 adults from the excavations of Schaeffer and Dikaios in Fischer 1986:28–9); and Lapithos *Vrysi tou Barba* (9 adults, Fischer 1986: Table 2; see also Fürst 1933:58) suggest a possible bias towards the inclusion of males more often than females, but if the individuals that could not be sexed with certainty were in fact female, the imbalance in numbers is relatively slight (see more detailed discussion in Chapter 4). A contrasting pattern is evident in a sample of 19 adult burials from MC tombs recently excavated in Kalavasos, in which females were more numerous than males among the nine individuals that could be sexed (Schulte-Campbell 1986). Another small sample of seven adult individuals recovered from excavations of probable Early Cypriot burials at Alambra *Mouttes* contained either four or five males and two or three females (cf. Coleman 1996:116–23; Domurad 1996:515–8), and a sample of twelve adult burials from the Philia–EC cemetery at Sotira *Kaminoudhia* contained three identified females and two males (Schulte-Campbell 2003:437). Much larger osteological samples were preserved from Late Cypriot tombs at Ayios Iakovos *Melia*, Enkomi *Ayios Iakovos* (for both sites see Fürst 1933; Hjörtsjö 1946–7; Fischer 1986), and Kourion *Bamboula* (Angel 1972), all of which reveal a very pronounced bias in favor of males (on the order of 2:1 at Enkomi and Kourion and as high as 4: 1 in the case of Ayios Iakovos *Melia*). However, it is likely that females outnumbered males in some Late Cypriot tombs, and perhaps especially in those used by very high status groups, as is evident in Kalavasos *Ayios Dhimitrios* Tomb 11 (South 2000:349–53). Given the present state of the evidence, it can only be concluded that gender biases in burial practices were probably prevalent throughout the Bronze Age, but these may have varied within and between communities based on a variety of social considerations.

Regarding the third issue, the question of whether burial samples provide a broad representation of various sub-groups of the community or a narrower view of particular (and in some cases relatively privileged) kin segments or other sub-groups, there is also no simple answer. It has already been noted that some mortuary samples are so small that any assessment of local variability in mortuary treatments is precluded. In the case of sites where relatively few tombs were encountered or excavated, it is indeed likely that we are looking at the practices of a single or a very small number of closely related burying groups. Larger cemeteries, such as those of EC–MC Bellapais *Vounous* and Lapithos *Vrysi tou Barba*, might ostensibly be interpreted to represent the practices of larger communities, but it could also be argued that the burial populations of even these sites may have accumulated from the deaths within a fairly limited number of kin groups reusing the same burial grounds over several generations. Archaeological surveys in the vicinity of Marki *Alonia* (Frankel and Webb 1996; Sneddon 2002) and the Episkopi area (Swiny 1981) suggest that the residents of particular settlements or settled areas probably made use of multiple cemeteries, some used contemporaneously, others in overlapping chronological succession. Unless several cemeteries have been sampled in a given region, the full range of mortuary and social variability cannot be reconstructed. However, at least until the end of the Late Bronze Age, there is no compelling evidence to suggest that 'formal' or preservable burial, and chamber tomb burial in particular, was the exclusive prerogative of a social elite from which certain kin groups were excluded on the basis of class, as Morris (1987) proposes for Iron Age Greece, and as Manning (1993:45) implies was also the case in Early–Middle Bronze Age Cyprus. While some segments of the population are undoubtedly underrepresented in the mortuary sample, the range of variation in grave assemblages from the larger cemeteries or burial complexes of both the EC–MC and LC periods implies the presence of individuals and groups of diverse wealth and prestige.

## Towards a methodology for the interpretation of collective burial practices

The study of collective burial groups presents many complex interpretive problems for the mortuary analyst, problems that are rendered even more difficult by the historical and intellectual context of previous archaeological investigations in Cyprus. The foregoing sections of this chapter have discussed some of the processes affecting the formation and recovery of the archaeological record, along with critical sampling issues. I turn now to the question of how it may be possible to interpret programs of ritual treatment, the relationship between tomb groups and social groups in the living community, and the significance of variability in tomb assemblages in what is undoubtedly a complicated and very imperfect dataset. The fundamental assumptions of this research methodology, informed by the ethnographic cases and theoretical issues discussed in Chapter 2, are outlined below.

## Reconstructing programs of mortuary treatment in collective burial groups

An understanding of the system of ritual practices through which the mortuary record was produced is both a subject of intrinsic interest and an essential prerequisite to any viable interpretation of mortuary variability and its cultural significance. After all, archaeologically observable differences in the treatment of the dead need not be exclusively related to horizontal or vertical status distinctions, expressions of prestige, or even the manner of an individual's death. They may also reflect a series of different phases in the processing of the deceased, phases which, for a variety of historically contingent reasons, are not always brought to completion (e.g. Miles 1965; Brown 1981). In any situation where pronounced differentiation in the modalities of contemporary burials are observed (e.g. single primary inhumations versus collective burials displaying evidence of secondary treatment), it is important to be able to distinguish mortuary variability relating to different phases of mortuary ritual from mortuary variability relating to differences in social status or cultural affiliations (Keswani 1989a:91).

The complex, overlapping depositional and post-depositional processes affecting collective burial groups make it difficult to distinguish mortuary programs involving primary or secondary treatment, but some general guidelines can be suggested. The principal lines of data needed to elucidate the problem include (a) the condition and relative completeness of the skeletal remains, (b) the evidence for sequential and simultaneous interment practices, which may sometimes be inferred from the stratification of burial deposits, and (c) evidence for functional variations in mortuary facilities. In some cases, the condition, disposition, and chronology of the burial goods in the terminal interment facility may afford further evidence for the depositional history of the burials in that mortuary unit.

In a ritual system involving the use of collective tombs for sequential primary interments, it might be predicted that the skeletal remains of the latest burial in each undisturbed tomb would be found in articulation, with the preservation of body parts consistent with statistical expectations based on relative bone density, the age and health of the individual at the time of death, and post-depositional climatic or geochemical conditions within the burial chamber. Waldron (1987) provides a useful discussion of the preservational propensities of various skeletal elements and bone types. Moreover, although some of the earlier burials in the tomb may have been disarranged by successive interments, others in unaffected parts of the tomb ought to be largely intact. Even some disturbed burials might be at least partly articulated. Additionally, if post-depositional conditions throughout the chamber were more or less constant, the body parts of each individual should be consistently represented. The stratification of the burial deposits ought to reflect sequential phases of reuse, and if goods were included with the burials, earlier types should in general be found in stratigraphically lower positions than later types (although prolonged curation of valued heirlooms may sometimes complicate this issue). Differences in the type or quality of burial features may be correlated with differences in age, sex, prestige (as signaled by associated status regalia) or type of death, but not with the presence of primary versus secondary, or articulated versus disarticulated burials. In other words, all burial units should represent the same, terminal phase of mortuary processing.

The prevalence of a ritual system involving 'dual obsequies' or the at least occasional exhumation and reburial of corpses may be inferred from several lines of evidence. The presence of notably incomplete and/or disarticulated skeletal remains whose condition is not readily explained in terms of susceptibility to decay or later human disturbance would be suggestive of practices involving secondary treatment, as would the occurrence of seemingly simultaneous burials of two or more individuals. Evidence for the use of temporary burial facilities may afford critical supporting evidence for such a ritual system, especially where the condition of the skeletal remains is ambiguous. Temporary burial features may be distinguished in part by the comparatively low level of energy expended in their construction relative to permanent collective burial units and in part by their apparently 'looted' or 'emptied' condition, although the completeness of their evacuation may differ, as Metcalf and Huntington (1991:101–2) have observed in various Indonesian societies. Direct indications of the redeposition of skeletal material, grave goods, and associated sediments from the temporary to the permanent burial facility would of course constitute the most conclusive evidence for complex ritual programs. Practices of exhumation and reburial may be signaled, for example, by the presence of damaged or incomplete vessels in terminal mortuary features, joins between ceramics from different mortuary features, and/or the mixing of grave goods of different chronological periods within the same stratigraphic level (as when several individuals of different generations are exhumed and reburied simultaneously, although here, once again, practices of heirloom curation must be weighed as an alternative explanation). Alternatively, it is conceivable that an entirely new complement of goods might be prepared for the deceased in the interval between primary and secondary burial, in which case there may be striking anomalies between the condition of the skeletal remains and the condition of the associated artifacts.

In tombs where both primary and secondary burials are indicated, the stratigraphic sequence of the burial types may be informative. If secondary burials are associated with the earliest use of the tomb, succeeded mainly by primary burials, then a mortuary system like that of the Berawan (Metcalf 1982; Metcalf and Huntington 1991), involving elaborate and long-term preparations for the memorialization of the deceased (as well as for feasting, gift-giving, and other costly displays among the living), may be indicated. If, on the other hand, secondary burials occur mainly in later strata, then it is possible that these represent delayed interments of low status individuals (children, victims of disease) and/or customary restrictions on the reopening of tombs within certain intervals. It should be cautioned, however, that both of these modes of secondary treatment may be prevalent within the same community, as Bloch's (1971) account of Merina mortuary practices indicates.

Finally, it is important to remember that mortuary programs involving the exhumation and reburial of individuals or groups are not the only forms of complex ritual processing that may be practiced. Stratigraphic evidence for deliberate post-interment disturbances of earlier burials on subsequent occasions of tomb reuse, the patterned rearrangement or removal of certain body parts, and the removal of goods associated with earlier burials, may be indicative of secondary or tertiary mortuary treatments undertaken within the permanent burial facility in conjunction with the introduction of new burials or other periodic ritual observances. In some instances—and here again the Merina serve as a case in point—skeletal remains may even be removed from an ostensibly final and permanent collective tomb in order to inaugurate the use of a new one, since an individual may not be placed alone in a tomb (Bloch 1982:213).

## Interpreting the relationship between tomb groups and social groups

In addition to reconstructing the system of mortuary ritual which configured the formation of the mortuary record, it is also important to develop some understanding of the relationship between tomb groups and social groups in the living community. A survey of the ethnographic literature on societies practicing collective burials suggests that the grouping of individuals together in a common tomb is more often an important expression of social identities and interests than an entirely random process (Keswani 1989a, Chapter 3). This does not mean that kinship structures and residential patterns can be inferred directly from mortuary data, since tomb group composition may also be affected by the expression of cross-cutting personal ties, prestige claims, and broader communal or pan-communal affiliations. However, under optimal circumstances, considerable insight into community social structure may be obtained from an analysis of the age-sex composition of collective tomb groups, determination of genetic similarities or other evidence for close biological relationships among groups of skeletons, and an assessment of the relative degrees of relatedness or homogeneity amongst males as one group and females as another (e.g. Lane and Sublett 1972; Gamble *et al.* 2001).

Unfortunately, the paucity of comprehensive osteological analyses of human skeletal remains from tombs of the Cypriot Bronze Age renders this one of the most problematic areas of mortuary analysis. However, the scale or inclusiveness of tomb groups (approximated by the number of burials per chamber or other mortuary unit), the timespan over which kin or other social relations are reaffirmed through the continuing use of the same tomb, and the relationship between tomb group size and the elaboration of burial treatments may also express important aspects of social structure and ideology, and these variables are more readily monitored in the Cypriot context. Tomb chronologies have consistently been a major focus of research, and most excavators have provided at least minimum counts of the number of burials present in a chamber when they were able to do so. Although these figures may often underestimate the actual number of individuals present in the burial group, they are useful in defining long-term changes in patterns of tomb use, especially in terms of group size and longevity. From a synchronic perspective, a consideration of the relative standardization (or, conversely, the range of variability) in tomb group size may also shed some light on the relative preeminence of large and long-established descent groups versus smaller and shorter-lived familial groupings whose membership may have been affected by social fissioning and/or social mobility.

## Interpreting variations in wealth and social prestige

As noted in the previous chapter, there are numerous channels of mortuary variation through which prestige differentials among individuals may be expressed or asserted, including the locational patterning and spatial associations of mortuary units, their relative elaboration and cost, the burial programs accorded to the deceased, and the range, quantity, and quality of associated grave goods. In general, variations in social status among collective burial groups may be monitored through some of the same dimensions of variability considered in studies of single interments, but a number of caveats apply. Factors of tomb location and architectural elaboration, for example, may be more closely related to the status and resources of the earliest users or builders of the tomb than to the status and socioeconomic resources of later users; reuse of the tomb may sometimes imply at most an expression of ancestral connections to the tomb. However, the redundancy and relative richness of status goods observed in successive chronological periods may give some indication of the extent to which wealth and social position were inherited.

Additionally, when the unit of analysis is shifted from the individual to the group, the comparative evaluation of grave assemblages is inevitably complicated by variations in burial group chronology, size, and age-sex composition, all of which are likely to have affected the range and quantity of artifact types present. Amongst all of these variables, the effects of differing age-sex composition on grave goods assemblages are by far the most difficult to establish in the Cypriot context. Nevertheless, as noted in Chapters 4 and 5, there are certain instances where it may at least be suggested that assemblage composition was influenced by gender, and future investigations may help to elucidate the dimensions of this variation. Meanwhile, differences in tomb chronology can be controlled for through the careful grouping of tombs for comparative analysis, and problems associated with differences in burial group size can be obviated to some extent, as discussed below, through the use of rough per capita wealth indices.

Regardless of whether single or collective interments were practiced, the prestige-related treatments accorded to the dead are likely to have been strongly influenced by the social aspirations of their survivors. The significance of synchronic variability in material expenditure must therefore be interpreted in the context of long-term variations in such expenditure. This should help to elucidate the relative importance of prestige competition in different periods. In drawing comparisons between tomb groups and assessing long-term trends in mortuary display, I have chosen to work with several quantitative indices of energetic and material expenditure or wealth. One is an estimate of chamber floor area, calculated by superimposing published tomb plans on graph paper, counting the number of squares enclosed and multiplying the total according to the scale of the drawing. Another is the gross value of 'pots per chamber', considered in conjunction with a per capita index of 'pots per burial', calculated for each chamber by dividing the total number of pots present by the estimated number of burials present. Clearly this need not reflect the actual number of pots associated with any one person in the tomb; however, it permits comparison of relative mortuary outlays in tomb groups with varying number of burials. Gross counts of 'copper-based items per chamber' and the per capita index of 'copper-based items per burial', similarly calculated, are especially valuable in assessing patterns of expenditure during the Early–Middle Cypriot periods. Total gold weight per chamber is generally more useful in evaluating patterns of wealth disposal in the Late Cypriot period, particularly in the large mortuary sample from Enkomi, since bronze consumption was relatively low among most tomb groups between LCIB and LCIIC.

It has also been suggested that the degree of structure or intercorrelation between variables such as tomb location, tomb construction, the richness and diversity of associated tomb goods, and the internal redundancies of tomb assemblages may be an important measure of the structure or hereditary rigidity of the status hierarchy and its associated system of mortuary distinctions. In order to assess the relationships between these variables, assemblages of grave goods, largely from intact and well-documented burial groups, have been charted with measures of tomb size and locational data wherever possible to allow comparison of specific distributional and associational patterns, and statistical measures of correlation and association have been calculated where appropriate.

The foregoing analyses ultimately permit a final, qualitative phase of discussion, in which the ideological symbolism of burial goods may be interpreted in conjunction with their distributional patterns, iconographic content, contextual associations, and the ongoing social changes observed in other aspects of the archaeological record.

## A note on chronology

Throughout this work I have chosen to use the traditional terminology for chronological subdivisions of the Cypriot Bronze Age, namely the 'Early', 'Middle', and 'Late Cypriot' periods (see Table 1.1 for specifics), defined and further subdivided on the basis of changing frequencies of ceramic wares or shapes and anchored, in terms of absolute dating, mostly by synchronisms with other Mediterranean regions suggested by associated imported goods. As Knapp (1990a; 1994a; 1994b) and others have observed, the divisions of the conventional ceramic chronology do not coincide neatly with broader complexes of cultural transformations, and it is in many respects more logical (and, indeed, safer, given the ongoing revisions of regional chronologies) to work with more general cultural-chronological divisions such as the prehistoric Bronze Age (with PreBA I encompassing the so-called 'Philia Culture' and the 'canonical' Early Cypriot I–II, and

PreBA II equivalent to the conventional Early Cypriot III–Middle Cypriot II) and the protohistoric Bronze Age (in which ProBA I equates to the transitional MCIII–LCI periods, ProBA II equates to Late Cypriot II, and ProBA III equates to Late Cypriot III). The merits of this schema notwithstanding, in the present context I suspect that the redesignation of hundreds of tombs currently known to specialists by the traditional chronology would only perpetuate confusion, and would also, in some cases, unnecessarily forego the fine-grained chronological controls which the nuances of the prevailing system permit.

In assigning tombs to specific periods, I have adhered largely to the datings proposed by those who excavated or published the tombs, taking into account subsequent revisions proposed by Stewart (1962a), Åström (1957; 1972a; 1972b), and Kehrberg (1995) among others. However, both the relative and absolute chronologies of the Cypriot Bronze Age are constantly being modified with the expansion of regional exploration, the publication of stratified settlement deposits, and the increasing use of radiocarbon dating (e.g. Manning *et al.* 2001). Therefore, I have tried to ensure that the groupings of tombs treated as contemporaneous are relatively broad in definition, so that minor revisions of their dating should not occasion major revisions of the interpretations advanced here concerning variations within and between periods.

# 4 The Early and Middle Bronze Age

## Introduction

The Early and Middle Bronze periods in Cyprus pose a major contrast to the preceding Chalcolithic era in almost every aspect of material culture and socioeconomic life. The settlements of discrete, monocellular 'roundhouses' or huts predominant throughout the Neolithic and Chalcolithic periods gave way to agglomerative villages of multi-roomed rectilinear houses (Swiny 1989) that were almost certainly characterized by different modes of household production, kin and community organization, and strategies of wealth accumulation (Flannery 1972). Cattle were reintroduced to the island early in the period, several thousand years after their initial appearance and demise (Peltenburg *et al.* 2000), augmenting the traditional faunal complement of domesticated sheep, goat, and pig. Ceramic models of plowing scenes (Dikaios 1940: Pls 9, 18) indicate that cattle were exploited as draft animals as well as sources of meat and milk in what Knapp (1990a, following Sherratt 1981; 1983) has interpreted as the Cypriot 'Secondary Products Revolution'. The donkey was also introduced during the Early Bronze Age, providing another source of animal traction and transport. The use of copper-based metal objects, scantily attested in the Chalcolithic, increased significantly as metallurgy came to be practiced more extensively. Further technological developments are evident in the manufacture of 'Philia' and other regional variants of the Red Polished ceramic tradition (Stewart 1962a; Bolger 1983; 1991; Swiny 1991), in the first use of mould-made mudbricks on the island, and in the appearance of distinctive artifacts associated with spinning, weaving, and food processing ( Frankel *et al.* 1996; Webb and Frankel 1999). In conjunction with these changes in technology and social life, the early prehistoric burial tradition of single inhumations in intra-settlement pit graves gave way to the use of extramural cemeteries of pit and rock-cut chamber tombs that often contained multiple burials. Funerary celebrations became more protracted and elaborate, sometimes involving rites of secondary treatment and reburial. Some tombs were very large, ornate, and labor-intensive in their construction, and many were filled with rich collections of metal and ceramic goods, contrasting with the more austere burial traditions of earlier periods.

The causes of these cultural transformations have long been a subject of debate. Early theories emphasizing the role of stimulus diffusion, linked with trade or population movements from Anatolia (e.g. Stewart 1962a; Watkins 1981; Swiny 1986a; Mellink 1991; Dikaios 1962; Catling 1971), were countered in the early 1990s by models of indigenous transformation driven by the dynamics of local power relations (Knapp 1990a) and the emulation of Anatolian elite prestige symbolism (Manning 1993; Manning and Swiny 1994). More recently, Frankel *et al.* (1996) and Webb and Frankel (1999) have renewed the case for migration, arguing that the wide range of innovations in technology, coincident with other cognitive and behavioral transformations in burial practices and the curation of portable goods, can best be accounted for in terms of colonization (see also Peltenburg 1996:27; cf. Knapp 1999:81; 2001). According to this model, occurrences of the so-called 'Philia Culture' or 'Philia facies' currently dated to 2500–2350 BC

(Webb and Frankel 1999:5) mark the presence of immigrants from Anatolia, primarily the southwestern region, although some eastern Anatolian cultural attributes have been noted as well. These colonists are thought to have established their settlements in the west, southwest, and central parts of the island, evidently targeting not only the copper-rich foothills of the northwestern Troodos but also promising agricultural zones and key locations along transport routes and coastal outlets (Webb and Frankel 1999:7, 40). Webb and Frankel postulate a largely peaceful mode of interaction based on exchange among indigenous communities and newcomers, with the latter initially displaying 'a high degree of identity maintenance' (1999:38) that eventually gave way to the assimilation and integration of local groups by the onset of the Early Cypriot period, *c.* 2300 BC (Table 1.1).

Both the new migration model and prior discussions of EC–MC cultural change inspire important questions for the interpretation of burial evidence. First of all, to what extent do the new forms of mortuary practice noted represent customs transplanted from the mainland as part of the immigrants' cultural baggage, and to what extent do they represent new ideologies and modes of practice developed and deployed by heterogeneous communities interacting within an increasingly complex social landscape? Second, what do these practices, in their initial form and as they evolved over time, tell us about the social structures, strategies, and belief systems of older and newer Cypriot communities? Do the architecturally elaborate and metal-rich tombs of the Philia Culture cemetery of Vasilia *Kafkallia* and the ostensibly later EC–MC cemetery of Lapithos *Vrysi tou Barba* mark the presence of an hereditary elite, one that was perhaps already established among newcomers and adopted by older indigenous communities as some groups began to exploit new forms of prestige symbolism and technology (Manning 1993; Manning and Swiny 1994)? Does the depiction of a male figure presiding over a ritual scene in the famous Vounous 'sacred enclosure' model represent the existence of an hierarchical social order dominated by a ritually sanctified leader, drawn from a status-conscious, weapon-bearing elite whose notions of the social and cosmological order had Mesopotamian antecedents (Peltenburg 1994)? Or was EC–MC society characterized by the absence of political and settlement hierarchies (Swiny 1989) and a relatively low level of socioeconomic differentiation, as other recent studies (Davies 1997; Baxevani 1997:65–66) imply?

These questions are addressed throughout this chapter as the spatial dimensions of mortuary practice, the ritual treatment of corpses, the grouping of the dead, and the changing patterns of mortuary expenditure and material symbolism expressed in tomb architecture and grave goods are examined in detail. Based on this analysis, I argue that EC–MC mortuary practices, while drawing to some extent on a lexicon of cultural practices and symbolic forms encountered in neighboring regions, evolved into a unique and locally rooted system that may initially have distinguished but ultimately united communities of recent immigrants and older inhabitants in a broadly shared complex of social representations and ancestral ideologies. Moreover, I propose that it is not so much the emergence of an hereditary elite that is represented in increasing levels of mortuary expenditure, but the increasing centrality of mortuary ritual as an occasion for prestige competition and the negotiation of social identities, statuses, and alliances at both the local and regional levels. One of the most far reaching consequences of the intensification of prestige competition in mortuary ritual was the concomitant intensification of metallurgical production and the expansion of interregional exchange networks during the Early–Middle Bronze periods, setting the stage for further socioeconomic change at the onset of the Late Cypriot Bronze Age, an argument which is further developed in Chapters 5 and 6.

## The spatial dimensions of Early–Middle Bronze Age mortuary practice

One of the most frequently noted contrasts between EC–MC mortuary practices and those of both earlier and later periods is the consistent use of extramural cemeteries, or burial grounds that were spatially separated from the habitation areas of the living community. During the Middle and Late Chalcolithic periods, numerous burials of adults and children were located adjacent to or within contemporaneous or abandoned buildings, as attested at the extensively investigated settlement sites of Lemba *Lakkous* and Kissonerga *Mosphilia* (Peltenburg *et al.* 1985; 1998). Nevertheless, extramural cemeteries of bottle-shaped, rock-cut shaft tombs were in use during the Middle Chalcolithic period at Souskiou *Vathyrkakas* and *Laonas* (Christou 1989), and Peltenburg suggests that many of the adults residing at Kissonerga *Mosphilia* in Period 3b (equated with the Middle Chalcolithic) may also have been buried off-site, whether at Souskiou, some 20 km to the east, or at other local cemeteries (Peltenburg *et al.* 1998:85). Therefore, the Philia Culture practice of extramural burial was not without local precedents, a circumstance that may ultimately have contributed to its more general adoption in the context of social interactions between locals and newcomers imbued with cultural traditions of extramural burial. The mortuary customs of western Anatolia in the Early Bronze Age typically entailed the use of extramural cemeteries, although the burials were usually placed in jars or pithoi rather than in the pit and chamber tombs characteristic of Cyprus (Wheeler 1974). In eastern Anatolia and Syria, many different types of burial have been observed, including extramural cemeteries of pithos burials, as well as pit and rock-cut shaft tombs located in both intramural and extramural contexts (Carter and Parker 1995). Interestingly, while the majority of known Philia sites are indeed extramural cemeteries (Webb and Frankel 1999:8), some of the earliest manifestations of the Philia mortuary complex are the chamber tombs found in intra-settlement contexts in Period 4 at Chalcolithic Kissonerga *Mosphilia*. It might be speculated that Philia 'frontierswomen' and men at *Mosphilia*, newcomers who may have been welcomed and integrated as members of the local community, chose to express both their distinctive identity and their claims of belonging through the use of chamber tombs inside their adopted settlement, where traditional intra-settlement burials seem to have resumed (Baxevani 1997:65; Peltenburg *et al.* 1998). Alternatively, assuming that local extramural cemeteries of Late Chalcolithic date existed but remain undiscovered, it is possible that immigrants were deemed ineligible for burial within the communal cemeteries. In other parts of the island where immigrants were sufficiently numerous to found entire communities by themselves, mainland traditions of extramural burial may have been more readily and immediately perpetuated.

The scattering of querns and other artifacts in some Early–Middle Cypriot cemetery sites suggests that not all domestic activities were completely segregated from burial zones, but the intermingling of tombs and houses characteristic of both earlier and later periods of Cypriot antiquity is not apparent (Swiny 1981:79; Coleman 1996:18). Most cemeteries were located within a few hundred meters of adjoining residential sites, although greater distances may have prevailed in cases where contemporary occupation areas have not been identified (Swiny 1981:79 and notes 91, 92). Tombs were cut in a variety of topographical settings, often along ridges and hillsides or in lower lying areas that would have been readily visible from nearby settlements. With the exception of the tumulus-covered site of Korovia *Paleoskoutella* (Gjerstad *et al.* 1934), none of the EC–MC cemeteries known to date can truly be described as 'monumental', but with or without above ground markers, they must have served as important local landmarks and persistent reminders of the dead.

Burial sites varied considerably in scale, from small clusters of tombs that were probably used by a few closely related kin or residential groups—such as Ayios Iakovos *Melia* and Korovia *Paleoskoutella*—to spatially extensive cemeteries such as the Philia Culture cemetery at Vasilia

*Kafkallia*, reportedly extending over at least one and a half kilometers (Hennessy *et al.* 1988:25), and the Philia–EC–MC site at Dhenia *Kafkalla*, described as 'a vast necropolis' whose size 'proclaims it the cemetery of a prominent settlement—unless, indeed, it was the sacred burial ground of a province' (Åström and Wright 1962:225). Many settlements seem to have made use of several different cemetery sites over the course of their occupation. Marki *Alonia*, for example, overlooked five distinct burial areas with an estimated total of 786 tombs, all located within an 800 m radius of the settlement site (Frankel and Webb 1996:11; Sneddon 2002:9–12). At Erimi *Kafkalla*, surveyors counted a total of 227 dromoi in two closely spaced groups over a stretch of approximately 500 m (Swiny 1981:66, Fig. 6), while at Evdhimou *Amolo*, 282 dromoi opened onto an estimated 750 chambers in three spatially discrete locations (Swiny 1981:70, Fig. 7). In some cases, different clusters of tombs appear to have had distinctly different chronological components, as is evident at Bellapais *Vounous* sites A and B (Dikaios 1940; Stewart and Stewart 1950). In other cases it might be hypothesized that different social segments (e.g. kin groups) of the larger community made use of specific burial areas. However, there may also have been instances where discontinuities in the distribution of tombs, as well as variability in tomb form, were more simply attributable to the discontinuities and limitations of the local terrain suitable for tomb construction (e.g. Coleman 1996:113).

Tomb orientations varied considerably from site to site. North-south orientations were common at the north coast cemeteries of Bellapais *Vounous* (Dikaios 1940: Fig. 1; Stewart and Stewart 1950: Figs 3, 6, 174; Dunn-Vaturi 2003:177) and Lapithos *Vrysi tou Barba* (Gjerstad *et al.* 1934: Plan V: 1; Herscher 1975), but most of the tombs at the geographically intermediate site of Karmi *Palealona* seem to have been oriented from east to west (Stewart 1962b: Fig. 1). At all three of these sites, a few tombs had dromoi that were perpendicular to the majority of other tombs, and at Ayios Iakovos *Melia* in northeastern Cyprus, there seems to have been a roughly equivalent proportion of tombs—all within the same immediate vicinity—oriented either approximately south-north or approximately east-west (Gjerstad *et al.* 1934: Plan V:3). Swiny (1981:79) has opined that tomb orientations in southwestern Cyprus and elsewhere were entirely random. To some extent, the directional placement of both chambers and dromoi orientations may have been determined by the peculiarities of the local topography, but it is also conceivable that some form of social symbolism, such as assertions of lineal identity, was expressed in opposing tomb orientations. However, the limitations of the present archaeological sample make it difficult to adduce any further evidence—in the form of distinctive architectural elements, artifact styles, or burial practices—that might support the latter hypothesis.

It is important to note that exceptions to the pattern of extramural burial during the EC–MC periods have recently been observed at Marki *Alonia*. At this settlement site, single interments of two adults, one identified as female, were discovered in burial pits in seemingly abandoned rooms of the settlement (Frankel and Webb 1997:88; 1999:90). Fragmentary skeletal remains of an adult female and three juveniles not associated with a burial pit were found in another abandoned habitation area (Moyer 1997:111–15), and the cranium, mandible, and a few postcranial bones of an additional individual were recently encountered on what may also have been an abandoned habitation floor (Frankel and Webb 2000b:70). Perhaps the circumstances associated with the abandonment or dissolution of the associated domestic units resulted in the suspension or indefinite postponement of standard mortuary rites involving chamber tomb burial, at least in the case of the adults represented. It is also possible that the human remains from these settlement units are rarely preserved examples of alternative mortuary treatments, that is, of modes of disposal accorded to individuals who were typically excluded from chamber tomb burial on the basis of age, sex, manner of death, or other social factors. This explanation may be especially relevant given the preponderance of juvenile and adult female remains (Frankel and Webb 1999: 90). Infants and children are notably underrepresented in EC–MC chamber tomb contexts, and there are slight, although as yet unsubstantiated, indications that females were also underrepresented, as discussed in more detail further on.

The identification of either local or foreign precedents for the use of formal cemeteries, as well as for other aspects of mortuary ritual, may help to elucidate the source of traditions that were modified and redeployed in Early–Middle Bronze Age Cyprus, but it does not answer the fundamental question of why particular practices were maintained and more generally adopted at this time. Most changes in mortuary practice, including the shift to extramural burial, must have been articulated with a much broader complex of ideological and socioeconomic developments underway simultaneously. The practice of plow agriculture and attendant changes in patterns of wealth accumulation, for example, would have had important ideological ramifications, particularly with regard to perceptions of the value, permanence, and heritability of movable and immovable property (Bogucki 1993; Goody 1976). Houses, land, and domestic animals were now probably passed down through multi-generational family groups, and, as Webb (1995) has noted, various types of portable household goods were also more often retained than discarded when settlements were abandoned, in contrast to Chalcolithic practice. EC–MC communities were thus becoming more akin to 'delayed' return societies (see Woodburn 1982 and discussion in Chapter 2), placing a greater reliance on the hereditary transmission of economic and social capital than the comparatively 'immediate' return societies of the early prehistoric period, and their mortuary rituals were probably elaborated concomitantly. Under these circumstances, the establishment of formal burial grounds for the disposal of the dead—spatially reserved precincts where the remains of the ancestors would not be disturbed in the course of long-term household and settlement occupation—may have been viewed as a logical necessity.

Meanwhile, rising population densities, coupled with the larger land areas required for plow cultivation and cattle pasturage, could have created a perception of land scarcity and a growing emphasis on social boundaries or group membership, a phenomenon perhaps symbolized in part by the development of diverse regional ceramic traditions (Frankel 1974; Herscher 1976; 1981; 1991). These concerns may have received further expression through the establishment of corporate burial areas, by reference to which both indigenous and immigrant communities could affirm a permanent connection between the ancestors, the land in which they were buried, and the territory, antiquity, and communal identity of their descendants. And concurrently, as mortuary ritual became an increasingly important forum for the assertion or negotiation of those connections (and of social status in general), major elaborations of mortuary celebrations, involving extensive communal and even multi-communal participation in funerary rites, would also have favored the establishment of open, public, and permanent spaces in which those rites could be conducted.

## Programs of mortuary ritual

Diverse and complex forms of mortuary treatment are represented in many Early–Middle Cypriot tombs. While some tombs were probably the repositories of sequential primary interments, as some researchers have assumed (see most recently Webb 1992:88), others contained one or more secondary burials, and some were apparently used for several simultaneous secondary burials. Secondary treatment of infants and children appears to have been practiced at Chalcolithic Lemba *Lakkous* (Niklasson 1991:186–7) and Kissonerga *Mosphilia* (Peltenburg *et al.* 1998.85), and at least during the Middle Chalcolithic, this treatment may have been 'honorific', given the other indications of special treatment for young individuals (e.g. 'libation hole graves' containing infant burials: Peltenburg *et al.* 1985:114–8, 327–8; also the inclusion of picrolite 'fertility' pendants and other exotica in various child burials: Peltenburg 1992; Peltenburg *et al.* 1998:91; Baxevani

1997; Bolger 2002). Adults may also have been subject to rites involving exhumation and reburial in more elaborate Chalcolithic tombs such as those of Souskiou *Vathyrkakas* (Christou 1989; Niklasson 1991:189; Peltenburg *et al.* 1998:85, 246), where adults significantly outnumbered children (Lunt 1994). Although multi-stage burial programs are certainly attested in adjacent regions of Anatolia (e.g. Carter and Parker 1995:110) and the Levant (Chesson 1999; Porter 2002; see also Byrd and Monahan 1995; Wright 1978 for much earlier examples), the development and widespread adoption of these practices in EC–MC Cyprus, as with the shift to extramural burial, may be explicable at least in part as an ongoing elaboration of local Chalcolithic practices.

The evidence for multi-stage ritual programs in EC–MC Cyprus derives from the observation of emptied mortuary features that may have served as temporary interment facilities, the disarticulated or incomplete condition of skeletal remains not attributable to post-depositional processes alone, the occasional occurrence of multiple simultaneous burials, and other supporting information relating to the stratification (or lack thereof) of burials and grave goods (see summary in Table 4.1). The following analysis begins with a discussion of the evidence from sites where human burial remains are best-documented, followed by a consideration of how the ritual system may have worked and its ideological significance.

## Bellapais *Vounous*

Despite the effects of repeated flooding, many well-preserved burials are described in the accounts of the *Vounous* cemeteries published by Dikaios (1940), Stewart and Stewart (1950), and Dunn-Vaturi (2003). Roughly two-thirds of the burials found *in situ* at the predominantly ECI site *Vounous* A were placed on their left sides with the legs flexed, often with their arms folded across the body and raised towards the face. Other burials were often found in dorsal positions with the legs flexed to varying degrees. At ECII–MCII *Vounous* B, all but a few of the articulated burials were laid out in dorsal positions, some with one or both legs flexed, others with both legs extended. Burial orientations and their relative frequencies are difficult to determine from the accounts of *Vounous* B, but at *Vounous* A it appears that most of the burials were oriented roughly north-south, although some instances of the opposite orientation were noted as well. Occasionally the deceased seem to have had small bowls placed in their hands (e.g. Tombs 17 and 19 at *Vounous* B, Dikaios 1940:39, 43), and in at least one instance a pair of burials were apparently arranged so that they were holding hands (*Vounous* B Tomb 19).

Many of the intact and apparently articulated burials observed in the *Vounous* cemeteries were probably primary interments, and many of the disarticulated burials may have been disarranged when they were pushed aside or relocated to make room for subsequent interments. There are, however, a number of other cases where practices involving secondary treatment would seem to be indicated. These include several tombs in which some of the individuals present were represented by skulls only, a few tombs in which simultaneous burials seem to have taken place, several intact tombs containing no articulated burials, and still other tombs in which body parts seem to have been deliberately rearranged after decomposition had occurred (Table 4.1). One notable example of the last-mentioned phenomenon comes from Tomb 164A at *Vounous* A, where two more or less complete skeletons were found missing their skulls, which had evidently been repositioned some distance away from the bodies within the same chamber (Stewart and Stewart 1950:227, Fig. 165). According to the excavators:

> Since the skeletons lacked skulls, and vice versa, and in the absence of further human remains, it is probable that the skulls belonged to the bodies but it is not clear how they attained their present positions, as symmetrical as the corpses, but in front of both. While it was possible that it was due to water action, it seems extraordinary that either or both should have successfully negotiated the

solid barrier of pottery by the head of B. Mr. Rix carefully examined the remains for evidence of decapitation, but none was forthcoming (Stewart and Stewart 1950:226).

It is possible that the corpses were interred with skulls intact and at some later date, probably after the flesh had decomposed, the tomb was reopened and the skulls were deliberately separated from the bodies. Alternatively, prior to interment the dead may have been subjected to some sort of treatment which involved the separation of the skulls from the remainder of the skeletons. Another instance of ritual practice involving the rearrangement of disarticulated bones may be represented in Dikaios' Tomb 31, where three skeletons were found, all reportedly decomposed and disorderly, one in the southwest, the others towards the front of the tomb. Their positions were unclear because, according to Dikaios (1940:65), 'most of the long bones were found grouped together'. This could indicate secondary burial after decomposition with deliberate grouping of the body parts, or some other type of post-interment manipulation.

Also important is the case of Dikaios' Tomb 36, which contained the remains of at least nine individuals, the largest number of burials recorded in a single tomb at *Vounous*. Dikaios reports a large heap of human bones representing several burials 'grouped anyhow' in the foreground of the chamber; these could not be related stratigraphically to the burials in the west 'as even these were in no order'. Most significant is Dikaios' observation that no single individual in the uppermost level of burial remains was found in an articulated condition. There in fact appears to have been only one articulated or *in situ* skeleton in the entire tomb, located in the lower stratum of finds beneath many pots. Despite the stratigraphic position of this individual, Dikaios concluded that this must have been the last burial in the tomb (1940:72–4). It is possible, alternatively, that Tomb 36 contained several secondary burials, some perhaps interred *en masse* after one or more primary burials had taken place.

The evidence from both *Vounous* A and B suggests that a program of secondary treatment placing special emphasis on the manipulation and curation of the skull was practiced for at least some individuals. These were presumably adults, as the excavators did not indicate otherwise. In some instances, secondary treatment seems to have been associated with individuals who were provided with large tombs and outstanding assemblages of grave goods, as in the case of Tomb 164A. This was the single largest chamber excavated at *Vounous* A, and it contained an impressive assemblage of grave goods including an unusual deep bowl or 'cult vessel' (no. 13, with modelled miniature scene depicting a trough, a jug, and a similar bowl), a ceramic horn, considerable quantities of ox bones and four copper knives (Stewart and Stewart 1950:228, 230–7). Similarly, Tomb 36 at *Vounous* B contained a diverse array of ceramic items including two horns, a ring vase, and several vessels with modelled bulls' heads, snakes, birds, horned animals etc., along with a large part of an ox skeleton and other faunal remains (Dikaios 1940:72–6). Tomb 143 at *Vounous* B, in which eight skulls and very scanty postcranial remains were found, yielded a large collection of eleven copper-based artifacts, remains of sheep or goat, and an unusual stone axe amulet and stone tablet (Stewart and Stewart 1950:324, 328, Fig. 242). However, as is apparent in the cases of Site A Tomb 82B and Site B Tombs 44, 125, 141, 31, 146 (see Tables 4.7a–c for the distribution of goods), not all cases of secondary treatment were associated with elaborate tomb construction or distinctive concentrations of valuable grave goods. It is possible that the ideological-ritual complex represented involved extended 'reverential' treatment for important kin group members, regardless of their wealth or prestige relative to their peers within the community.

## Lapithos *Vrysi tou Barba*

Accounts of the Swedish Cyprus Expedition excavations at the ECII–MCII/III cemetery at Lapithos *Vrysi tou Barba* in 1927 (Gjerstad *et al.* 1934) and the University of Pennsylvania Expedition excavations conducted in 1931–32 (Grace 1940; Herscher 1978) reveal a considerable diversity of burial practices. Among the burials whose original postures could be determined, lateral flexed or contracted positions predominated, with roughly equal numbers of individuals placed on their right and left sides, but dorsal burials with legs flexed (e.g. Gjerstad *et al*. 1934:72) and individuals placed in squatting positions were also found (Herscher 1978:786 and note 23; Myres 1940–45:79–80). Numerous individuals are described with distinctive arm positions, such as the person in T302A said to have had the hands placed under the head grasping two knives, a scraper and a tweezer (Gjerstad *et al*. 1934:42), the skeleton of a child in T836B with hands placed under the head (Herscher 1978: 675), the burial in T837A holding a bowl in the right hand (Herscher 1978:685), and the skeleton in T829A whose hands appeared to have been tied between the legs (Herscher 1978:572–3). Some of the other individuals just described may also have had their body parts bound into certain positions.

Most reports of the burial remains at Lapithos explain the disarticulation or scanty preservation of skeletal remains either in terms of flood damage or the disarrangement of earlier burials in the course of sequential primary interments. While this may often be correct, there are nevertheless, as at *Vounous*, some cases in which more complex programs of mortuary ritual would seem to be attested. These include tombs such as Swedish Tomb 302B, where four burials found in different parts of the chamber, all having the appearance of disarticulated or partly disarticulated bone piles (Gjerstad *et al*. 1934:42), and others such as Swedish Tomb 306A where a single individual represented only by fragments of the head, neck, and arms was observed (Gjerstad *et al*. 1934:60) and Swedish Tomb 312A, in which one of two individuals present was represented by fragments of the vertebral column only (Gjerstad *et al*. 1934:84, 77, Fig.37:7). Other tombs were described as containing the skeletal remains of two or more individuals in extremely variable states of preservation, perhaps indicating that those which were incompletely preserved had been subjected to some form of secondary treatment (Table 4.1).

Some of the mortuary features recorded at Lapithos may illustrate forms of primary interment that preceded the exhumation and reburial of the dead. As enumerated in Table 4.1, there are numerous chamber tombs where fragmentary skeletal remains were found lying on top of or mixed up with pithos sherds that were sometimes embedded in the chamber floor or propped up with small stones. Two of the 'pithos' burials were clearly identified as children (one individual in Tomb 311B, Gjerstad *et al*. 1934:79–80; another in Tomb 823, Herscher 1978:468); the ages of the others are unknown. It is possible that pithoi were sometimes used as containers for primary burials of young individuals, then broken to allow removal of the remains in subsequent ceremonies of reburial, possibly occurring in conjunction with other adult burials. Pithos burials were most commonly associated with Early Cypriot ceramics, and may represent the continuation of Philia practices observed at Kissonerga *Mosphilia* and Marki *Alonia* (Peltenburg *et al*. 1998:72; Frankel and Webb 2000:764, both involving children) and others alluded to at Philia *Laxia tou Kasinou* (originally known as Philia *Vasiliko*) and Vasilia *Kafkallia* (Dikaios 1946; Hennessy *et al*. 1988:29). Traditions of pithos burial also have Anatolian antecedents (Wheeler 1974; Carter and Parker 1995).

Other features that could have served as temporary burial facilities for infants and very young children are dromos 'cupboards' or very small chambers cut in the dromos walls of many tombs. Although the bones of children may have a lower aptitude for archaeological preservation than those of adults, it is curious that so many of these features were devoid of skeletal remains (e.g. Pennsylvania Tombs 804 with 17 of 18 cupboards containing finds but no bones, Tomb 829d, e, f, and Tomb 835c as described in Herscher 1978:91–3, 571–2, 661; Swedish Tombs 307, 315, 316, 318, 322C as described in Gjerstad *et al*. 1934:63, 106, 115, 126, 144). It is possible that the burials originally placed in the dromos cupboards were sometimes removed after the original interment and redeposited in the principal tomb chambers when these were reopened for the burial of adults. There is some direct evidence for this practice in Late Bronze Age tombs at Myrtou *Stephania* (Tombs 3 and 5; Tomb 14 side and main chambers, Hennessy 1964) and Kalavasos *Ayios Dhimitrios* (Tomb 11; South 2000:349–53).

Pit graves may also have been used as temporary or primary interment facilities at Lapithos, possibly for adults as well as children. In the appendix of her dissertation, Herscher noted that a series of shallow rectangles, averaging 1 m in length and 0.5 m in width, with a preserved depth of *c.* 20 cm, were found over the entire exposed area of bedrock, arranged in fairly regular rows (illustrated in Herscher 1975). These pits were sometimes cut by the dromoi of the chamber tombs, and they would seem to have been dug prior to the construction of many, if not all, of the chamber tombs. Deposits on the rock surfaces of the rectangular features suggested that they had been covered by a layer of soil for at least a few years before the chamber tombs were cut. Herscher observed that the cuttings were large enough to have served as shallow pit graves for contracted burials, but as they were all found empty, she concluded at the time of writing that they were agricultural features (Herscher 1978:818–9). However, in light of the total corpus of mortuary evidence from Lapithos and other Early and Middle Cypriot sites, it is quite possible that they were in fact mortuary features from which all of the burial remains had been removed and redeposited in adjacent chamber tomb complexes. The emptying of earlier burial features on occasions of collective secondary burial rites would help to explain the chronological anomalies observed in some chamber tombs, such as 812A, dated architecturally to ECIII, in which the long duration of use suggested by the pottery (ECI–MCI) is at odds with the absence of evidence for stratigraphic separations or disturbances among the seven burials found (Herscher 1978:296–7, 311–7).

As indicated in Table 4.1, there are several other chamber tombs, in addition to Tomb 812A, in which the disposition of the burials and tomb deposits suggests that simultaneous and probably secondary interments had taken place. Some cases of synchronous interments were associated with small to medium-sized tombs that were fairly poor in metal artifacts (e.g. Pennsylvania Tombs 812A, 826A, 832A; see Tables 4.11a and b), suggesting once again that this practice was not exclusively the prerogative of groups of great wealth or high social status. One of the most outstanding examples of collective secondary burial, however, comes from the unusually large and rich Swedish Tomb 313, yielding a total of 134 copper-based objects, 80 from Chamber A alone. This tomb also contained the largest number of burials documented for a single complex, with an estimated total of 26. The 15 burials in chamber 313A were very carefully arranged in the niches and in a circle upon the floor of the chamber, with no evident stratigraphic separation or disturbance that might have resulted from reuse (Gjerstad *et al.* 1934:87, Fig. 43). The excavators postulated lateral contracted or squatting positions for most of the burials, although the preservation of the bones appears to have been very fragmentary. The extraordinary size, form, and wealth of this tomb all suggest that it was used by a group of considerable social prominence and resources, and it is conceivable that the 15 burials in chamber A represent one or two extremely ostentatious mortuary events, involving the exhumation and reburial of several individuals at the same time, in a celebration staged to demonstrate the wealth and prestige of the kin group while honoring its ancestors.

Finally, other manifestations of postmortem rituals and complex physical processing of the deceased may be attested in Swedish Tomb 322A, an unlooted chamber in which some scanty human bone remains were found on the floor, and a few finger bones, which have a low aptitude for archaeological preservation and recovery, were found in the niches. This was the single largest chamber found at Lapithos, with a floor area well over 20 sq m and a ceiling height of perhaps as much as 2 m. It was also part of one of the richest tomb complexes discovered at the site (see Table 4.11c), yielding 82 copper-based artifacts, ornaments of gold and silver, cult vessels, idols, and bones of cattle and equid (allegedly horse; see Stewart 1962a:287; Reese 1995:38). As Gjerstad writes:

> Splendid funeral ceremonies must, surely, have taken place in connection with many of the burials. The great quantity of animal bones (horse and bull) in Chamber B and E, the fine collection of cult vessels, the fetish stone and the marble idol…in Chamber D, bear eloquent witness of the cult of the deceased. The presence of gold and the many bronzes in Chamber A, as well as the magnificent

construction of Chambers A and B give evidence of the wealth of the burials (Gjerstad *et al.* 1934:147).

It is possible that pre-interment processing (e.g. primary burial and exhumation) and/or subsequent secondary or even tertiary post-interment ritual affected the representation of bones in this tomb. Chamber 322A and several other tombs at Lapithos were large enough to allow several living individuals to move around inside and perhaps to engage in ceremonies that hastened the destruction of previously interred skeletal remains, conceivably on the occasions of successive primary or secondary burials.

The variable condition of the skeletal remains, the diversity of mortuary features, and the depositional histories of the Lapithos tombs suggest that secondary treatment of the dead was a common practice at this site. It is possible that many individuals were not interred in chamber tombs immediately following their demise, but were stored in other temporary facilities—pithoi, dromos cupboards, and pit graves—for a certain period of time. This would have given groups of both greater and lesser resources a chance to accumulate the necessary supplies to stage a proper funeral celebration, and/or to construct a new chamber tomb. Some groups may then have initiated the use of a new tomb by exhuming and reinterring not just one but several of their previously deceased members collectively. Still other post-interment rituals involving the handling and rearrangement of human bones may sometimes have been carried on within the tomb chamber subsequently, as suggested in the case of Swedish Tomb 322A.

## Middle Cypriot tombs at Ayios Iakovos *Melia*

There are strong indications of secondary treatment and post-interment manipulations of skeletal remains in tombs of both Middle and Late Cypriot date at Ayios Iakovos *Melia*, excavated by the Swedish Cyprus Expedition (Gjerstad *et al.* 1934). The most striking example from the Middle Cypriot period occurs in Tomb 6, which may have been used for a large-scale mortuary celebration involving multiple secondary interments. Fifteen burials were discerned in this tomb, with associated finds dating from the end of MCI to early MCIII (Åström 1972a:179, 189). Eleven skeletons were found near the walls of the chamber (I–XI) and four more (XII–XV) were placed in the central area, probably later in the sequence of interment as they obstructed access to those around the walls (Gjerstad *et al.* 1934: Fig. 124:2). The excavators believed all of the skeletons had been seated in crouching positions, as otherwise there would have been very little room for the bodies; the space available for each individual amounted to only 70 × 90 cm per burial. Skeleton VII, according to the excavators, had its knees drawn up at a sharp angle, the tibia partly covered. Its spinal column was preserved in two pieces; the skull was found near the wall. Skeletons VIII and IX may have been placed in similar positions, but their remains were less well-preserved. The burials were surrounded by tomb goods which had apparently been left *in situ*. There is little evidence for the disturbance of the deposits in the course of reuse, or for any stratigraphic separation between them, nor did the dromos stratification show any evidence for multiple episodes of clearance and refilling (Gjerstad *et al.* 1934:315–7). The systematic placement of the skeletal remains in a circle around the walls of the chamber and in the center is rather similar to the orderly arrangement of the burials already noted in Lapithos Tomb 313A and also in Korovia *Paleoskoutella* Tomb 7, discussed further on. The wide chronological range of the finds might be explicable in terms of the exhumation and reburial of several persons of different generations.

Burial remains in Tombs 1 and 4 would seem to attest to other types of post-interment ritual and/or episodes of ancient looting. In Tomb 1, masses of skeletal remains representing 10–12 individuals were gathered along the south wall on top of a silted layer, partly covered by rock debris (Gjerstad *et al.* 1934: Fig. 119:1) The only finds from the tomb were some pots of small size hidden by the thick deposit of silt on the chamber floor. The stratification of the dromos indicated that the tomb was deliberately resealed after a possible episode of disturbance, which the excavators proposed was not a typical looting incident, but rather a clearing in preparation for future burials that never occurred (Gjerstad *et al*. 1934:304). The same explanation was proposed for Tomb 4, where both the principal chamber and the side-chamber contained very few finds and only fragmentary skeletal remains (Gjerstad *et al*. 1934: Fig. 124:1). Another possible explanation might be that portions of the burials and some of the associated goods were removed to inaugurate the use of a later chamber tomb whose users wished to unite their ancestors (as in Bloch's 1982 discussion of the Merina). Such practices could account for the highly variable condition of the four quasi-contemporaneous burials in Ayios Iakovos Tomb 7, where two fragmentary and two relatively complete burials were found (Gjerstad *et al*. 1934:322–5, Fig. 125:5). In the course of such rituals, it is possible that the original grave goods were selectively transferred to new tombs, left behind in the older tombs, discarded, and perhaps sometimes returned to use by the living.

## Korovia *Paleoskoutella*

The site of Korovia *Paleoskoutella*, with its diverse array of mortuary features surmounted by tumuli made of earth and stones, is unique amongst all of the Cypriot Bronze Age cemetery sites found to date. Although the construction of tumuli over stone cist graves seems to have been fairly common in late third millennium Palestine (Gophna 1992:141 and Fig. 5.7), such features have rarely been observed elsewhere on Cyprus, with a few examples from the Late Cypriot town of Enkomi, (see discussion in Chapter 5), and others at EC–MC Alambra *Mouttes*, where some tombs were reportedly covered by mounds of earth and stones (Decker and Barlow 1983:83). Possibly, as Webb (1992:94) suggests, erosion and plowing have often removed the evidence for such features at other sites. The *Paleoskoutella* cemetery appears remarkable nonetheless, dominated as it was by a very large tumulus (Tumulus 7) surrounded by several smaller tumuli set at regular distances from each other. Some of the tumuli were badly damaged or had almost vanished, but it was estimated that the entire cemetery originally contained about 20 such features (Gjerstad *et al*. 1934:416). The expedition investigated seven of the tumuli, including the largest one in the center, revealing both simple and elaborate chamber tombs along with other complexes of cuttings that may have been created in related mortuary practices. The associated finds have been variably dated to MCIII and LCIA (cf. Gjerstad *et al*. 1934:437–8; Catling 1963:157, no. 99; Merrillees 1971; Åström 1972a:194; Webb 1992:92–3), but given the compelling similarities between the mortuary practices observed here and at some of the EC–MC cemeteries just discussed, the site is treated in this chapter rather than in the next.

The diverse features encountered in the *Paleoskoutella* cemetery attest to a complex, multi-stage program of mortuary rituals. Tombs 2 and 5, both relatively small, crudely cut chamber tombs, had been cleared of all burial remains and goods except for a few sherds in antiquity. Their overlying tumuli were partially (in the case of Tomb 2, Gjerstad *et al*. 1934: Fig. 163:9) or entirely displaced (in the case of Tomb 5, Gjerstad *et al*. 1934: Fig. 166:3) when the tombs were emptied, and the chamber and dromos of Tomb 5 had been systematically packed with rubble. This filling, along with the complete removal of bones and artifactual material, supports the excavators' interpretation (Gjerstad *et al*. 1934:421–2, 426–8, 438) that the emptying of the tombs was a deliberate ritual act rather than an instance of recent looting. It is probable that Tombs 2 and 5 served as primary interment facilities for burials that were later exhumed and reinterred in more elaborate chamber tombs such as Tombs 4 and 7. Tomb 4 contained two very poorly preserved burials, thought to have been interred simultaneously (Gjerstad *et al*. 1934:423–5; 427, Fig. 166:1). Tomb 7 seems to have contained fourteen burials interred in a single event. Sjöqvist described their arrangement as follows:

In three of the niches lay two skeletons in lateral, contracted positions, and below the niches seven other skeletons were found; one niche was occupied by a single burial only. All had been deposited, so far as could be stated, in the same positions. Two of the skeletons in the niches and three of those on the floor were, however, very fragmentary. A striking feature of the deposit is that no skeleton had been removed in order to make place for another. Also, the fact that two skeletons were placed in each of the three niches is very unusual. As a matter of fact, all the fourteen bodies make the impression of having been brought into the chamber simultaneously (Gjerstad *et al*. 1934:431).

He also offered this further interpretation:

To my mind, it is inevitable to connect the heterogeneous burial of the large tumulus with the emptying of the smaller tombs, viz., their contents having been brought over to a common resting place below the central tumulus. Several of the burial remains in Tomb 7 give also the impression of having been brought into the chamber as mouldered corpses, or perhaps only as skeletal remains. This is the case, especially, with three of the bodies on the floor which, had we to deal with successive burials, ought to have been the latest, and best preserved (Gjerstad *et al*. 1934:438).

There is some direct evidence for the transferral of burial deposits, inasmuch as one of the Plain White fragments found in Tomb 2 came from the same jar as a sherd found in the main trench dug through Tumulus 7 (Gjerstad *et al*. 1934:438).

The features associated with Tumuli 1, 3, and possibly 6, may represent other activities carried on in conjunction with primary and secondary burial rites. Each of these tumuli covered a series of cuttings of variable extent. The complex beneath Tumulus 1 was the largest and most intricate, extending over an area of 9.10 × 5.50 m. It was divided into two sections into which a number of pits, some of which could have been postholes, were dug. Several of the features in the area were reported to contain strata of 'mouldered organic material' and/or bits of burned animal bone (Gjerstad *et al*. 1934:416–8). There were no complete finds from the complexes in either section except for a Black Slip II jug and a spindle whorl. Some 725 sherds, derived mainly from shallow bowls, were also recovered (Gjerstad *et al*. 1934:419). The complex of features beneath Tumulus 3 was roughly similar but smaller in scale (Gjerstad *et al*. 1934:423, 420, Fig. 163:11–13). Tumulus 6 covered an approximately rectangular shaft with an apsidal niche at the southeast corner, a circular pit in the northwest corner, and a low north-south rubble wall two courses high dividing the shaft into two unequal portions (Gjerstad *et al*. 1934:427, Fig. 166:4–6). The presence of burned animal bone and the large quantity of shallow bowls from beneath Tumulus 1 suggest that food preparation and feasting occurred in this area, lending support to Sjöqvist's interpretation of the Tumulus 1, 3, and 6 features as 'cult places' where sacrifices and other religious ceremonies were carried out (Gjerstad *et al*. 1934:419, 423). It is possible that these areas were used, in addition, for some phase of mortuary treatment involving exposure and defleshing of the deceased on structures such as platforms, scaffolding, or hammocks for which postholes were required. Some of the pits said to contain mouldered organic material could have been used to collect and/or burn the secretions of decaying corpses. It is also conceivable that the Tumulus 6 feature may have been a temporary grave, different in form from Tombs 2 and 5, with the low wall being used to separate different burials. It bears some resemblance to Korovia *Nitovikla* Tomb 3, also a rectangular shaft (with a small buttress/divider) found emptied of its original contents.

Sjöqvist believed that the features associated with Tumuli 1, 3, and 6 had been 'desecrated' prior to being covered with mounds, resulting in the removal of 'votive' gifts and covering of the area with thick layers of white clayey earth and rock matter (Gjerstad *et al*. 1934:437–8). However, if these features were also used as processing facilities for human remains, the white deposits may instead represent ritual 'purification' rather than 'desecration' of the area; many cultures which practice secondary treatment or 'bone-picking' regard the decaying flesh as unclean or profane (Metcalf 1982:178), and if this was the case at *Paleoskoutella*, a cleansing ritual may have been practiced. It has also been argued that the exhumation and reburials associated with Tomb 7 and its tumulus may have been undertaken in response to external danger (Gjerstad *et al*. 1934:438; Webb 1992:94), but given the widespread evidence for secondary treatment at other EC–MC sites and the close similarities between this tomb and others of varying date at Lapithos and Ayios Iakovos, it seems reasonable to conclude that the *Paleoskoutella* mortuary complex was instead a locally distinctive manifestation of a widely shared system of collective, multi-stage mortuary observances.

## Evidence for secondary treatment of the dead at other sites

Although the best documented examples of secondary treatment and reburial come from the sites just discussed—*Vounous* and Lapithos in the northwest, along with Ayios Iakovos and *Paleoskoutella* in the northeast—other likely occurrences may be discerned at various localities throughout the island. The cemeteries of Dhenia *Kafkalla* and Dhenia *Mali*, for example, although extensively looted over the years, nonetheless display some interesting variability in mortuary features that may be indicative of complex ritual processing as well as differences in wealth between groups. The majority of the ECIII–MC features reported by Nicolaou and Nicolaou (1988) were rather small tombs, seldom exceeding 2 m in length, and sometimes quite shallow. These features contained few finds or human bones *in situ* (e.g. *Kafkalla* Tombs 49, 163, 167–9; *Mali* Tombs 24–6). They contrast strikingly with three much larger tombs at *Kafkalla*, including Tomb 48, which contained at least 10 skulls and over 277 objects of ECIII–MCIII date (Nicolaou and Nicolaou 1988:72–93), Tomb G.W.1, which measured approximately 5 × 4 m and was reported to have been used for several burials of MCII date (Åström and Wright 1962: Fig. 6), and Tomb 6, an extremely large (9 × 4 m) niched complex representing the fusion of two tombs and containing approximately 175 objects of MCI–MCIII date (Åström and Wright 1962: Fig. 8). It is possible that the emptying and 'looting' of the smaller, poorer features noted above may have taken place in antiquity in conjunction with rites of exhumation and reburial in these larger and more elaborate chamber tombs. Possible evidence for other activities that may have been carried on in conjunction with ritual celebrations come from the recent discovery of *senet* and *mehen* type gaming boards cut in the bedrock at Dhenia *Kafkalla* (Herscher 1998:320), supporting the proposition that funerals were indeed occasions for social activity and entertainment as well as the mourning of the deceased.

Other instances of multi-stage ritual treatment may be represented in Politiko *Lambertis* Tomb 18, where the occurrence of a so-called 'mass burial' of 12–15 individuals, possibly an instance of collective secondary reinterment, was noted (Gjerstad 1926:81). This was one of the larger tombs found at *Lambertis*, measuring 3.5 × 3.5 × 0.75 m in size, and it was also one of the few tombs in which silver objects were reported at this site (Åström 1972a:151–2). Still other 'mass burials' may be represented in Nicosia *Ayia Paraskevi* Tomb 13, a small (2 × 1.6 m) feature in which Stewart observed nine skulls and long bones along with a few small pots and metal ornaments attributed to MCIII (Hennessy *et al*. 1988:20–21), and in Kalopsidha Tomb 28, where Myres (1897:146) noted a thick layer of bones including three skulls. Along the south coast, mortuary practices involving secondary treatment may be attested at the Philia–EC cemetery at Sotira *Kaminoudhia*, where only a few of the 21 tombs excavated had articulated skeletal remains and small cist graves containing artifacts but no human bones were found (cf. Swiny *et al*. 2003:143–4). The intentional manipulation of human bones in settlement contexts may perhaps be best attested by the triangular arrangement of two humeri and a femur on the floor of Unit 16 and the presence of a femur and a pelvis in Area 6 (Schulte-Campbell 2003:436; Swiny *et al*. 2003:27). In the ECIII–MCIII cemetery in Kalavasos village, several tombs with burials represented mainly by skulls were excavated by the Department of Antiquities (Karageorghis 1958; detailed here in Table 4.1) and other intact tombs containing only disarticulated human remains were excavated by the Vasilikos Valley Project (e.g. Tombs 39–40, Todd 1986:28–30).

## Ritual practices and ideology

The mortuary evidence discussed above suggests that rituals involving secondary treatment were widely practiced throughout the island during the EC–MC period. While some degree of regional variation may have been associated with the processing of the physical remains of the deceased (for example, treatment emphasizing the curation and reburial of the skull, sometimes to the exclusion of other body parts, is more apparent at *Vounous* and Kalavasos than elsewhere) and the types of primary interment facilities may have differed from place to place, there are nonetheless some marked resemblances in the overall systems of mortuary ritual practiced in diverse areas. This is perhaps best exemplified in the extraordinary examples of collective secondary interments associated with Lapithos Tomb 313A, Ayios Iakovos Tomb 6, and *Paleoskoutella* Tomb 7, each with 14–15 carefully spaced burials arranged either in the chamber niches or in circles around the perimeter of the chambers. Other instances of so-called 'mass burials' observed at Politiko, Nicosia *Ayia Paraskevi*, and Kalopsidha may represent similar large-scale, collective secondary interment rituals. The frequency of such events appears to have been on the rise during the Middle Cypriot period, occurring in conjunction with other aspects of mortuary elaboration evident in tomb architecture and assemblages of grave goods, as discussed further on.

Although collective secondary burial rituals appear to have been widely practiced, they were not necessarily the norm for all burials. There are many more tombs at *Vounous*, Lapithos, and other sites in which single or sequential primary interments occurred, or in which mixed primary and secondary burials are probably represented, than there are tombs in which simultaneous secondary interments are apparent. The explanation for this variation in patterns of ritual treatment does not seem reducible to variations in group status alone, for while some secondary burials were exceptionally rich, there are a number of examples of relatively modest secondary and/or collective burials at *Vounous* (Site A Tomb 82B, Site B Tombs 31, 44, 125, 141, and 146), Lapithos (e.g. Tombs 812A, 826A, and 832A), and *Paleoskoutella* (e.g. Tomb 4). Moreover, not all 'rich' burials were necessarily secondary and/or collective (see Tables 4.7a–c and 4.11a–c). As in the case of the Berawan *nulang* (Metcalf and Huntington 1991; Metcalf 1982), the selection of a particular mortuary program was probably linked to a combination of variables—the status of the deceased, the social aspirations of his or her survivors, and the timing or practical exigencies surrounding the death.

In the Berawan system, as discussed in Chapter 2, dual obsequies entailing a primary funeral followed by secondary rites of exhumation and reburial are in effect the cultural 'ideal' based on cosmological beliefs, but they are not consistently practiced. *Nulang* is undertaken mainly when the survivors decide to construct a new and elaborate mausoleum to 'ennoble' themselves as much as the deceased, or when seasonal and economic constraints prevent the family from staging what they would consider an appropriate celebration at the time of an individual's death. When a suitable tomb already exists and the family has adequate resources to stage a primary 'grand' funeral immediately, secondary treatment of the dead is not practiced. Such a system might be postulated as a working model of Cypriot Bronze Age mortuary practice, with the qualification that in Cyprus secondary mortuary rituals were sometimes performed for groups of ancestors as well as for individuals, an issue to which I will return in the next section.

The cosmological and meaningful, as opposed to the practical, dimensions of Early–Middle Cypriot mortuary rituals are more difficult to reconstruct. In Hertz's general formulation, the separation of the flesh from the bones, commemorated in rituals of secondary treatment and collective burial, marks the separation of the soul of the deceased from the society of the living and its incorporation in a spiritual world of the ancestors. However, beliefs about the journey of the soul and the afterlife are variably developed in different cultures that practice secondary treatment. As Metcalf and Huntington have observed, Hertz's model works well for the

Indonesian societies upon which it was based, but not for Madagascar, where notions of the soul and the afterlife are seldom articulated. Amongst the Bara communities studied by Huntington, the completion of the journey from 'mother's womb to father's tomb' via reburial of the deceased in the collective tomb of the lineage is more importantly a means of restoring the imbalance between the sterile order associated with death and the vitality essential to natural fertility and human reproduction (Metcalf and Huntington 1991:111–3). Cypriot communities, far removed both culturally and geographically from Indonesia and Madagascar, undoubtedly cherished unique beliefs of their own. However, the placement of ceramic 'genre' scenes depicting agricultural and food-processing activities, sometimes juxtaposed with images of human fertility (women holding babies, couples embracing etc., see Morris 1985:264–90; Herscher 1997:31–5), or rituals such as those portrayed in the *Vounous* 'sacred enclosure' model (discussed below) does indeed suggest an instrumental link between the dead and the control of fertility in Cypriot mortuary ideology.

It might be speculated that EC–MC mortuary rituals were both elaborations and transformations of earlier Cypriot practices in which infants and children were viewed as important intercessors or supplicants between the living community and the 'Earth Mother' or fertility 'goddess' represented in stone and terracotta figurines. Over the course of the Chalcolithic, practices of secondary treatment for adults as well as children seem to have expanded, as the reproduction of kin group identities and status competition took on increasing importance in some communities. This trend may have been reinforced by the traditions of 'Philia' immigrants, and it was further intensified during the EC–MC period. At this time, adult ancestors seem to have been repositioned as the principal intermediaries between the human community and the natural/supernatural worlds, bestowing fertility and in turn legitimate authority upon the descendants who performed their obsequies and presented them with the gifts, possessions, food offerings, and libations to which they were entitled. Some of these offerings may have continued on an ongoing basis even after a funeral, as suggested by the occasional placement of goods in tomb dromoi (Stewart 1962a:294; Dunn-Vaturi 2003:178) and the iconography of tomb carvings and ceramic models (Frankel and Tamvaki 1973; Åström 1988; see further discussion below in the context of tomb architecture).

## The grouping of the dead

The practice of collective burial, through either sequential or simultaneous interments, became standard in Bronze Age Cyprus with the onset of the Philia phase or facies. As in other ethnographically known societies, it may have been integrally related to a broader system of multi-stage mortuary rituals through which the deceased were gradually transformed from individuals with distinct personal identities to members of a more abstract and collective society of the ancestors. There are local antecedents for collective burial practices in the Middle Chalcolithic cemeteries of Souskiou (Christou 1989; Lunt 1994), and examples of similar practices in the mainland societies of Syria and Palestine, although it is currently difficult to pinpoint close parallels from southwestern Anatolia, the proposed homeland of the Philia immigrants. It is most likely, therefore, that traditions of tomb reuse and the simultaneous grouping of individuals in common tombs evolved in a local context as mortuary ritual became a focal arena for the expression of ancestral ideologies and group identities.

As discussed in Chapter 3, the reconstruction of tomb group demography and the social dimensions of tomb group membership in Bronze Age Cyprus is extremely difficult because of the dearth of comprehensive analyses of human skeletal material. However, close consideration of the mortuary samples from the larger EC–MC cemeteries at Bellapais *Vounous* and Lapithos *Vrysi tou Barba* reveals pronounced biases in age and possibly in gender representation, along with notable increases in burial group size over time, trends which undoubtedly had important social and ideological implications. Similar patterns can be discerned in other less extensively investigated cemeteries, but significant variability, particularly with regard to gender-based practices, may also have existed between communities.

## Regional and diachronic patterns in burial group composition

One of the most noteworthy departures from Chalcolithic traditions in the EC–MC periods was the tendency to exclude infants and children from the burial chambers used for adults. Infant remains were noted in only one tomb at *Vounous*, Tomb 121 at Site B, where the rib of a neonate or stillborn infant was found inside a pot. Admittedly, the preservation of infant bones is apt to have been affected by post-depositional flooding, erosion, and recovery factors, but their nearly complete absence suggests that they may have been disposed of in some other manner than chamber tomb interment. Similarly, the number of individuals loosely classified as 'children' is much smaller than might be expected; two such burials were observed at *Vounous* A (an 11–13 year old from Tomb 107 and a 6–12 year old from Tomb 120, Stewart and Stewart 1950), and seven or eight were remarked at *Vounous* B (Tombs 20, 23A, 38B, 40, and 41B, Dikaios 1940; Tombs 50b and 77, Dunn-Vaturi 2003), including some foot bones from a person of small stature in a pot from Tomb 131a (Stubbings in Stewart and Stewart 1950:378). In other pre-industrial populations infants and young children may account for as much as 50% of all deaths (Acsádi and Nemeskéri 1970; Weiss 1972), whereas at *Vounous* they amount to roughly 6% of the minimum number of burials observed by the excavators. The contrast with the high proportion of infant and other sub-adult burials at the Chalcolithic site of Lemba *Lakkous* in southwestern Cyprus (Niklasson 1985:241) is remarkable.

Similarly, infants and children were very much underrepresented at Lapithos. Approximately a dozen instances of infant and child burials are referred to in the excavation reports of the Pennsylvania and Swedish Cyprus expeditions (e.g. in Tombs 311B, 313A, 313C, 319, 820, 823, 832B, 836B, and possibly Tombs 819A, 829B, and 833B), amounting to a very small fraction of the entire mortuary sample. However, it is possible that infants and children were accorded formal burial rites somewhat more often at Lapithos than at *Vounous*, given the frequency of the so-called dromos cupboards, which occur in at least 13 tomb complexes or 23% of the Pennsylvania and SCE tombs. Pennsylvania Tomb 804, which was unusually rich in metal and imported goods, had a total of 18 such cupboards cut into the sides of its dromos. During the MC period in particular, it is possible that wealthier kin groups at Lapithos may have begun to invest more energy in the disposal of the young, including some of them at least in an expanding society of the ancestors (see also Ribeiro 2002:207). Elsewhere (e.g. Alambra *Mouttes* and Kalavasos), however, the absence of infants and the relative scarcity of children in chamber and pit tomb contexts attests to the broad exclusion of the very young from the mortuary rites accorded to adults.

The evidence for differential treatment of the dead on the basis of gender is more equivocal. Among the 16 crania examined by Rix from the Stewarts' excavations at *Vounous* A and B, four were identified as female, nine as male, and three as possibly male (Stewart and Stewart 1950:374). Among the four skulls retained from Schaeffer's and Dikaios' excavations at *Vounous* B, restudied by Fisher (1986:28–9), three were identified as adult males and one as an adult female. Information on the relative preponderance of adult males and females at Lapithos derives only from the very limited analyses of Fürst (1933:58), who identified four males and one female among the specimens which he studied, and the more recent work of Fischer (1986: Table 2), who identified five males, one female, three or four adults of uncertain sex, and one child within the sample available to him. While both of these samples suggest that females were accorded chamber tomb burial less frequently than males, it must be stressed that the sample sizes are very small, Rix's identifications at *Vounous* remain uncorroborated, and at both sites, if the 'possible' males were in fact female, the sex imbalance would appear much less dramatic. Moreover, it is conceivable that the original analysts deliberately selected male specimens for publication, further biasing the sample.

It should be noted that different patterns of sex representation are indicated elsewhere, particularly at Kalavasos, where Schulte-Campbell identified one definite and one probable male, five definite and two probable females among a total of nineteen adults in the tombs excavated by the VVP (Schulte-Campbell 1986:168–76). The small sample of sexed adults from Alambra *Mouttes* is also ambiguous, with four or five males and two or three females (cf. Coleman 1996:116–23; Domurad 1996:515–8). The suggestion that adult females were sometimes excluded from burial in chamber tombs in certain communities, perhaps being subject to less elaborate forms of interment (e.g. the intramural burials at Marki *Alonia*, Frankel and Webb 1999:90) must therefore be regarded with caution.

The foregoing data suggest that EC–MC chamber tomb groups were not broadly inclusive representations of local community composition, as children certainly, and women possibly, were not consistently buried there. However, there is some evidence from *Vounous* indicating that the size and inclusiveness of the burying groups that contributed to the cemeteries' population increased over time. Based on the total number of tomb chambers recorded and the mean number of burials per chamber, estimates of approximately 100 burials at *Vounous* A and 350 at *Vounous* B may be calculated. While *Vounous* B was probably in use for a longer timespan than *Vounous* A, it is possible that there was a slight increase in the size of the local population contributing to the cemetery at the later site, or that a larger number of individuals were entitled to, or deemed eligible for, chamber tomb burial at *Vounous* B. Unfortunately, in the absence of related settlement excavations and survey data, it is difficult to determine which of these hypotheses is correct.

In addition to the increasing number of burials overall, small but perhaps highly significant diachronic changes in patterns of tomb use are evident at both *Vounous* and Lapithos. As is apparent from Table 4.2, the mean number of burials per chamber at *Vounous* B (2.68) is significantly higher than the mean of 1.59 at *Vounous* A, and the range of burials per chamber is also greater at the later site, increasing from 1–6 to 1–9 per chamber. Considering this in conjunction with the increased proportion of multi-chambered tomb complexes at *Vounous* B (see discussion in the following section of this chapter), it would seem that tomb complexes were becoming more permanently associated with family groups of some type, in contrast to the situation at *Vounous* A, where they were used mainly for one or two individuals. The same pattern is discernible at Lapithos, where the mean number of burials per chamber is approximately the same as at *Vounous* B initially, but rises to 4.19 in tombs of predominantly MC date (Table 4.2). A few particularly rich tombs such as Markides' Tomb 204 and Swedish Tomb 313A had unusually large numbers of interments, 10 and 15 respectively. As was the case at *Vounous*, chamber tombs seem to have become more permanently identified with larger kin groups, rather than being used for only one or two individuals. The prevalence of multi-chambered complexes at Lapithos once again suggests the grouping of related lines of individuals.

Beyond the north coast cemeteries of *Vounous* and Lapithos, synchronic and diachronic trends in burial group size are difficult to gauge. Very small burial groups or single burials seem to have been common at Alambra in the EC and early MC periods, based on the evidence of Tombs 101–106 (Coleman 1996). The Middle Cypriot burial group sizes reported at Ayios Iakovos *Melia* (4 individuals in Tomb 7 of MCII date, 10–12 in Tomb 1 of MCIII date, and 15 in Tomb 6 of MCI/II–MCII/III date) and at Korovia *Paleoskoutella* (the 14 burials in Tomb 7, of MCIII or slightly later date) suggest that the use of tombs for larger burial groups was becoming prevalent throughout the northern region by the later part of the Middle Cypriot period. This phenomenon is further attested by the 'mass burials' at Politiko and Nicosia *Ayia Paraskevi* and by the extremely large tombs at Dhenia in the central part of the island. In the south, comparably large burial groups have yet to be found. Within the 18 tombs with preserved bone material excavated by the Department and the VVP in Kalavasos Village, the mean number of burials per chamber was 2.33 (standard deviation = 2.18, n=9), with a range of 1–7 burials per tomb (although, in light of the disturbance of the chambers, these should perhaps be viewed as minimum figures). 'Mass' burials like those observed in the north and the east are not attested to date, although the seven burials in the incompletely excavated Tomb 36 might conceivably have taken place in the context of a collective secondary burial celebration. Still, there are some indications of a more protracted reuse of mortuary facilities; Herscher's (1976) analysis of the material from the Department of Antiquities Tomb 5, for example, suggests that this tomb continued in use from ECIII–MCIII. And further to the west, at Episkopi *Phaneromeni*, the construction of multi-chambered tomb complexes with large trench-type dromoi bespeaks an emphasis on common descent or 'lineality' in mortuary practice also evident at *Vounous* and Lapithos in the north.

## Tomb groups and social groups in EC–MC Cyprus

In the absence of complete and detailed osteological studies of all of the skeletal material from EC–MC tomb groups, the nature of the social and biological relationships pertaining among the individuals interred in a common tomb remains largely a matter for conjecture. In tombs with pairs of burials, for example, it is usually difficult even to say whether male-female couples or same sex pairs—whether siblings, parent and child kin units, or other social groupings—are represented; at least one male-female couple was allegedly found in *Vounous* A Tomb 89 (Stewart and Stewart 1950:274), but other pairs might have varied. Similarly, in the case of larger burial groups, we do not know whether the individuals buried together were related by common descent, or by a combination of descent and marital ties, or whether there were perhaps closer biological relationships amongst the males or the females included in the same tomb chambers.

Nevertheless, it is possible to infer at least one important aspect of social structure from EC–MC tomb groups. This is the growing ideological significance of social affiliation and identity, which received frequent ritual and material expression through the reuse of tombs for multiple sequential burials, sometimes over several generations, and through the occasional collective, simultaneous reburial of several individuals in the same tomb. It is extremely likely that the persons buried together within a particular tomb were indeed linked by kinship ties rather than being random collections of unrelated individuals who simply died at the same time (cf. Webb 1992:88). The construction of ornately niched chambers such as Lapithos Tomb 313A and *Paleoskoutella* Tomb 7 suggests prior architectural planning, undertaken with the intention of accommodating specific individuals in a carefully structured fashion, with forebears perhaps assigned to the niches and their descendants arranged below on the chamber floor. The largest tomb groups may well have represented the cultural ideal, if not the statistical norm, for kinship groupings within the living society.

It is striking that the most lavish EC–MC mortuary celebrations were invariably focused on groups of burials rather than on individuals. Regardless of the extent to which individuals, or individual households, might achieve and augment their status through the production of agricultural surplus, the manipulation of alliance and exchange networks, and the elaboration of their personal paraphernalia (weaponry, ornamentation), status claims were ultimately founded on membership in a group rather than on individual identity. A number of factors may have enhanced the prominence of broader kin affiliations and descent group structure: (1) the importance of inheritance or intergenerational transfers of wealth in establishing and maintaining individual household groups; (2) the need for cooperation in agricultural labor, especially in harvest season and other periods of intensive activity; (3) the need to meet ritual responsibilities (such as the staging of large-scale mortuary celebrations and the fulfillment of marriage payments), and (4) the need for households to pool resources and borrow from one another, similarly, in order to participate in regional exchange alliances that sustained essential supplies of metal, livestock, marriage partners, and other goods. The kinship ties which helped to reproduce basic household groups and their broader social connections may thus have been celebrated in mortuary ritual, with periodic affirmations of their significance concurrently reinforcing the consciousness of lineage or descent group identity.

## The origins and elaboration of the EC–MC chamber tomb tradition

The use of rock-cut chamber tombs became widespread in Cyprus with the onset of the Philia Culture, although this like other mortuary phenomena cannot be simplistically characterized as a 'Philia' innovation. The bottle-shaped rock-cut shaft tombs of Middle Chalcolithic Souskiou attest to local precedents that could have influenced both the use and form of early chamber tombs at nearby Sotira *Kaminoudhia* (Swiny 1985:123) and perhaps also at Kissonerga *Mosphilia*. It is difficult to identify definitive antecedents in southwestern Anatolia, whence most of the Philia immigrants are thought to have originated, but earth or rock-cut shaft tombs similar in principle to Cypriot chamber tombs were used at Aleppo Ansari in northern Syria and at Middle Euphrates sites such as Selenkahiye and Halawa between 2450–2350 BC, at the same time as they were becoming common in Cyprus (Carter and Parker 1995:106–4). As discussed further on, the similarities between the Halawa tombs and the most elaborate EC–MC tombs from Lapithos *Vrysi tou Barba* and Korovia *Paleoskoutella* in Cyprus are striking. Collective burials in multi-chambered rock–cut tombs were also common in the EBA–MBA Palestine (Gophna 1992:138–40), although Stiebing (1971) has argued that the Palestinian chamber tombs were as much influenced by interactions with Cyprus as vice versa. On the basis of current evidence, the best hypothesis may be that some Philia and later EC–MC tomb forms were influenced by familiarity with and/or emulation of mainland burial traditions, but the broad adoption of chamber tombs within Cypriot communities came about because they already had local precedents that were amenable to revival, reinterpretation, and elaboration in the context of ongoing social change.

A regional overview of EC–MC tomb architecture reveals the emergence of both localized peculiarities in tomb form and of cross-cutting inter-regional similarities. There was in addition considerable differentiation in tomb forms and sizes within communities, variation which may have been linked to differences in ritual usage (i.e. position within the ritual cycle) and differences in social resources between groups. From a diachronic perspective, it is also possible to discern a tendency towards increasing energy expenditure and competitive elaboration in tomb construction. This trend was broadly based within communities such as Lapithos and *Vounous*, and may also have characterized other settlements whose mortuary records are less extensively documented. The following sections examine the evidence for these trends and possible explanations.

### Local and regional diversity in EC–MC tomb architecture

The earliest known chamber tombs in Cyprus are associated with the Period 4 or late Chalcolithic occupation levels at Kissonerga *Mosphilia* in southwestern Cyprus. Thirteen examples were excavated there, all generally described as having subcircular chambers offset from vertical or oblique entrance shafts (Peltenburg *et al*. 1998:70). Some of the Philia and EC tombs of Sotira *Kaminoudhia* were similar, with small, irregularly rounded chambers and short dromoi, while others were simply undercut pits without dromoi or irregular bottle-shaped shafts like those of Chalcolithic Souskiou

(Swiny 1985:122; Swiny *et al.* 2003; Christou 1989). Early Philia Culture tombs from the northwestern part of the island included a pit tomb at Kyra *Kaminia* and both pit and chamber tombs at Philia *Laxia tou Kasinou* (Fig. 4.1a, 4.1b). The chamber tombs at the latter site seem to have been undercut from natural hollows along the edge of a plateau, and were entered via shallow oblique or horizontal dromoi (Dikaios 1962: Figs 73, 75–79; Toumazou 1987:171–3). Some of the tombs (e.g. 3 and 4) had fairly simple semicircular chambers, but others, such as Tomb 1 with its three rounded niches and Tomb 2 with a long narrow chamber leading down into a smaller, rounded inner chamber, had more complex plans. In the vicinity of Vasilia *Kafkallia*, both pit graves and five chamber tombs of extraordinary construction have been reported. The chamber tombs, all of which had been looted, had rectangular chambers, pillar-like buttresses, rectangular wall niches, and lime-plastered dromoi up to eight meters in length (Stewart 1957; 1962a; Hennessy *et al.* 1988: Figs 33, 81, 37, 40; see also Fig. 4.1c here). The chamber façades were decorated with rectangular pillars (Stewart 1962a:216), and unique stone transverse walls were built in the dromoi, apparently to impede access to the burials. Some scholars have argued that the complex architecture of the Vasilia chamber tombs, in conjunction with the sophisticated metal finds from the dromos of Tomb 1, must indicate their relatively late construction, perhaps in ECIII (Stewart 1962a:270, 275; Hennessy *et al.* 1988:41; see also Swiny 1985:116), but Webb and Frankel (1999:8) contend that the ceramic finds place them securely within the Philia facies preceding the Early Cypriot period. If they are correct, then extensive variation in chamber tomb forms was prevalent from the very onset of the EC period.

Many subsequent EC–MC chamber tombs had relatively simple plans. The tombs of *Vounous* A usually had single circular to ovoid chamber plans with few internal features and were accessed via short, rectangular dromoi (Fig. 4.1d); dual chambered complexes (Fig. 4.1e) were comparatively rare. More than one-third of the tombs at *Vounous* B, however, were dual- or multi-chambered complexes, a phenomemon that may have expressed a more prolonged and inclusive emphasis on familial identity. They were also generally larger than the tombs of *Vounous* A (Table 4.3). The dromoi of two or more tombs may have been deliberately joined to create one particularly elaborate complex at *Vounous* B (Schaeffer 1936: Fig. 10, Tombs 78, 75, 70, 70a, 70b, 73; Dunn-Vaturi 2003: Fig. 23; see also Fig. 4.1f here). Tombs at Site B were sometimes equipped with rock cut niches or recesses in the chamber walls, frequently located to the right of the entrance (e.g. Dikaios 1940: Tombs 3, 5, 6, 13, 15, 17, 29, 37). Further to the west, at the site of Karmi *Palealona* located midway between the cemeteries of *Vounous* and Lapithos, tombs were similar in plan to those of *Vounous*, with one to three round or ovoid chambers of varying size, entered via rectangular or bath-tub shaped dromoi (Fig. 4.1g). Some of the smaller chambers may have been 'dromos cupboards'. Tomb 3A (Stewart 1962b: Fig. 1; Fig. 4.1h here) had a comparatively elaborate plan, with four semicircular burial niches similar to those seen at Lapithos to the west and at Korovia *Paleoskoutella* and Ayios Iakovos *Melia* in northeastern Cyprus.

Despite their general simplicity, both the *Vounous* and the Karmi tombs display a few notable architectural embellishments. Tombs 114, 116, and 117 at *Vounous* Site A were unusual in having entrances with carved façades depicting vertical uprights and horizontal lintels (Stewart 1939a; Stewart and Stewart 1950:152, Figs 111 and 112; 158, Figs 120, 124b; 162–3, Fig. 123). It is possible that the decorated doorways were intended to mark the tombs as shrines like those depicted in un-provenienced ceramic models from Kótchati and Kalopsidha in central and eastern Cyprus (Frankel and Tamvaki 1973; Åström 1988). The two well-preserved models from Kotchati (Karageorghis 1970; 1991:142–3, Pl. CII:2–3, Pl. CIII:1–2) portray a woman standing by an amphora in front of a structure with similar architectural features and mounted bulls' heads, perhaps intended to represent a libation scene associated with the ritual veneration of ancestors. It is unclear as to whether these architectural embellishments served to distinguish groups of superior social or ritual position within

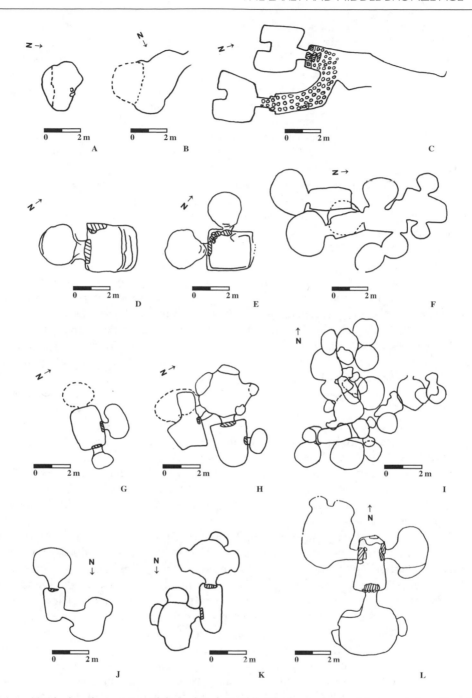

**Figure 4.1** Tomb plans from various Philia and Early–Middle Cypriot cemeteries: (a) pit tomb from Kyra *Kaminia* (after Dikaios 1962: Fig. 73), (b) Philia *Vasiliko/Laxia tou Kasinou* Tomb 1 (after Dikaios 1962: Fig. 75), (c) Vasilia *Kafkallia* Tombs 1 and 2 (after Stewart 1962a: Fig. 86:1), (d) Bellapais *Vounous* Site A Tomb 90 (after Stewart 1962a: Fig. 88:2), (e) Bellapais *Vounous* Site A Tomb 87 (after Stewart 1962a: Fig. 88:1), (f) Bellapais *Vounous* Site B Tomb 70 complex (after Schaeffer 1936: Fig. 10), (g) Karmi *Palealona* Tomb 11 (after Stewart 1962b: Fig. 1), (h) Karmi *Palealona* Tombs 8 and 3 (after Stewart 1962b: Fig. 1), (i) Episkopi *Phaneromeni* tombs in Area C (after Carpenter 1981: Fig. 3-3), (j) Lapithos *Vrysi tou Barba* Tomb 312 (after Gjerstad *et al*. 1934: Fig. 37:7), (k) Lapithos *Vrysi tou Barba* Tomb 309 (after Gjerstad *et al*. 1934: Fig. 34:1), (l) Lapithos *Vrysi tou Barba* Tomb 806 (after Grace 1940: Fig. 8).

the community; Tombs 114 and 117 were found looted, and none of the three contained any copper-based artifacts, although they did yield certain distinctive items of pottery such as the dagger and sheath models from Tomb 114 and remains of ceremonial or cult vessels from Tombs 114 and 116. Yet these objects were by no means unique throughout the site as a whole. At Karmi *Palealona*, the most unusual architectural feature was a unique sculpture of a human figure in bas-relief cut in the dromos of Tomb 6 (Stewart 1962b:197), which might be interpreted as a representation of the deceased or of a person paying homage to the ancestors (Frankel and Tamvaki 1973; cf. J. Kara-georghis 1977:44). Other examples of carved doorways or other dromos sculptures have been observed at Lapithos *Vrysi tou Barba* (Herscher 1978:705 and note 14) and Vasilia *Kafkallia* (Stewart 1962a:216).

The Lapithos tombs display considerably more variation and elaboration in their architectural plans and features than was characteristic of the tombs at either Bellapais *Vounous* or Karmi *Palealona*. The majority of the tombs were multi-chambered complexes, with a range of 1–5 chambers opening from a common dromos. Chamber size variation was considerable, as indicated in Table 4.4 and Fig. 4.1j–l (4.1j illustrates a small tomb of relatively early ECIIIA date, while 4.1k and 4.1l illustrate size variation between two ECIIIB/MCI tombs). Numerous examples of recessed doorframes and at least two examples of decorated tomb façades were observed (Herscher 1978: 705). While some burial chambers had simple ovoid or circular plans, others were provided with more intricate structural features such as semicircular raised burial niches cut back from the chamber walls (e.g. Swedish Tombs 322 and 313; Fig. 4.2a and 4.2b). Sometimes the form of the chamber was rendered bilobate by carving projecting 'piers' along the rear wall of the chamber, features which have alternately been interpreted as roof supports and cult symbols (e.g. Herscher 1978:706). The similarities between these tombs and the descriptions and illustrations of late third millennium tombs from Halawa in the Middle Euphrates area (Fig. 4.2h and 4.2i) are worth considering. The Halawa tomb chambers had vaulted roofs, wall niches and cuttings reminiscent of windows, and benches that were sometimes provided with built-in 'pillows' thus 'giving the burial chamber or chambers the appearance of underground bedrooms or houses' (Carter and Parker 1995:108, see also Orthmann 1980:101–104; 1981, Pls 29–35, 42). The disposition of the burials in Halawa Tomb H-37 also resembles the arrangement of bodies in Lapithos 313A, suggesting more than a superficial relationship between these distant regions. It may be speculated that the builders of the Lapithos tombs were familiar with earlier burial traditions of the Middle Euphrates, whether by cultural heritage or ongoing cultural interactions.

The Middle Cypriot tombs (Tombs 1, 4, 6, 7, 12 and 13) of Ayios Iakovos *Melia* (Fig. 4.2c–e) usually had one large single chamber, although a small sidechamber was found along the side wall of the dromos of Tomb 4. The tombs were entered by long, tapering dromoi, sometimes stepped, a characteristic which seems to have been peculiar to the northeastern part of the island. Tomb 6, which may have been used for one or more episodes of collective secondary burial, was roughly circular in plan, but most of the other chambers were provided with one or more bays or raised niches that were cut in varying locations, possibly to accommodate specific burials (Gjerstad *et al.* 1934: Figs 119, 124, 125, 130). Circular pits of uncertain function were found in several tombs, sometimes to the left or the right of the door or opposite the entrance in the right rear corner of the chamber. There is no evidence that the pits were cut to hold special goods or offerings, but they may have been used as ossuaries for earlier burials swept aside in the course of reuse or for fragmentary secondary burials; the pit in Tomb 4, for example, was reported to contain dark 'culture earth' and fragmentary burial remains (Gjerstad *et al.* 1934:310). The mean chamber floor area of tombs cut in the Middle Cypriot period was 12.13 sq m, comparable to the latest tombs at Lapithos, but variations in floor area were considerable, ranging from 5.6–20 sq m.

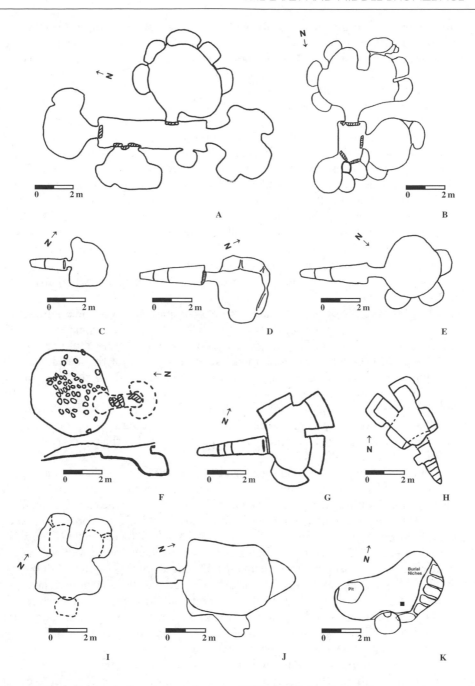

**Figure 4.2**    Niched tombs and other mortuary features of ECIII–MCIII date along with comparative examples from Halawa in Syria: (a) Lapithos *Vrysi tou Barba* Tomb 322 complex (after Gjerstad *et al.* 1934: Fig. 53:5), (b) Lapithos *Vrysi tou Barba* Tomb 313 complex (after Gjerstad *et al.* 1934: Fig. 43:1), (c) Ayios Iakovos *Melia* Tomb 6 (after Gjerstad *et al.* 1934: Fig. 125:1), (d) Ayios Iakovos *Melia* Tomb 1 (after Gjerstad *et al.* 1934: Fig. 119), (e) Ayios Iakovos *Melia* Tomb 12 (after Gjerstad *et al.* 1934: Fig. 130:6), (f) Korovia *Paleoskoutella* Tumulus/Tomb 5 plan and section (after Gjerstad *et al.* 1934: Fig. 166:2–3), (g) Korovia *Paleoskoutella* Tomb 7 (after Gjerstad *et al.* 1934: Fig. 166:8), (h) Halawa Tomb H-21 (after Orthmann 1981: Pl. 29), (i) Halawa Tomb H-37 (after Orthmann 1981: Pl. 37), (j) Dhenia Kafkalla Tomb G.W. 1 (after Åström and Wright 1962: Fig. 6), (k) Dhenia Kafkalla Tomb 6 (after Åström and Wright 1962: Fig. 8).

The varied mortuary features and unusual tumuli found at the cemetery of Korovia *Paleoskoutella* have already been mentioned above in the discussion of the evidence for multi-stage ritual programs. Only four chamber tombs (Tombs 2, 4, 5, and 7) were encountered beneath the seven tumuli investigated by the SCE (Gjerstad *et al*. 1934: Figs 163, 166). The simple construction and small sizes of Tombs 2 and 5 (Fig. 4.2f), with floor areas roughly half those of Tombs 4 and 7, are consistent with their use as temporary, primary burial facilities, and both in fact seem to have been emptied. Tombs 4 and 7, with their long, narrow stepped dromoi, large, elaborately niched and buttressed chambers, and floor areas of 12.8 sq m and 15.1 sq m respectively, attest to a level of planning and labor input perhaps not readily achieved in the immediate aftermath of a family member's death, but possibly executed instead in the context of delayed, collective secondary obsequies. The form of Tomb 7 (Fig. 4.2g), with each of its two lobes subdivided into two distinct angular niches raised approximately 0.40 m above the floor level of the chamber, bears a strong resemblance to Lapithos Tomb 313A, also interpreted as the locus of collective secondary interments, and to the Halawa tombs previously noted. The tumulus constructed above Tomb 7, measuring 22 × 17.5 × 3.1 m, was three to four times larger than the others and, according to Sjöqvist, had been built up in multiple, oblique layers of clay and stone surmounted by concentric casings of rubble (Gjerstad *et al*. 1934: 429–30). Its construction might have taken place over a period of several years, perhaps in the course of periodic ritual observances.

In central Cyprus, the only site where chamber tombs comparable in size and elaboration to those of Lapithos, Ayios Iakovos, and *Paleoskoutella* have been found to date is Dhenia *Kafkalla*. The most notable of these is Tomb 6, an extremely large complex of 9 × 4 m, possibly representing the fusion of two tombs, located near other imposing tombs at the north end of the plateau. Tomb 6 was entered through the roof via a deep cylindrical shaft-type dromos, and the chamber was provided with six small burial niches or cavities on the east side of the chamber (Fig. 4.2k). It is possible that Tomb 6 and other large tombs at Dhenia such as G.W.1 (Fig. 4.2j) and Tomb 48 (Nicolaou and Nicolaou 1988:72–93) were also used for elaborate mortuary celebrations, in the course of which some individuals previously interred in the smaller pit tombs noted at the same site were exhumed and reburied.

Construction of such impressive tombs may have been feasible in the chalk plateau environs of Dhenia, but more problematic in the alluvial and conglomerate soils found elsewhere in the Mesaoria plain and the Troodos foothills. This may explain, at least in part, the prevalence of smaller pit and chamber tombs with simple circular, ovoid, and occasionally 'beehive' forms observed in the vicinity of Katydhata, Politiko, Marki *Davari*, and Nicosia *Ayia Paraskevi* (Gjerstad 1926; Åström 1989; Webb and Frankel 1999:8). It is also conceivable that the full diversity (formal, function, and chronological) of the mortuary features employed in many localities is not represented within the existing samples. Both small pit tombs and larger chamber tombs seem to have been used at sites such as Alambra *Mouttes*, Marki *Davari*, and Dhenia *Kafkalla* and *Mali*. In some cases, tomb form may have been dictated by the constraints of geomorphology (e.g., Coleman 1996:133; Sneddon 2002:102, 114), while in other cases, factors such as diachronic changes in mortuary practice, different phases of mortuary ritual, or status and wealth differentials may have been involved.

Architectural elaboration is poorly attested in southeastern Cyprus, where the elaborate niched tomb plans and extremely large floor areas observed at some northern sites have yet to be reported. The tombs of Arpera *Mosphilos* are described as small pit or cave-shaped rock-cut chambers (Gjerstad 1926:54–57). Tomb 1, dated to the MCIII period (Merrillees 1974), consisted of four chambers opening from a central dromos, similar, perhaps to the MCIII/LCIA tombs of Maroni *Kapsaloudhia* (Cadogan 1984: Fig. 4; Herscher 1984). At the more extensively investigated cemetery of Kalavasos Village, most of the tombs appear to have had single, small, rock-cut chambers measuring 2–3 m in diameter, with entrances through apertures in their roofs (Karageorghis 1958; Todd 1986). Once again, a variety of factors may be adduced to explain the contrast with the north: peculiarities of local geology, sampling, and perhaps a lesser predilection for mortuary display, particularly in tomb architecture, within this region during the EC–MC period.

A somewhat different picture emerges from southwestern Cyprus. Tombs of probable EC date observed in surveys at Evdhimou *Amolo* and *Ambelovounos* consist of small oval dromoi opening onto single tomb chambers, not unlike the plans of EC tombs at *Vounous* in the north. More elaborate multi-chambered tomb complexes have been found, however, at some sites with MC ceramics, including examples with trench-type dromoi at Erimi *Kafkalla* (Swiny 1981:82), and others with smaller and deeper shaft-like dromoi at Evdhimou *Shilles* and Paramali *Mandra* and *Pharkonia* (Swiny 1981:83), reminiscent of Tomb 1 at Arpera *Mosphilos* and others at Maroni *Kapsaloudhia*. Similarly, tomb complexes observed at the EC–MC site of Episkopi *Phaneromeni* (Weinberg 1956; Carpenter 1981) seem to have had either pit-like or trench-type dromoi, often opening onto multiple chambers, which tended to be rounded or ovoid in shape with domed roofs, their diameters ranging between 1–3 m (Fig. 4.1i). The most extraordinary of the trench-type dromoi (Weinberg's Trench 4), measuring 7–8 m in length, opened on to a minimum of nine and possibly as many as twelve chambers. On the long sides of the dromos, the chambers were arranged in four pairs, their doors almost opposite each other (Weinberg 1956:121, Fig. 13). Unfortunately, none of these were ever excavated. However, the construction of such a complex would presumably have entailed a considerable level of energy expenditure, and it may represent the emergence of groups with a strong interest in expressing lineal or other kin ties among their members.

## Variations, elaborations, and convergences in tomb architecture

From the onset of the Philia facies and throughout the EC–MC period, there was considerable diversity in the form and scale of tomb architecture within local communities. In some areas, as in the vicinities of Philia *Laxia tou Kasinou*, Vasilia *Kafkallia*, Lapithos *Vrysi tou Barba*, and Dhenia, there are dichotomies between pit tombs or pit graves and chamber tombs, as well as variations in chamber tomb elaboration. Manning, contrasting the Vasilia chamber tombs and a nearby cemetery of simple pit graves, has proposed that the chamber tombs were 'the burial places of the few elite lineages or families' (1993:45). In light of the relationship between pit graves and chamber tombs observed at Lapithos, however, it is possible that the different burial types were associated with different phases of mortuary processing rather than with class differences. Similarly, the tremendous size variation observed among the *Paleoskoutella* chamber tombs also seems to have been related to protracted sequences of ritual, with burials from the smaller tombs eventually being removed to the larger ones. Still other variations in chamber tomb size and elaboration evident in the large samples from *Vounous* and Lapithos (Tables 4.3, 4.4) may well have been linked to competitive displays and differences in social resources between groups—especially in terms of the number of family members, friends, and allies that could be drafted for tomb construction activities.

While the synchronic variations in energy expenditure are considerable, the diachronic patterns are even more noteworthy. At *Vounous*, the means for most of the major dimensions of tomb size underwent statistically significant increases over time, with mean chamber floor area rising from 3.68 sq m at Site A to 7.16 sq m at Site B (Table 4.3). At Lapithos, chamber floor area rose from a mean of 5.25 sq m in tombs dating between ECII–ECIIIA to 11.18 sq m in tombs dating between MCI–MCIII (Table 4.4). These increases in tomb size may reflect the increasing importance of mortuary rites as occasions for status competition among the living over the course of the late third and early second millennia. In later tombs at both *Vounous* and Lapithos, the values for chamber floor area were more strongly correlated with other wealth measures such as the number of pots and/or copper-bronze objects per chamber than with the number of burials per tomb

(Tables 4.5, 4.6), suggesting that the increase was part of a general trend towards increasing mortuary expenditure, rather than resulting simply from increased burial group size. Moreover, not only the means but the median chamber floor areas rose at each site, indicating a trend towards increasing energy expenditure that was broadly based within the community, not merely skewed by the ostentation of a few very high status tomb groups (Fig. 4.3). The construction of extremely large and elaborate tombs at sites such as Dhenia, Ayios Iakovos and *Paleoskoutella* during the MC period may have been part of the same trend towards increasing mortuary investment and competitive elaboration.

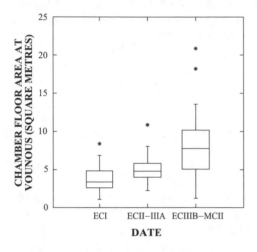 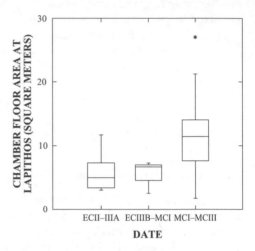

**Figure 4.3**     Box plots showing the increase in chamber floor areas at Bellapais *Vounous* and Lapithos *Vrysi tou Barba* through time. Note the rising medians, represented by the horizontal line inside each rectangle. The box itself represents the 'hspread' or absolute value of the 'hinges' which delimit half of the values from below the median and half of those above it (i.e. the midrange or interquartile range). The vertical lines or 'whiskers' at either end of the boxes represent the range of values within 1.5 hspreads of the hinges, where hspread is the absolute value of the difference of the two hinges. Values plotted by asterisks are those which fall outside the 'inner fences' (i.e. beyond plus or minus 1.5 hspreads of the upper and lower hinges respectively). See Wilkinson (1990:166–7) for further details on this type of plot.

Geographically distinctive traditions of tomb cutting and embellishment are evident at sites such as Vasilia, Lapithos, Dhenia, Ayios Iakovos and *Paleoskoutella*, yet at the same time there are some notable cross-cutting regional similarities in overall tomb form. The multi-chambered tomb complexes of *Vounous*, Karmi, and Lapithos in the north, for example, have analogues at Episkopi *Phaneromeni* and other southwestern sites (Fig. 4.1). The elaborately niched chamber types of Lapithos in the northwest can be seen much further east at Ayios Iakovos and *Paleo-skoutella*. The explanation for these similarities may lie in extensive social ties based on kinship, marriage, alliance, and exchange relationships that would have brought members of different and sometimes distant communities together periodically, perhaps in the context of large-scale mortuary celebrations and other ritual occasions. Such reunions and visitations may have stimulated the emulation of practices and designs observed in other communities, resulting in widespread parallelisms and an overarching *koinē* of ritual and material symbolism. Concurrently, architectural forms would also have expressed ongoing transformations in social structure and ideology—including the increasing emphasis on kin group identity discussed earlier—that were underway in many different parts of the island.

## Early–Middle Cypriot grave goods and the symbolism of prestige

The practice of burying the dead with personal ornaments and other types of grave goods was not uncommon throughout the course of Cypriot prehistory, but it seems to have undergone a major expansion and transformation in its social and ideological significance between the Chalcolithic and Bronze periods. Prior to the Late Chalcolithic, infants and small children appear to have been the primary recipients of grave goods such as picrolite pendants and dentalium beads at the sites of Lemba *Lakkous* and Kissonerga *Mosphilia*, while adult burials at these sites were seldom accompanied by any goods amenable to archaeological preservation. However, the rock-cut shaft tombs of Middle Chalcolithic Souskiou, which contained a preponderance of adult burials, yielded impressive arrays of ceramics and ornaments made of bone, shell, copper, and picrolite (Christou 1989), and it is possible that the disposal of material wealth with the dead was a practice closely associated with rites of secondary treatment and collective burial emerging in local communities at this time.

Subsequently, with the onset of the so-called Philia facies, the bestowal of grave goods seems to have become a standard practice throughout the island. A small number of pots and other objects such as stone and faience necklaces may be noted in the Late Chalcolithic, transitional Philia chamber tombs at Kissonerga *Mosphilia* (Peltenburg 1991b:30; Peltenburg *et al.* 1998:90–92), and thereafter 'Philia Culture' burials were often equipped with larger and more diverse arrays of goods that included pottery, spindle whorls, shell pendants, flint blades, various small stone objects, and copper-based artifacts such as knives, toggle pins, and spiral earrings. The three or more burials in Philia *Laxia tou Kasinou* Tomb 1, for example, were accompanied by at least 68 objects, eight of which were metal (Dikaios 1962:176). The looted 'Philia Culture' tombs of Vasilia *Kafkallia* yielded a 'magnificent' deposit of copper objects including three heavy armbands, a unique dagger, two knives and two toggle pins, along with other items of copper, alabaster vases, gold earrings, and large quantities of smashed pottery (Hennessy *et al.* 1988).

Given the exceptional concentrations of wealth in tombs at Vasilia *Kafkallia* and in the later or partially contemporaneous cemeteries of Bellapais *Vounous* and Lapithos *Vrysi tou Barba*, some researchers have argued for the emergence of hereditary, status conscious elites during the EC–MC period (Manning 1993; Peltenburg 1994). However, a closer examination of both the distributional patterns of prestige goods among contemporaneous tomb groups and the diachronic increases in mortuary expenditure at *Vounous* and Lapithos suggests that EC–MC mortuary ritual bears witness not to the emergence of a rigidly defined status hierarchy, but rather to the development of an ongoing dynamic of prestige competition within and between communities. It is also apparent that patterns of mortuary consumption varied regionally, and yet, as observed in the case of tomb architecture, there are broadly cross-cutting similarities in the complements of prestige goods deployed in distant localities. These similarities may reflect the development of a widely shared complex of prestige symbolism and ideology that was created as the scale of the social networks participating in mortuary celebrations became ever more extensive.

## Burial assemblages at Bellapais *Vounous*

The *Vounous* tombs are perhaps most renowned for their lavish arrays of Red Polished pottery, ceramic models, and coroplastic genre scenes. Types from *Vounous* Λ included a variety of jugs and bowls of diverse sizes, and many unusual pots referred to by the excavators as 'cult vessels' (Stewart and Stewart 1950, *passim*). These were most commonly tulip-shaped bowls with flat bases, or stemmed bowls referred to as chalices. Many were decorated with pairs of modelled animals' heads, ear-like projections, and vertical hilt-shaped handles on the rim. Some were embellished with minia-ture models of ceramic vessels and troughs (e.g. Tomb 111 nos. 1 and 8, Tomb 160A nos. 13 and 17, Tomb 164A no. 13). Other unusual ceramic items found in the *Vounous* A tombs included models of spindles, daggers, and sheaths, a bellows nozzle from Tomb 92, a ceramic horn from Tomb 164A, and a 'finger guard' from Tomb 164B. The *Vounous* B tombs contained an even more diverse array of ceramic types such as composite vessels comprised of multiple jugs and bowls, other jugs and bowls with modelled and incised zoomorphic representations, along with double-necked and triple-necked jugs, ring vases, kernoi, pyxides, and askoi. In addition to the earlier dagger, sheath, and horn models, some other rather unusual ceramic items found in the *Vounous* B tombs included brush models, plank idols, a ceramic 'offering table,' a plowing scene model, and the truly extraordinary model of a 'sacred enclosure' found in Tomb 22 (Dikaios 1940: Pls VII, VIII). This famous bowl depicts a circular building with a number of human figures seemingly engaged in activities of a ritual nature, cattle penned in stalls along the inner periphery of the room, and bulls' heads mounted atop the walls of the building (reminiscent of the shrine models from Kotchati noted earlier; cf. Karageorghis 1970; Frankel and Tamvaki 1973; Åström 1988).

While some of these objects are truly remarkable in form and representational content, it is interesting to note that they were rather broadly distributed throughout both cemeteries. The so-called 'cult vessels' and other unusual, possibly ceremonial forms found at *Vounous* A occurred in nearly one half of all the tombs excavated at that site (Table 4.7a). Similarly, at *Vounous* B, although the frequencies of particular 'distinctive' types were fairly low, more than one-third of the earlier tombs contained at least one of these unusual or elaborate forms (see Table 4.7b), and the proportion of the later tombs in which such items were found was more than two-thirds (see Table 4.7c). They are sometimes present, but often absent from the richest tomb assemblages, and in some cases, most notably that of the extremely complex 'sacred enclosure' model from *Vounous* B Tomb 22, they occur in otherwise unimpressive contexts (note, however, that this tomb had been disturbed by looting). Consequently, it may be more appropriate to characterize them as part of a widely shared complex of ceremonial or sacred regalia rather than as the exclusive perquisites of high ranking individuals.

As Tables 4.8 and 4.9 illustrate, the quantities of pottery disposed of per chamber and 'per burial' varied considerably at both cemeteries, with both the ranges and standard deviations increasing at *Vounous* B. Such disparities between tombs may indicate considerable differences in wealth between kin groups, although the possibility that mortuary outlays were conditioned by age and gender variables must also be kept in mind. It is also notable that both the mean and the median values (see Figs. 4.7a and 4.8a) of these wealth measures increased diachronically. The mean number of pots per burial, for example, rose from 14.4 at *Vounous* A to 18.7 at *Vounous* B. Thus even though expenditures differed widely between groups, they were increasing substantially throughout the community at large, as might be expected in the context of generalized prestige competition.

Copper-based objects occurred in just under half (49%) of the intact or substantially preserved tombs at *Vounous* A (Table 4.7a). The great majority of the finds were either knives, axes, or hook-tang weapons (most recently interpreted as spearheads, see Philip 1991). The number of copper artifacts per tomb was generally quite small, with a mean of 1.2 items per chamber, and an even lower mean of 0.7 items per burial (Table 4.8). Only three tombs were notably rich in metal: Tomb 164A with two burials (possibly subjected to secondary manipulation of the skulls, as discussed above) provided with a total of four knives, Tomb 161 with a single burial equipped with two knives, two axes, and one hook-tang weapon, and Tomb 105 with two burials accompanied by four knives, two axes, one hook-tang weapon, a chisel, an awl, and another unidentified object.

As suggested in the previous discussion of pottery, it is possible that the occurrence of metal objects in the *Vounous* A tombs was also significantly conditioned by the age and gender of the associated burials, but the lack of expert osteological analyses of the skeletal remains precludes any definitive conclusions in this regard. It may be significant that seven of the tombs with no copper at *Vounous* A contained one or more spindle whorls, which have sometimes been interpreted as female gender markers. However, the use of artifact types such as knives, weaponry, jewelry or tools such as spinning and weaving equipment as indicators of sex is sometimes problematic. Tombs 110b and 155 were both reported to have single burials that were provided with knives as well as spindle whorls; Tomb 110b contained a whetstone as well (Stewart and Stewart 1950). A knife and a spindle whorl, along with a bronze pin and numerous ceramic artifacts, may possibly have been associated with a single burial in Schaeffer's Tomb 69 at *Vounous* B (Fischer 1986:28–9). In light of recent discussions (Knapp and Meskell 1997; Talalay and Cullen 2002) of potentially multivalent gender constructions in Chalcolithic and Bronze Age Cyprus, the lack of well-documented osteological sex determinations is all the more regrettable.

At *Vounous* B there was a marked decline in the consumption (and thus, presumably, in the availability) of metal objects in the earlier (ECII–ECIIIA) phase of the cemetery's use, when they occurred in only 27% or 7 out of 26 chambers (Table 4.7b). Interestingly, ceramic models of daggers, sometimes accompanied by sheaths, were deposited in five tombs of this date, perhaps as substitutes for scarce and valuable metal 'originals'. However, the later tombs of *Vounous* B attest to a dramatic increase in the consumption of metal wealth. Of the tombs dating to ECIIIB–MCII 75% contained metal artifacts, and the mean number of copper-based objects per chamber rose to 4.5 (versus 0.6 for the earlier *Vounous* B tombs and 1.2 at *Vounous* A). The mean number of copper items per burial was also notably higher in the later tombs of *Vounous* B, averaging 1.2 as compared with 0.2 for the ECII–ECIIIA tombs and 0.7 at *Vounous* A (cf. Tables 4.8, 4.9).

This increase in metal consumption was accompanied by a change in the frequencies of types present. Hook-tang weapons, which were rare at *Vounous* A, occurred in 23 out of 39 of the later *Vounous* B tombs containing copper-based artifacts, a level of ubiquity which would seemingly preclude their interpretation as exclusive 'elite' status symbols (see also Philip 1991:69). Meanwhile, the overall proportion of ostensibly 'male' metal types declined; whereas at *Vounous* A knives, hook-tang weapons, and axes made up for 81% of all metal types, at *Vounous* B they accounted for less than 48%, with pins representing the single most frequently occurring type, making up 34% of the copper-bronze assemblage. 'Toilet' articles such as scrapers, razors, and tweezers were also more common at *Vounous* B than at *Vounous* A (cf. Tables 4.7a, 4.7c). The increasing preponderance of pins could reflect an increased emphasis on ornamentation and perhaps on female prestige, but as there is no osteological evidence to document any sex-related associations of artifacts, the possibility that such items were also worn by males must be kept in mind.

The relatively discontinuous distribution of copper artifacts at *Vounous* A (Fig. 4.4 top) is suggestive of a social system with limited but relatively well-defined status categories, perhaps based on seniority and authority within the hierarchy of the local kin group structure, augmented by individual wealth and accomplishments, rather than membership in a superordinate 'elite'. The more continuous distribution of copper-based items in the later tombs of *Vounous* B (Fig. 4.4 bottom), by contrast, may have been associated with the development of increasing prestige competition, with the display of wealth objects gradually becoming an instrument in the active creation of status distinctions (Shennan 1986; Knapp 1990a). This trend may also be attested in the wide range of size variation observed in the category of hook-tang weapons at *Vounous* B, with lengths ranging from 12 to 47.5 cm (mean 31.6 cm, standard deviation 8.5, n=42).

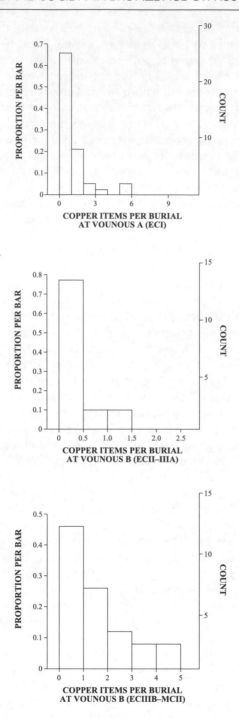

**Figure 4.4** Copper items per burial at Bellapais *Vounous*. 'Proportion per bar' refers to the proportion of the entire sample represented by each bar. Each observation represents the value obtained when the total number of copper items in a tomb chamber is divided by the total (estimated) number of burials in that chamber. This 'per capita' calculation permits the comparison of wealth among tombs with differing numbers of burials. The earlier (ECII–IIIA) *Vounous* B tombs display an extreme scarcity of metal. Note that the distribution of copper wealth appears more continuous in the later (ECIIIB–MCI) group than in the ECI group.

One form of mortuary expenditure that seems to have remained fairly constant at both sites was the sacrifice of domestic livestock, including cattle, sheep, and goat, whose bones were recorded in roughly one-third of the *Vounous* A and B tombs. In some cases (e.g. Dikaios' Tombs 13 and 36 at Site B), substantial portions of an ox or calf were said to be present (Dikaios 1940: 30, 72–6). In other cases, the deceased may have received offerings of 'joints of meat' with the flesh intact, according to Stubbings, who observed cut marks on the shafts of the humeri related to butchering practices and also the preservation of bony material inside the bone shafts, suggesting that the marrow had not been removed (Stewart and Stewart 1950:376). Presumably the remainder of the animals sacrificed to the dead were consumed by the living guests at the funeral.

Items of non-local origin were extremely rare in the *Vounous* tombs. Four fragments of sheet gold and an imported jug of Syro-Palestinian origin (cf. Stewart 1939b; Wright 1940; Amiran 1971; 1973; Ross 1994) were recovered from *Vounous* A Tomb 164B, the chamber adjoining 164A, in which possible evidence of secondary treatment was observed. At *Vounous* B, imported goods included a number of faience beads from Tombs 2, 15, 19, 20A, 23, 33, 59, 64, and 72, a gold earring in Tomb 59 (Dunn-Vaturi 2003:61) and daggers of probable Minoan origin (Catling and Karageorghis 1960; Philip 1991: 85) from Tombs 19 and 143. Two pots of Syro-Palestinian origin were also recovered from Tombs 64 and 68 (Ross 1994). It is noteworthy that these exotic goods occur mainly in tombs which had rich and diversified assemblages of local goods as well. This may imply that groups of greater wealth were more likely to participate in long distance exchange transactions.

As illustrated in Tables 4.7a–c, it appears that a small number of tombs at *Vounous* A and *Vounous* B contained exceptional arrays of goods, comprising several distinctive ceramic types, metal goods and animal bones. These were among the largest tombs overall in terms of floor area, and at *Vounous* A, most were located along the highest contour elevation of the cemetery. This is certainly suggestive of differences in wealth and social status within the community. Yet at the same time, statistical patterns of association between the copper-based objects, ceremonial vessels, ceramic models, and fauna are not particularly strong at either site (Table 4.10), and ceremonial regalia seems to have been rather broadly distributed. It is interesting to note that at *Vounous* B, only one of the very rich tombs (Tomb 143) was placed at a distinctively high elevation vis-à-vis the others, suggesting a breakdown of spatial exclusivity that is consistent with the more general availability of metal goods at the same time. It seems likely that any ascribed differences in social status that may have been established early in the history of *Vounous* gave way as prestige competition increased within the community, with status differentials perpetually being created and revised in the context of ritual displays.

## Burial assemblages at Lapithos *Vrysi tou Barba*

Long-term trends in the disposal of material wealth at Lapithos are broadly similar to those observed at *Vounous*, but distinctive local tendencies are also evident. Ceramic goods, primarily Red Polished and White Painted wares, were disposed of in smaller quantities throughout the site's history in comparison with *Vounous* (see also Davies 1997:18). Unusual ceramics and probable ceremonial vessels (e.g. composite vases, askoi, ring vases, pyxides, pots with anthropomorphic or zoomorphic modelling or engraving, stemmed and conical vessels) were also less common at Lapithos. Although the mean and median numbers of pots per burial rose through time, the increasing consumption of material wealth is much more dramatically illustrated by the rising disposal of metal goods in the mortuary context (see Tables 4.11a, 4.11b, 4.11c, and 4.12).

The earliest phase of the Lapithos cemetery (Table 4.11a), dating to ECII–IIIA, was by far the poorest period in terms of the consumption of metal wealth, but metal was nonetheless more plentiful than in the contemporaneous tombs of *Vounous* B, occurring in approximately 40% of all chambers, or 14 out of 35 cases. Unfortunately, as was also the case at *Vounous*, it is difficult to determine to what extent the occurrence of metal was related to differences in the age-sex compositions of the burial groups in question. Knives, hook-tang weapons, and pins were the most common metal types. Copper axes do not appear to have been deposited in tombs of this date, although three stone axes were found in Tomb 825 and four more were discovered in the dromos of Tomb 826. These tombs were otherwise unexceptional in their contents. The two 'richest' chambers were Tombs 314B—in which a total of six burials were provided with one knife, 2 hook-tang weapons, one pair of tweezers, and two pins—and 829C, in which one burial was provided with two pins, the other with a knife and tweezers, and two other pins were also found in the wall of the chamber. It is perhaps significant that these tombs, along with Tomb 813a, the third richest chamber overall, were the only tombs containing elaborate ceremonial vessels in this period.

Among the tombs dating to the ECIIIB/MCI transitional period (Table 4.11b), roughly contemporaneous with the later phase of the cemetery at *Vounous* B, copper was much more widely distributed, occurring in approximately 74% of all chambers (28 out of 38). The distribution of hook-tang weapons and knives broadened considerably. Two tomb assemblages (Tombs 309A and 806A) had unusually high concentrations of pins, along with a dearth of weaponry, perhaps indicative of a concentration of rich female burials. One notably rich single burial in Tomb 301C, possibly male, was equipped with three knives, two hook-tang weapons (one of unusual length, 63.6 cm), one razor, and two tweezers. It is noteworthy that this person was not distinguished by the use of any symbols which differed *in kind* from those available to others, but rather by the number and the elaboration of the goods with which he or she was provided. Also of interest are the two burials in Tomb 827B, less ostentatiously equipped with metal items that included one knife, one hook-tang weapon, one razor, and one pair of tweezers. The hook-tang weapon was by far the smallest which has been found to date, measuring only 9.3 cm in length, perhaps a deliberate miniature or 'votive' item (Philip 1991:69). This tomb contained an unusually large number of pots, 76 in all, one a double-necked jug of possible ceremonial function. Ox bones were also recovered from the chamber, along with two stone axes. This assemblage is suggestive of a fairly costly mortuary celebration in honor of one or more persons of senior or traditional status, but of lesser wealth than, for example, the burial in Tomb 301C.

The two most outstanding collections of grave goods attributable to this period were found in Tombs 806A and 322B. Tomb 806A, used for at least 3–4 burials, contained a massive accumulation of 110 pots, among them several composite vessels, and a total of 14 copper-based artifacts, including ten pins, two knives, one hook-tang weapon, and one pair of tweezers. This tomb is most renowned for the occurrence of a Minoan bridge-spouted jar (Grace 1940) and several small ornaments of gold, silver, and faience, all of which appear to have been extremely rare, imported valuables. It also yielded cattle bones, perhaps the remains not only of sacrificial offerings to the deceased but also of the mortuary feasts staged in conjunction with the burials. Tomb 322B, located very nearby Tomb 806, contained a total of nine copper-based objects, including two knives, two hook-tang weapons (one of which was unusually large, measuring 58.6 cm), one razor and three pairs of tweezers. Although only 20 pots were recovered from the tomb, two of them were cult vessels, one a coupled vessel of two bird-shaped vases. A fragmentary plank figurine and small ornaments of imported silver and faience were also present. Of particular interest is the occurrence of equid bones in the tomb, which, along with the equid bones from Politiko *Chomazoudhia* Tomb 3 (Buchholz 1973:304, note 29) and Kalopsidha Tomb 9 (Myres 1897: 138), may represent the earliest evidence for the presence of horses in Cyprus (see also Reese 1995). Because the skeletal remains from the only preserved burial in this chamber belonged to an adult female aged 18–24 years, Fischer (1986:29) has referred to Tomb 322B as the tomb of a female warrior or 'Amazon'. However, inasmuch as the chamber seems to have been looted, it is likely that other burials, some possibly male,were present originally in the chamber's four niches and/or on the floor (Gjerstad *et al.* 1934:143).

In the latest (MCI–MCIII) phase of the Lapithos cemetery (Table 4.11c), 22 out of 23 or 96% of all chambers contained copper-based artifacts, and the gross, community-wide level of metal expenditure increased significantly (Table 4.12). The number of copper items per chamber was highly correlated with the number of burials per chamber (Table 4.6, Pearson's r=0.858). At the same time, variation in the distribution of metal wealth appears to have been more continuous than in earlier periods; in other words, even though some tombs appear extraordinarily rich, when variation in the number of burials is controlled, the differences between groups appear graded rather than discrete, a pattern suggestive of competitive rather than rigidly ascribed status differentials (Fig. 4.5). Meanwhile, hook-tang weapons were broadly distributed, occurring in 65% of all chambers with metal (15 tombs), reinforcing their interpretation as a form of widely shared status insignia rather than as esoteric symbols of rank. Their ubiquity seems to have provoked a tremendous outburst of competitive elaboration, nascent in the previous period, with rising variation in size (Fig. 4.6) and numerous impractically large examples. The most extreme examples occur in Tomb 313A (the single richest chamber in the entire cemetery) which yielded 10 hook-tang weapons, with 5 ranging between 50–60 cm in length and two measuring over 60 cm (Balthazar 1990: Tables 64, 66). Other evidence for competitive display comes from the remarkable proliferation of copper-based pins and rings, presumably associated with a heightened emphasis on personal ornamentation. Status competition and innovation may also be represented by the appearance of a broader range of copper-based implements such as needles, awls, spatulas, and chisels in the richest tombs overall, perhaps included in an attempt to 'embroider' the core complement of prestige goods.

Copper-based axes were scarcely represented in pre-MC tombs at Lapithos yet were relatively numerous in tombs of MC date. A total of 27 examples occurred in 9 out of the 23 chambers containing metal. The most extraordinary concentration—eight axes in all—was found in Tomb 322A of MCI date. This chamber is remarkable in terms of a number of other criteria: its elaborate architecture, the probable practice of post-interment rituals which contributed to the disappearance of most of the skeletal remains, the wealth of copper-based objects present (53 items), and the occurrence of exotic materials such as gold, silver, faience, and iron. Surprisingly, however, hook-tang weapons are entirely absent from the copper assemblage in Tomb 322A. Possibly they were deliberately removed in the course of secondary or tertiary post-interment rituals in the chamber and reused by living descendants in their own prestige displays. Alternatively, it is possible that in the context of the competitive elaboration of hook-tang weapons, their 'cachet' was somewhat diminished, and the use of an archaic status symbol, the axe, was revived as a distinctive symbol of seniority or traditional authority in the context of widespread political competition and mobility.

Faunal remains belonging to cattle, sheep, goat, dog, and other unidentified species were reported in roughly 25% of the earliest tombs at Lapithos and in a slightly smaller percentage of tombs assigned to the intermediate chronological group. They appear to have been absent (or were not identified) in any of the latest tomb chambers, except for the nearly complete dog skeleton found in Tomb 322d. Considering the increasing scale of expenditure in other aspects of mortuary ritual, this decline seems peculiar. Possibly, however, an increasing proportion of animal sacrifices were diverted to the consumption of funeral participants rather than to the provisioning of the deceased.

Imported goods were absent from the earliest Lapithos tombs, rare in all but the richest tombs of intermediate (ECIIIB–MCI) date, and more broadly distributed subsequently (Tables 4.11a–c).

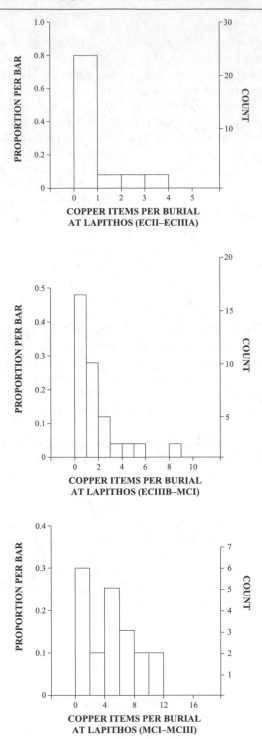

**Figure 4.5**    Copper items per burial at Lapithos *Vrysi tou Barba*. Note the partially discrete distribution of values in the ECIIIB–MCI group, suggesting that a few groups may have been substantially richer than others, and the more continuous distribution in the MCI–MCIII group.

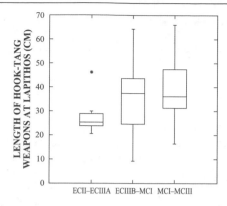

**Figure 4.6**    Box plots showing the distribution of the lengths of hook-tang weapons from Lapithos *Vrysi tou Barba* in different periods. Note the increase in the median values (horizontal lines inside the boxes) and interquartile ranges (span of the boxes) in the later groups relative to the ECII–ECIIIA group, as well as the extension of the upper hinges and whiskers in the latest group. (For an explanation of box plot symbols and terminology, see Figure 4.3 caption.)

Silver and faience ornaments were the most commonly available exotic goods, sometimes occurring in tombs of low to moderate overall wealth (e.g. Tomb 828a, Table 4.11b). Among the more notable arrays of valuables were the gold, silver, and faience items from Tomb 806A (Table 4.11b, see also Grace 1940), which also contained the famous Minoan bridge-spouted jar, along with the finds from Tomb 322a (Table 4.11c), which contained gold, silver, and faience ornaments as well as a dagger and a razor of apparently Minoan origin (Catling and Karageorghis 1960:111). Another dagger of Minoan type occurred in Tomb 313C–D, a chamber adjacent to the extraordinary chamber 313A (Catling and Karageorghis 1960:111; see Branigan 1966; 1967 for daggers possibly of Syrian origin in other, less well-documented tombs). These tombs were among the richest in copper within their respective chronological periods. As suggested in the case of *Vounous*, it is possible that groups with the greatest access to local copper wealth were most likely to participate in exchanges with foreigners.

While some rich tombs such as Pennsylvania Tomb 806 and Swedish Tomb 322 seem to have been clustered together in certain parts of the cemetery (see Herscher 1978:3), possibly reflecting close kin relationships between those groups, there is little evidence to indicate that rich tombs were deliberately segregated from smaller, poorer tombs. Swedish Tomb 313, for example, was set amidst a cluster of tombs varying considerably in size and wealth (Gjerstad *et al.* 1934: Plan V:1). Nor does there seem to have been any complex of emblematic goods or iconography depicting cosmic or supra-local symbols which would qualitatively distinguish 'elite' from 'non-elite' individuals or groups, symbolism which is often found in other societies with ascribed social rank (Peebles 1971; Wright 1984). The display and disposal of large quantities of metal wealth seem to have constituted the principal means of negotiating and asserting social status, but the types employed in this competition were broadly available throughout the community, with differentials in the distribution of prestige goods appearing quantitative and graded, rather than qualitative and discontinuous, and measures of association between various categories of status goods appearing rather weak (Table 4.13). However, the display of exotic goods whose value, at least initially, could be maintained because of their scarcity and restricted availability may certainly have enhanced prestige differentials.

## Burial assemblages at other sites

At sites other than *Vounous* and Lapithos the published samples of mortuary data do not support analyses of wealth and status variations between tomb groups or changes in mortuary consumption over time. However, they do contribute to an understanding of regional similarities and variations in mortuary expenditure and symbolism. For example, an examination of Åström's (1989) catalogue of finds from Katydhata and Flourentzos' (1989) account of more recent excavations at nearby Linou *Alonia* and *Ayii Saranta* suggest that the numbers of tombs with and without copper-bronze objects may have been approximately equal, and that the range of metal types and the quantities in which they occurred (0–14 items per chamber) were comparable to the distributions observed at *Vounous* B and the middle group of tombs from Lapithos (Tables 4.7c, 4.8, 4.11b, 4.12). The Katydhata community may therefore have shared general similarities in prestige symbolism and patterns of mortuary consumption with these sites, although considering Katydhata's proximity to the rich copper mines of the Skouriotissa region, it is perhaps surprising that metal objects were not in fact more plentiful.

Further to the east, in the vicinity of Politiko, the contents of most of the 50 EC–MC tombs reportedly opened by Ohnefalsch-Richter between 1885–1895 (Gjerstad 1926:5; Buchholz and Untiedt 1996:72) have never been completely reported, yet it is apparent that some of the tombs were very rich. *Lambertis* Tombs 16 and 18 contained silver bracelets, and small gold ornaments were recovered from *Lambertis* Tombs 16 and 21 (Åström 1972a:151–2). Politiko *Chomazoudhia* Tomb 3, a relatively small (1.6 × 1.4 × 1.4 m) tomb of MCII date used for at least ten burials, yielded approximately 60 copper-based objects including daggers, flat axes, tweezers, and one tin-bronze shafthole axe of possible foreign workmanship (cf. Courtois 1986a:73; Buchholz 1973:304–309; Buchholz 1979). Bones of horse or other equid and dog were also reported in this tomb (Gjerstad 1926:81; Buchholz 1973:304, note 29). The level of mortuary expenditure evinced in *Chomazoudhia* Tomb 3, along with the probable evidence for collective burial in *Lambertis* Tomb 18, suggests that lavish mortuary festivities comparable to those observed at Lapithos were also conducted in this part of the island during the MC period. The very rich hoard of copper and bronze objects (including two shafthole axes) allegedly originating from nearby Pera, and possibly deriving from one or more looted tombs (Åström 1977), may be a further attestation of these practices.

Similarly, despite the depredations of tomb looting and the scanty recording of many early excavations, it appears that some extremely rich mortuary complexes were also present at Nicosia *Ayia Paraskevi*. Two tin-bronze shafthole axes, deemed on both technological and stylistic criteria to be the products of foreign craftsmen, were reportedly recovered from tombs found in the nineteenth century (Buchholz 1979:78–9; see also Swiny 1982:73–4; Courtois 1986a:73–4; Philip 1991:80–83), and an imported Syro-Cilician jug with painted decoration was attributed to Tomb 9, discovered in 1949 (Merrillees and Tubb 1979; Kromholz 1982:18–9). Other tombs such as Tomb 8, of MCII–III date, although fairly poor in copper (7 items), appear to have been quite rich in ceramic finds (more than 300 in its principal chamber), lead (20 spiral ring fragments), and faience (608 beads ) (Kromholz 1982:16–8, 32–9). Horse teeth were found in Tomb 14, excavated in 1894 (Myres 1897:135, 138).

Mortuary complements from the eastern Troodos foothills and the southeastern region of the island are scarcely better attested, but several tombs afford useful information. Among the 31 tombs of later Early and Middle Cypriot date investigated by Myres (1897) at Kalopsidha, Tomb 9 is notable for the alleged occurrence of horse bones and teeth, and Tomb 11 contained a number of copper-based items along with probable imports of silver and faience. The seemingly intact Alambra Tomb 102, which contained one disarticulated and one *in situ* burial, along with 14 pots, a copper-based axe, a hook-tang weapon, tweezers, a scraper or razor, ten fragments of a tin-bronze spiral ring, a whetstone, two bone needles, and cattle bones is interesting inasmuch as the finds seem to have been primarily associated with a single, male individual (the later, intact burial; see Decker and Barlow 1983:83; Coleman 1996:118–9). The range of metal types represented is once again very similar to the metal complements found in the tombs of *Vounous* and Lapithos.

Other instances of elaborate mortuary celebrations may be inferred from occurrences of shaft-hole axes in various localities of eastern Cyprus, including one from the Alambra region (Cesnola 1877: Pl. 5; Buchholz 1979:79–80), another from the looted Dhali *Kafkallia* Tomb G (where the axe was accompanied by an undecorated bronze belt, Overbeck and Swiny 1972), a further un-provenienced example from the Larnaca area (Buchholz 1979:82), and fasteners apparently derived from a bronze 'warrior' belt alleged to have been found in the vicinity of Klavdhia (Philip 1991:85; Courtois 1986a:75 and note 33). Similar pairs of finds dating to *c.* 1700 BC have been observed in Palestinian tombs such as Jericho Tomb J3 and Tell el Farah Tomb A, and also in Tomb LVII at Ras Shamra (Philip 1991:85; 1995; see also Courtois 1986a: 74–79). Both Philip and Courtois have interpreted the occurrence of these objects, singly and in pairs, as an indication of the Cypriot adoption of 'the standard symbols of power of the Levantine MBA' (Philip 1991:85; 1995; see also Courtois 1986a:74–9; Keswani 1989a:512–4).

By contrast, the burial assemblages known from the northeastern cemeteries of Korovia *Paleo-skoutella* and Ayios Iakovos *Melia*, both sites at which large and elaborate tomb complexes were constructed and collective secondary burial celebrations were evidently staged, appear remarkably poor relative to those of the northern and eastern sites just described. At *Paleoskoutella*, the two burials in Tomb 4 were provided with only 14 pots and two spindle whorls, with no metal items at all. The 14 burials in Tomb 7 were provided with 102 pots of local origin, 11 small copper-based ornaments, one knife, a lead whorl, and other non-exotic small finds. Among the Middle Cypriot tombs of Ayios Iakovos, the quantities of valuables disposed of were also limited. Tomb 6 appears to have been the richest, with 134 pots (or roughly 9 'pots per burial') and 26 copper-based artifacts (1.7 'copper items per burial') associated with its 15 burials (cf. the mean values for the latest tombs at Lapithos: *c.* 14 and 4.9 respectively as indicated in Table 4.13). Interestingly, however, the metal finds included three circular metal discs with relief decoration which Philip (1991:84; 1995:145) has identified as part of another Levantine type 'warrior' belt. An unusually large concentration of faience beads (*c.* 238) was also found in this tomb, although exotic goods were otherwise extremely scarce throughout the cemetery. In general, it would seem that the remote locations of *Paleoskoutella* and Ayios Iakovos served to isolate these communities from the major sources of copper and from long distance exchange contacts, limiting the availability of metal and exotic goods and in turn the propensity to dispose of them in the context of mortuary rituals.

Variable patterns of mortuary consumption would seem to be attested along the south coast of the island. Impressive collections of Red Polished pottery were found in the ECIII tombs exca-vated at Kition *Chrysopolitissa* and *Kathari* (Karageorghis 1974) and in the MCI–II tombs found at Kition *Ayios Prodromos* (Herscher 1988), but as Herscher has noted, metal objects were rela-tively scarce in these tombs in comparison to those of Lapithos. The question of whether this is attributable to looting or to regional differences in wealth and mortuary customs remains un-resolved. Further to the west, at Kalavasos, the tremendous accumulations of wealth characteristic of some of the Lapithos tombs are also absent, but basic measures of material expenditure (Table 4.14) such as the mean numbers of pots and copper items per chamber and mean pots and copper items per burial compare favorably, and in some cases exceed the figures reported for Lapithos and *Vounous*. The two or more burials in Tomb 37, dated to the MCII period, were provided with a total of 26 pots and 16 copper-based objects, including three knives, three hook-tang weapons, two flat axes, one razor/scraper, two pairs of tweezers, one pin or needle, two bracelets, and three unidentified objects, as well as two whetstones and 111 faience beads (Todd 1986:141–2, 163).

An even more impressive MCII assemblage was found in Tomb 36, which, although only half excavated, yielded 20 copper-based artifacts, including three knives, two hook-tang weapons, three razor/scrapers, one awl, one pair of tweezers, two pins, one clothing ornament and seven other items of jewellery. Other finds included 2,608 faience beads, a macehead and a whetstone (Todd 1986:139–41, 163). The exceptionally large number of pots from Tomb 36 (100) may be linked to the unusually large number of persons buried there (at least seven). One of these ceramic artifacts was a remarkable Red Polished III mottled ware bowl with modelled figures, possibly depicting scenes of bread-making or wine-making (Todd 1986:151–4, Fig. 25:2, Pls XIX:3–4, XX–XXIII). These finds suggests that some southern sites participated in a complex of mortuary celebrations and display similar to that observed in the north.

Variations in wealth and the propensity for mortuary consumption are also apparent in south-western Cyprus. Two Philia phase tombs at Sotira *Kaminoudhia* (Cemetery A Tombs 6 and 15) which had burials provided with numerous copper objects and, in one case, a pair of electrum earrings, were succeeded by several ECI–ECII tombs with burials unaccompanied by any preserved metal finds (Swiny *et al.* 2003). The assemblages of metal and other grave goods reported in the multi-chambered, trench-type dromos complexes of Episkopi *Phaneromeni* were also relatively modest. Weinberg recovered a total of roughly 80 pots from the 12 tombs which he investigated, and discounted the effects of looting on the fairly sparse quantity of metal items recovered (1956:121; see also Swiny 1986b:88). In contrast, the 13 tombs rescued by the VVP in Kalavasos in 1978 (Todd 1986) yielded a total of 465 registered objects, of which over 50 were metal artifacts. Mortuary expenditure at Episkopi may therefore have been focused more intensively on tomb architecture than on the disposal of material wealth. A somewhat different picture emerges from the Limassol area, however, where communities may have enjoyed more favorable access to metal sources. Among the four tombs excavated at the locality of Limassol *Ayios Nikolaos* in the 1940s, Tomb 1 had a fairly rich array of finds, with 21 pots, three spindle whorls, four copper-based pins, two needles, a knife, a dagger, a pair of tweezers, and an elaborate necklace made of blue paste beads and cylindrical spiral bronze beads (Karageorghis 1958:143–6).

More recently, an impressive collection of copper-based artifacts, including two axes, three knives, two pairs of tweezers, two chisels, an awl, a scraper, a necklace, a bracelet, and a ring, was recovered from a substantially preserved, albeit water-damaged, tomb in the large EC–MC cemetery near Pyrgos *Mavrorachi* (variably referred to as *Maroraka*, *Mavrorakia* and *Mavroracha*), also in the Limassol region (Belgiorno 1997). Among the other finds from this tomb (Tomb 21) were approximately 70 pots, a faience necklace, several whetstones and maceheads, and a grinding stone. These objects were reportedly associated with a single individual, whom Belgiorno suggests may have been a coppersmith, given the presence of so many ground stone tools that could have been used in finishing metal objects, and the seemingly perfect condition of some of the metal finds, interpreted as 'a set of copper samples coming from the same coppersmith shop' (Belgiorno 1997:139). The Pyrgos site was centered between the Pareklishia, Mazokampos, and Monagroulli mines, and evidence of metallurgical activity has recently been found at the nearby settlement site (Belgiorno 1999; 2000; Giardino 2000). It is interesting to note that the 'coppersmith' was not provided with any hook-tang weapons, suggesting that such items were not an integral part of a craftsman's burial complement. However, this rich complement of goods would seem to illustrate both the special prestige associated with metallurgical expertise and the expression of a distinctive social identity (see also Giardino 2000:28–9).

## Mortuary display and social structure

It is noteworthy that virtually all intact tombs, if not all burials, included some grave goods during the EC–MC periods, suggesting that they were not merely accorded to a privileged minority of the dead. It is difficult to determine whether the objects deposited were the personal possessions of the deceased or gifts accumulated by their survivors, but most likely they were a combination of both. A recent analysis of ceramic types and use-wear, which included substantial collections of tomb pots from Kition and Kalavasos and smaller samples from Sotira *Kaminoudhia* and Bellapais *Vounous*, revealed no significant differences in frequencies of shapes or of used and unused vessels between these mortuary assemblages and those from settlement strata at Marki *Alonia*, suggesting that tomb pottery was not specifically commissioned for mortuary use (Dugay 1996). Sneddon (2002:97, 113) also noted the occurrence of similar types in settlement and cemetery contexts at Marki, although shapes associated with food-processing, as opposed to serving and storage, were infrequent in cemetery contexts.

In contrast, the copper types observed in funerary contexts far surpass the finds from contemporaneous settlement strata in both quantity and diversity. This need not imply that metal goods were not used, worn, or otherwise displayed by the living, but it does suggest that they were deliberately accumulated for mortuary consumption, which may have enhanced their value and stimulated increased copper production. Swiny (1986b: 88) has noted that some knives and daggers found in tombs display little evidence of use, and according to Giardino *et al.* (2002:44), two tin-bronze knives or daggers from Pyrgos *Mavrorachi* Tomb 21 shows casting defects, suggesting that these were probably made for symbolic purposes rather than actual use. The presentation of gifts to the dead may have been regarded as an essential means of ensuring the favor of the ancestors, and it would also presumably have bestowed considerable prestige on the heirs who ostentatiously removed costly goods from circulation (see also Nakou 1995:23). In a few instances at *Vounous*, Lapithos, and Ayios Iakovos, valuable weapons even appear to have been ceremonially damaged or 'killed' by bending their tangs and/or points (Stewart 1962a:294; Åström 1972a:274–5; Dunn-Vaturi 2003:181), rendering them useless to the living unless they were subsequently melted and reworked into new objects. The infrequent use of alloys prior to the Late Cypriot period (Swiny 1982; Balthazar 1990:430–32) may further attest to the relatively greater importance of copper in symbolic displays, as opposed to utilitarian consumption, during the EC–MC periods.

As discussed in Chapter 3, it is extremely difficult to assess differences in wealth and status treatments among individuals in tombs with multiple burials because the associations between particular burials and the goods interred with them have often been disturbed or obscured in the course of tomb reuse. The interpretation of mortuary variability is further complicated by the dearth of proper osteological studies of the human skeletal remains from most tombs, analyses which are essential to the definition of age- and gender-based differences in mortuary treatment, and even to the interpretation of expressions of individual identity that some researchers have recently advocated (e.g. Knapp and Meskell 1997; Meskell 2000). Therefore, only a few general statements are offered below on this subject.

With regard to age-based treatments, it appears that children were generally provided with fewer objects than adults, or with no goods at all. However, some did receive a number of items of pottery and occasionally other valuables as well. The child burial in *Vounous* B Tomb 41 received 23 pots, an unusually large number, while another juvenile in Tomb 23A was arranged with seven paste beads around the head, a copper-based pin to the left, and a group of small vases to the right of the skull (Dikaios 1940:79, 52). Collections of pottery numbering from 3–21 vessels were found with child burials in Swedish Tombs 311B, 313C, 319A, 319C and Pennsylvania Tombs 820, 832B, and 836B at Lapithos (the highest number came from 311B with three child burials), and a few received other items; a copper pin was found in Lapithos Tomb 319A; two faience beads were recovered from Tomb 319C, and goat remains were present in Tomb 836B (Gjerstad *et al.* 1934; Herscher 1978).

Distinguishing gender-based differences among adults is especially problematic, for in the absence of skeletal evidence we are reduced to speculating about what might have been 'appropriate' female goods (e.g. jewelry, dress ornaments, spindle whorls and spindles, needles) versus male goods (e.g. hook-tang weapons and certain metal tools such as axes), making assumptions and imposing binary categories strongly influenced by our own cultural biases (Knapp and Meskell 1997). Many ceramic goods and metal types such as knives, razor/scrapers, tweezers, awls, pins, and rings could well have been 'gender neutral', making even the interpretation of intact single burials difficult (see also Baxevani 1997:65–66).

However, assuming that hook-tang weapons were at least a consistent, if not an invariable appurtenance of males, a consideration of single burials equipped with these items helps to define what may have been the full or 'ideal' complement of status goods for some men. (It should be stressed that all of the associated goods need not have been exclusively 'male', and some women may have been presented in the same fashion; osteological data are essential for resolving these complexities.) The intact male burial in Alambra Tomb 102 serves as a point of departure; this individual, as already noted, was equipped with 14 pots, a copper-based axe, a hook-tang weapon, tweezers, a scraper or razor, ring fragments, a whetstone, two bone needles, and cattle bones (Decker and Barlow 1983:83; Coleman 1996:118–9). Among the ECI tombs of *Vounous* A, a similar complement may be seen in Tomb 161, where a particularly rich single burial was accompanied by a hook-tang weapon, two knives, two axes, a whetstone, 44 pots including four cult vessels and cattle remains (Stewart and Stewart 1950). Another example of a 'full complement' of later (ECIIIB–MCI) date may be observed in Lapithos Tomb 837, where the deceased was equipped with one knife, one hook-tang weapon, one razor/scraper, one pair of tweezers, an awl, a whetstone and 12 pots (Herscher 1978). A much richer, contemporaneous burial occurred in Lapithos Tomb 301C, which contained three knives, two hook-tang weapons, one razor/scraper, two pairs of tweezers and seven pots—in effect, the same types observed in Tomb 837 but with a greater quantity of metal goods and less pottery (Gjerstad *et al.* 1934). These tomb finds suggest a recurring complement of status goods, which, once again, may or may not have been gender specific, but in at least one instance was associated with a male.

Other single burials were provided with only portions of this complement. The burial of the 'seafarer' in Karmi *Palealona* Tomb 11 B, for instance, was equipped with one knife, one hook-tang weapon, a faience bead and seven pots, including a Minoan cup (Stewart 1962b). The presence of these metal artifacts and imports in association with a relatively simple burial in a very small (*c.* 1 × 1.4 m) chamber would tend to support the argument that such goods were not the exclusive perquisites of an exalted elite.

In two recent discussions of EC–MC sociopolitical structure and ideology, it has been postulated that the mortuary evidence from sites such as Vasilia, *Vounous*, and Lapithos attests to the emergence of 'status-conscious' and 'weapon-bearing' elites (Peltenburg 1994:158–9) with 'sustained and inheritable power' (Manning 1993:45) during the EC–MC periods. Manning has also suggested that the diversification and elaboration of the ceramic assemblage in the EC–MC periods may be viewed as evidence for the development of a 'drinking complex', in which alcoholic beverages were produced under elite supervision and consumed on occasions of elite sponsored hospitality and largesse (1993:45; following Sherratt 1987). Both Manning and Peltenburg have further argued for a highly stratified social order based on their interpretations of the symbolism of the 'sacred enclosure' model from *Vounous* Tomb 22. According to Manning, the representation of a male seated on a raised chair and evidently presiding over a ritual scene attests to 'an institutionalized, secular, form of power concentrated in, or expressed by, a key individual' (1993:45). Based on his interpretation of the bowl, he concludes that 'one may recognize an hereditary elite and all the ingredients for further development into entities like polities or states' (1993:46). Peltenburg likewise interprets the *Vounous* bowl as the expression of an hierarchical social order dominated by a ritually sanctified leader, drawing structural comparisons between the deployment of animal (bull), female, and male figures within the bowl and the depiction of nature, mankind, and the king and goddess on the famous alabaster relief vase from the E-anna precinct of Meso-potamian Uruk (Moortgart 1969:13).

There can be no doubt that differences in wealth and social status developed within and between communities in Cyprus during the EC–MC periods. Yet the complements of weaponry, ceramics, and associated symbolism which have been portrayed as evidence for the emergence of an hereditary elite dominating a rigidly stratified social structure are amenable to alternative inter-pretations. Some EC–MC tomb assemblages are indeed outstanding in the quantity and diversity of their metal, ceramic, faunal, and other artifactual remains. However, they appear so mainly because they contain the full array of items which can be identified as valuable, exotic, or 'cere-monial' in nature, not because of particular elements which set them apart from all the rest. Detailed quantitative and qualitative analyses of the *Vounous* and Lapithos cemeteries reveal that the most exceptional copper assemblages, allegedly associated with elites at these sites, are distinguished primarily by the numerical replication of metal types (significantly affected by the number of burials in the tomb chamber as well as by relative 'wealth'), along with the presence of the most diversified arrays of types, rather than by the presence of objects that were unattainable by other members of the community. Qualitative, as opposed to quantitative, differentiation in status paraphernalia between groups is difficult to discern, with most of the items present in 'full complements' broadly distributed throughout the cemetery, especially in the later tombs (Table 4.15). Elaborate ceramic containers, drinking, and pouring vessels were also widely distributed, especially in the cemeteries of *Vounous*, and they do not seem to have been exclusively associated with 'weapon-bearers'.

It is also evident that mortuary expenditure, as measured both by energy investment in tomb construction and material outlays in the form of grave goods, increased over time. The means for chamber tomb floor area, 'pots per burial', and 'copper items per burial' rose dramatically between the earlier and later periods at both *Vounous* and Lapithos. As the boxplots in Figs. 4.7 and 4.8 illustrate, the medians and interquartile ranges, as well as the means, were also rising, suggesting that the propensity for increasing mortuary expenditure was broadly based within the community. It is not simply the case that a few 'elite' groups were expending a great deal more in material display in the later periods; rather, a large number of kin groups were expending more over time. These gross increases in mortuary expenditure, like the increasing elaboration of mor-tuary ritual in general, would seem to reflect rising levels of prestige competition within the com-munity at large and the growing centrality of mortuary ritual as an arena for the construction of social status.

The ongoing processes of prestige competition and competitive elaboration are perhaps most vividly illustrated by the distributional and size patterning of the so-called hook-tang weapons, which were both the heaviest, most impressive items of copper found in EC–MC tombs and the most ubiquitous. Not only did the frequencies of these weapons increase through time at both *Vounous* and Lapithos, but concurrently, the mean size of the weapons and the variation in size also increased, as illustrated in Table 4.15 and Fig. 4.6. Some of the weapons were so large (up to 66 cm in length at Lapithos) as to be very difficult to haft or use, rendering them effectively 'baroque,' as Philip (1991:68) has dubbed them. The size patterning within this artifact category is highly suggestive of a contest in 'one upmanship' amongst individuals with access to the same symbolic paraphernalia. The wide distribution of the hook-tang weapons also makes it more likely that they were the basic prestige symbols to which most (presumably male) adults aspired (see also Philip 1991:69) rather than the private paraphernalia of an hereditary 'weapon bearing elite'.

In light of the foregoing discussion, the question of whether the *Vounous* sacred enclosure model represents any 'larger unit of society than an extended household' (posed by Coleman 1996:329) merits close consideration. The mortuary evidence strongly suggests that EC–MC

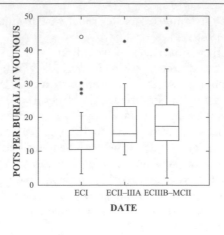

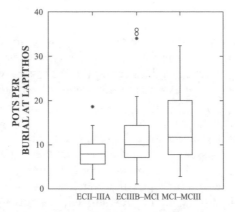

**Figure 4.7** Box plots showing the increase in median and upper ranges of pots per burial at Bellapais *Vounous* and Lapithos *Vrysi tou Barba* through time. Open circles seen in Figures 4.7 and 4.8 represent outlying values more than 3 hspreads from the upper hinges of the boxes. (For an explanation of terminology and other box plot symbols, see Figure 4.3 caption.)

prestige differentials were graded and negotiable, rather than being rigidly linked to a system of social hierarchy in which members of the elite differentiated themselves through the use of a qualitatively distinctive complement of prestige goods. The archaeological evidence from settlement contexts also offers little or no corroborating evidence for sociopolitical hierarchy. As Coleman and others (Swiny 1989:25; Frankel 2002:173) have noted, neither regional surveys nor excavations of EC–MC settlements have so far revealed any clear indications of the development of settlement hierarchies, tributary networks, elite residential architecture, or monumental public buildings of religious or administrative function, material correlates of political complexity that are generally associated with hereditary elites in other cultural settings (e.g. Peebles and Kus 1977; Earle 1977; 1987; Steponaitas 1978; 1981; Wright 1984). Hence it might be suggested that the status differentials portrayed in EC–MC mortuary rituals and in the iconography of the *Vounous* bowl had not yet become institutionalized relations of a sociopolitical hierarchy that structured the experience of everyday life. However, the dynamics of mortuary competition, its stimulative effect on copper production, and the demand created for exotic goods in the context of prestige displays may well have set the stage for the transformations in local economic and power relations that characterized the succeeding Late Cypriot period.

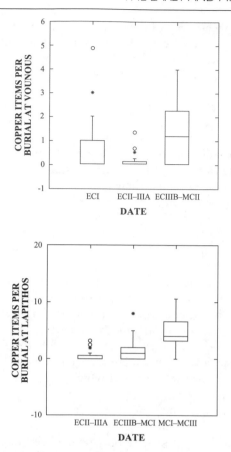

**Figure 4.8**     Box plots showing the changing medians and ranges of copper items per burial at Bellapais *Vounous* and Lapithos *Vrysi tou Barba* through time. (For an explanation of box plot symbols, see Figure 4.3 and 4.7 captions.)

## Foreign valuables and the changing construction of prestige

An examination of the finds tables from *Vounous* (Tables 4.7a–c) and Lapithos (Tables 4.11a–c) reveals that imported goods, for the most part, were found mainly in the richest tomb complexes overall at these sites. It is perhaps not surprising that groups with the greatest access to locally produced metal goods were also most successful in making exchanges with foreign traders or other overseas contacts. Yet the continuing scarcity and irregular occurrence of imported objects in even the richest tombs of the period makes it difficult to interpret them as essential components of an elite symbolic complex—no matter how great their prestige value—or as commodities which were strategically controlled and redistributed. Rather, they may represent the seeds of a gradual transformation in the symbolism of prestige.

In many societies, goods of non-local origin may be endowed with extraordinary value relative to their cost and symbolic significance where they were manufactured. Their value may derive in part from the intrinsic allure of esoterica, described by Brumfiel and Earle as 'the excitement of the exotic, the pleasure of the beautiful, and the significance of the symbolic' (1987:8; see also Helms 1988). To the extent that foreign goods are also associated with the elite of a more complex society whose material culture and ideology is perceived as more prestigious, their value may be further enhanced (Leach 1954; Flannery 1968; Wheatley 1975; Renfrew 1986). Such 'symbolic entrainment', as it is termed by Renfrew, generally serves to accentuate the degree of social differentiation in the borrowing society. In identifying itself with the values and ideology of the elite in the more complex society, the elite of the less developed polity may distance itself from the commoners of its own society and promote the development of a new hierarchical order which would have been incompatible with traditional ideology and social orderings (e.g. Wheatley 1975:238–9).

The unfolding of this process in Cyprus may be exemplified in part by the occurrence of shafthole axes, of which 14 have been recovered in a band of sites near the copper-mining zones of the Troodos and slightly farther to the south in the Larnaca/Klavdhia area (Buchholz 1979). Only two examples, from Dhali *Kafkallia* Tomb G and Politiko *Chomazoudhia* Tomb 3, have exact provenience. Swiny (1982:73–74) has observed that both the tin content and the high craftsmanship of the shafthole axes render them anomalous in comparison to contemporaneous copper-based objects of Cypriot manufacture, and it is likely that they were made by Near Eastern craftsmen, possibly working in Cyprus. Shafthole axes appear to have been an important compo-nent of contemporary, high status 'warrior' burials in Syria-Palestine, with the best, albeit inexact, parallels coming from MBII tombs at Ras Shamra (Philip 1991:80–83). The examples from Dhali and (allegedly) Klavdhia were associated with bronze 'warrior belts', items which co-occur with shafthole axes at the sites of Jericho, Ras Shamra, and Tell el-Farah in Syria-Palestine (Courtois 1986a:74; Philip 1991:84–5; 1995). The axe, perhaps a traditional symbol of authority in Cyprus, was now elaborated in form and quality, and it may have been further envalued through associations with more powerful and prestigious foreign elites. These associations may have been conveyed through personal interactions with foreigners at home and abroad and through the ob-servation of Near Eastern glyptics, which sometimes depicted warriors and gods with axes, other weapons, and regal paraphernalia such as maces which were also deposited in Middle Cypriot tombs.

Indeed, Syrian and Old Babylonian cylinder seals seem to have made their first appearance in Cyprus during the later Middle Cypriot period, sometimes remaining in circulation for several generations, presumably as highly valued heirlooms, before being deposited in Late Cypriot tombs at Enkomi and other sites (Courtois 1986a; Merrillees 1986). These too may have served as im-portant emblems of status. Further imports included silver and faience ornaments, ceramics of Syro-Palestinian and Egyptian origin, and possibly horses (as suggested by the bones found in Politiko *Chomazoudhia* Tomb 3, Kalopsidha Tomb 9, and Lapithos Tomb 322B). The 'sacrifice' of domesticated livestock had been a long established practice in Cypriot mortuary ritual; now it was expanded to include horses, which were important status symbols for Syrian and Babylonian elites (Sasson 1966; Courtois 1986a:74) and may have been similarly valued by the Cypriots as well.

The display of axes, maces, cylinder seals, horses, and other valuables on ceremonial occasions including mortuary ritual had the potential not only to augment but to differentiate qualitatively the status of their bearers with a symbolism that was lacking in locally produced goods. Political leaders controlling access to foreign prestige goods may have further enhanced their authority by the strategic distribution of certain items to supporters and allies. However, the effective manage-ment of this new symbolic capital required a monopolistic control over incoming goods, the move-ment of traders, and, concomitantly, the production of goods for exchange, conditions that were not realized before the Late Bronze Age.

## Summary and conclusions

The questions posed at the beginning of this chapter concerning the significance of developments in EC–MC mortuary practice are now revisited.

1. *To what extent do the new forms of mortuary practice noted represent customs transplanted from the mainland as part of the immigrants' cultural baggage, and to what extent do they represent new ideologies and modes of practice developed and deployed by heterogeneous communities interacting within an increasingly complex social landscape?*

This question presupposes some level of colonization from Anatolia during the Philia 'facies' preceding and perhaps partially contemporaneous with the EC period, as hypothesized in a number of recent articles (Webb and Frankel 1999; Frankel *et al.* 1996; Peltenburg 1996; see Knapp 2001 for a dissenting view). Several hallmarks of Philia and EC–MC mortuary practice, including the use of extramural cemeteries, practices of secondary treatment and collective burial, chamber tomb forms, and the provisioning of the deceased with grave goods, have parallels throughout mainland Anatolia and the Levant in the Early Bronze Age. However, the entire complement of practices that emerged in late third millennium BC Cyprus is not as yet readily discernible within any specific region of western Anatolia, the proposed homeland of the immigrants (Wheeler 1974; Carter and Parker 1995), whereas local precedents are clearly evident in the Middle Chalcolithic cemeteries of Souskiou in Cyprus. Therefore it seems likely that Cypriot mortuary traditions represent an evolving fusion of mainland and local practices, elaborated by indigenous and immigrant communities in the context of ongoing social competition and gradual cultural assimilation. The striking similarities between later Early and Middle Cypriot mortuary complexes (e.g. Lapithos Tomb 313A and *Paleoskoutella* Tomb 7) and Syrian tombs in the Halawa region and perhaps elsewhere suggest that social connections and cultural interactions with the mainland were far flung, and persisted throughout the EC–MC period.

2. *What do EC–MC mortuary practices, in their initial form and as they developed over time, tell us about the social structures, strategies, and belief systems of older and newer Cypriot communities?*

The widespread adoption of extramural cemeteries, along with other transformations in mortuary ritual, may be indicative of changing ideological constructions of the relationship between the living and the dead and of changing social dynamics within and between communities. The establishment of permanent, spatially reserved burial areas for the ancestors most likely had multiple meanings in the context of ongoing social and economic changes. With the expansion of settlement in conjunction with renewed colonization and the spread of plow agriculture, the affirmation of inter-community boundaries and territorial rights may have found expression in the creation and prolonged reuse of formal cemeteries. At the intra-communal or household level, the use of familial tombs could have served as a reference point for validating local rights of successorship, inheritance of cultivable land, livestock, and other forms of property, along with the perpetuation (or redefinition) of social status. Simultaneously, the stricter demarcation of habitation spaces for the living and the dead may reflect a new characterization of the ancestors as entities that empowered and/or exerted power over the living in their daily lives, and as such necessitated a more formalized, 'distanced' set of treatments. The prolonged reuse of houses and adjacent residential areas, in turn, may have favored the reservation of burial spaces that would not be inadvertently violated in the course of mundane activities. The need for dedicated, communal staging grounds would also have been reinforced over time as mortuary rites became ever more costly productions and the scale of social participation in mortuary celebrations widened.

Mortuary rituals involving secondary treatment and collective burial seem to have emerged in Chalcolithic Cyprus and were greatly elaborated during the EC–MC periods in conjunction with overall increases in mortuary expenditure and display. The ethnographic literature considered in Chapter 2 demonstrates that secondary funerals may have diverse social and ideological ramifications, but there are several consistent themes which may be relevant to the Cypriot case. For example, practices of secondary treatment may effect the transformation of the deceased from decaying corpses to purified ancestors with the power to bestow a variety of material and social benefits such as fertility, prosperity, and social prestige upon the heirs who venerate them. With the passage of time and the depersonalization that accompanies the separation of bones from flesh, the ancestors also become susceptible to social manipulation by their successors, who may use the often grand occasions of secondary ritual to magnify, if not totally redefine, the social importance that their forebears enjoyed while they were alive. Moreover, unlike death, secondary funerals may be scheduled and prepared for long in advance of their occurrence, allowing the sponsors to invite guests from far and wide while accumulating sufficient resources for their entertainment and amazement. The resulting displays of largesse, both in the form of hospitality extended to the living and commemorative expenditures on behalf of the dead, permit the sponsors to assert and sometimes renegotiate their social status and to cultivate, reaffirm and broaden social alliances. The quantities of metal and other grave goods disposed of at sites such as *Vounous* and Lapithos, the extensive sacrifice and presumable consumption of livestock at those sites as well, the evidence for funeral feasts and other specialized ritual activities at *Paleoskoutella*, and the general material elaboration (evident in tomb construction and grave goods) at many other sites all tend to suggest that mortuary rituals had a similar importance in EC–MC Cyprus.

The collective aspects of EC–MC mortuary practice, specifically the sequential reuse of tombs for one or more burials, and the occasional use of tombs for multiple secondary burials, offer further support for the interpretations of extramural burial and secondary treatment advanced above. The grouping of the dead in tombs that were permanent monuments rather than ephemeral or expedient burial locations would have permitted the living to make temporally extended references to their forebears, emphasizing and perpetuating a sense of kin group identity as well as social legitimacy. Given the increasing importance of heritable property associated with plow agriculture, the attendant labor requirements, and the high costs of mortuary and other prestige displays, the maintenance of kinship ties must have been essential to household reproduction. Consequently groups, or individual connections with particular groups, were memorialized in collective, as opposed to 'individualizing' burial chambers and ritual celebrations.

Dramatic increases in mortuary expenditure, as attested by the energy invested in tomb construction and the wealth consumed in the form of grave goods, are evident over the course of the EC–MC periods. This phenomenon, along with the establishment of formal burial grounds, the reuse of particular tombs, and the elaboration of ritual practices in general, bespeaks the increasing centrality of mortuary observances within the social life of the community. Funerary rites must have been occasions *par excellence* for competitive status displays and conspicuous consumption that are not readily evident in contemporary domestic architecture or other aspects of material culture. As more and more groups began to engage in these competitive displays, the attendant expenditures progressively increased, and it is possible that an economy once centered primarily upon subsistence was transformed, in some communities at least, into an economy with a large funerary sector, stimulating increasing copper production in response to ritual requirements. The unintended consequences of this development in turn would have been the increased visibility of Cypriot copper resources, increasing overseas trade aimed at acquiring those resources, and an increasing local demand for foreign valuables acquired in exchange for copper. Thus even as mortuary ritual was becoming the principal arena for the creation of social prestige and social alliances, it may also have set the stage for transformations in systems of production and power relations fundamental to Late Cypriot political and economic developments.

3. *Do the architecturally elaborate and metal-rich tombs of the Philia Culture cemetery of Vasilia* Kafkallia *and the ostensibly later EC–MC cemetery of Lapithos* Vrysi tou Barba *mark the presence of an hereditary elite, one that was perhaps already established among newcomers,*

*and spread rapidly to older indigenous communities as some groups began to exploit new forms of prestige symbolism and technology (e.g. Manning 1993; Manning and Swiny 1994)? Does the depiction of a male figure presiding over a ritual scene in the famous Vounous 'sacred enclosure' model represent the existence of an hierarchical social order dominated by a ritually sanctified leader, drawn from a status-conscious, weapon-bearing elite whose notions of the social and cosmological order had Mesopotamian antecedents (Peltenburg 1994)? Or was EC–MC society characterized by the absence of political and settlement hierarchies (Swiny 1989) and a relatively low level of socioeconomic differentiation, as other recent studies (Baxevani 1997; Davies 1997) imply?*

If the cemetery of Vasilia *Kafkallia*, with its monumental tombs and rich deposits of metal and other valuables, truly dates to a phase of the Philia facies preceding the EC period as Webb and Frankel (1999:8) have recently suggested, then it appears that the early Philia colonists practiced very elaborate prestige displays, either sustaining or creating some degree of elevated social status for themselves in their new Cypriot homeland. Interestingly, too, the discrete patterns of metal distribution in the earlier tombs of both *Vounous* and Lapithos are also consistent with a system of ascribed social statuses. However, much of the remaining mortuary evidence suggests that any 'hereditary' status differentials that may have been present at the onset of the EC–MC era soon gave way to a highly competitive and fluid system of status negotiation. Some individuals and groups were indeed endowed with remarkable collections of metal and ceramic prestige goods, and yet the individual types that comprised these assemblages were often widely distributed rather than exclusively accorded to a privileged few. Distinctions between richer and poorer individuals and groups alike tended to be quantitative rather than qualitative, and certain key prestige goods such as copper hook-tang weapons were clearly subject to competitive one-upmanship in manufacture and display. Consequently, the answer to the foregoing questions may be that while there were considerable differences in wealth and social status within and between some communities, these distinctions were not translated into rigid, insurmountable class barriers or political hierarchies that dominated and regulated social life. The origin of the social elites that emerged subsequently in Late Cypriot urban centers may be better traced to the dynamics of prestige competition unfolding in the EC–MC period, rather than to imported traditions of Mesopotamian cosmological hierarchy and sacred leadership. However, during the Middle Cypriot period, escalating social and economic interaction with adjacent mainland regions may well have introduced new complements of foreign prestige symbolism that would ultimately be adopted and deployed by Cypriot elites in the ensuing Late Bronze Age.

# 5 The Late Bronze Age

## Introduction

During the Late Bronze Age (*c.* 1650–1050 BC) a number of towns arose in Cyprus, first at Morphou *Toumba tou Skourou* in the northwest and Enkomi *Ayios Iakovos* in the east, and later at several localities along the southern shores of the island. The residents of these settlements engaged to varying degrees in copper working and other specialized craft activities, and they participated in extensive trade networks linking distant regions of the eastern Mediterranean (Muhly 1982; 1986; 1989; 1991; Stech 1982; Knapp and Cherry 1994; Manning and De Mita 1997). The role of trade in stimulating urban development is evinced by the location of copper producing centers in coastal areas that were sometimes relatively far from the principal sources of copper ore, the steadily increasing array of imported valuables in settlement and mortuary contexts, and the use of Near Eastern systems of weight metrology as early as Late Cypriot I (Courtois 1983; 1984a, 1984b; Petruso 1984). Textual references from Mari and Babylon (Dossin 1939; 1965; Millard 1973; Knapp 1979:179–81; 1996b:17–20) suggest that the land of *Alashiya*, which is generally equated with Cyprus or some part of the island (Dussaud 1952; Holmes 1971; 1975; Muhly 1972; Georgiou 1979; Knapp 1979:152–78; 1985, 1996b), was recognized as a copper source by these Near Eastern polities no later than the eighteenth century BC, or the later part of the Middle Cypriot period. Further texts from Alalakh, Ugarit, Amarna, and Boghazkoy indicate that *Alashiya, Isy,* or *'Irsa* as it is sometimes designated, had become a major participant in eastern Mediterranean trade and politics between the fifteenth–thirteenth centuries BC or the Late Cypriot II period (Knapp 1979; 1990b; 1996b; Malbran-Labat 1999; Yon 1999; Goren *et al.* 2003). These historical allusions receive further support from the increasing corpus of Cypriot artifacts recovered from archaeological sites throughout the eastern Mediterranean (e.g. Åström 1980; Gittlen 1981; Kiliian 1978; 1988; Merrillees 1968; Oren 1969; Portugali and Knapp 1985; Shaw 1984; Vagnetti and LoSchiavo 1989; White 1986) and from the late fourteenth-century shipwreck found off the coast of Uluburun, Turkey (Bass 1986; Pulak 1988; 1997; 1998).

The intensification of copper production, the expansion of long distance trade, and the rise of urban centers were accompanied by major transformations in the political and social structure of Cypriot communities (Knapp 1986a; 1994b; 1996a; 1997). The appearance of monumental ashlar residences, administrative buildings, and temples in the LCII–III periods bears witness to the development of power differentials and social stratification within the major centers. Egyptian and Near Eastern texts refer to *Isy/Alashiya/'Irsa* as a land governed by a king, and some researchers have argued for the existence of a single island-wide state centered on the site of Enkomi in LCI and the earlier part of LCII (Muhly 1989:299; Knapp 1994a:428; 1994b:291; Peltenburg 1996: 27–37; Webb 1999:305–307). While the importance of Enkomi as a political, industrial, and trading center during this timeframe is indisputable, there are nonetheless indications of emergent, contemporaneous 'peer' polities or independent town centers at Morphou *Toumba tou Skourou* (Vermeule and Wolsky 1990) and Maroni (Cadogan 1996; Herscher 1998:324–6) where LCI–IIB settlement strata and other interesting deposits (e.g. a group of large LCI sherds found in the under-water survey at Maroni; Manning and De Mita 1997:128–9) have been excavated. Moreover, recent petrographic studies of several clay tablets from Amarna and Ugarit containing letters written by rulers of *Alashiya* suggest that the political center or centers from which these letters were sent in the fourteenth and thirteenth centuries BC may have been located on the south coast of the island, possibly in the vicinity of Kalavasos *Ayios Dhimitrios* or Alassa *Paliotaverna* (Goren *et al.* 2003). The question of whether either of these centers or Enkomi ever exerted island-wide hegemony remains problematic at present. It is possible, however, that the titular center or capital of *Alashiya* shifted from time to time and that its ruler was, ideologically, a *primus inter pares* rather than the leader of a tightly integrated political system (see also Goren *et al.* 2003:252).

For the later LCII and LCIII periods, the archaeological evidence is certainly more consistent with the coexistence of a number of autonomous regional polities that varied in territorial scale and internal organization (Keswani 1996). Some of these polities, such as those which grew up in the vicinities of Kalavasos, Maroni, and Alassa, seem to have had relatively centralized, hierarchical structures, with political power and economic resources highly concentrated in readily identifiable elite administrative complexes. Others, such as Enkomi, Hala Sultan Tekke, and perhaps Kition, appear to have been dominated by a multiplicity of elite groups and institutions evincing a more 'heterarchical' pattern of local hegemony. Yet it is probable that there were fluctuations between hierarchical and heterarchical modes of integration in most urban centers over time, along with shifting patterns of collaboration and competition among them, such that regional political land-scapes were periodically reshaped. The relationships between urban centers and the hinterlands that supplied them with copper and agricultural produce were probably also diverse, based on variable systems of wealth and staple finance, direct domination, and more informal networks of alliance and cooperation (Keswani 1993; Webb and Frankel 1994; Knapp 1997; Manning and De Mita 1997).

In conjunction with these developments, the mortuary record of the Late Cypriot period clearly attests to the emergence of elite groups employing a distinctive complement of status symbolism (Keswani 1989a, 1989b). This prestige goods complex was comprised to a large extent of valuables acquired through trade and other, locally produced goods whose iconography was often strongly influenced by foreign cosmologies and ideologies of status that were adapted into the local culture. Meanwhile, traditional modes of mortuary practice underwent a variety of transformations. LC tombs were often located within settlement contexts, sometimes associated with private residential, administrative and workshop complexes, rather than in the extramural communal spaces favored in the past. Older practices of secondary treatment and collective burial persisted through much of the period, with several dramatic instances of large-scale collective burial taking place during LCI, and a propensity throughout the era for long-term tomb reuse far exceeding the timespans represented in EC–MC tombs. Yet towards the end of the Late Cypriot period there are signs of discontinuity, marked by the appearance of shaft graves used for single or relatively small numbers of burials, and a probable shortening of the traditional cycle of mor-tuary obsequies. Furthermore, an analysis of Late Cypriot patterns of material consumption reveals that even as large quantities of gold jewelry and other exotic wealth objects were disposed of in the course of elite funerals, there was a concurrent decrease in other forms of mortuary expen-diture such as tomb construction, and by the end of the Late Cypriot period, the disposal of valuable goods in mortuary contexts had also declined.

This chapter addresses a number of questions regarding (1) the significance of changes in the spatial disposition, ritual treatment and grouping of the dead, (2) long-term changes and syn-chronic variation in tomb architecture, (3) the evidence for social hierarchy and prestige symbol-ism in LC burial assemblages, and (4) the implications of declining mortuary consumption over the course of the Late Cypriot period. I will argue that these developments overall reflect the changing significance of mortuary ritual in stratified, socially heterogeneous urban communities, where status differentials were no longer primarily created through periodic, ritualized exhibitions among competitive kin groups, but were instead increasingly based upon differential access to copper, trade goods, and positions attained within a variety of court and temple institutions. In this altered sociopolitical context, mortuary ritual would have remained a crucial arena for the expression and reproduction of status differentials, but it most probably ceased to be the principal venue for their production.

## The spatial dimensions of Late Cypriot mortuary practice

Significant differences in the spatial dimensions of mortuary practice appear to have developed in rural and urban contexts during the Late Cypriot period. The persistence of the EC–MC tradition of burial in extramural cemeteries is evident at many localities which, based on current archaeological evidence, were not in close proximity to major towns. Some of these cemeteries, such as Ayios Iakovos *Melia* (Gjerstad *et al*. 1934), Dhenia *Kafkalla* and *Mali* (Åström and Wright 1962; Hadjisavvas 1985; Webb and Frankel 2001), Politiko *Ayios Iraklidhios* (Karageorghis 1965a), and Katydhata (Åström 1989), had already been in use for hundreds of years prior to the onset of the Late Cypriot period, while others including Ayia Irini *Palaeokastro* (Pecorella 1977), Myrtou *Stephania* (Hennessy 1964), Pendayia *Mandres* (Karageorghis 1965c), Milia *Vikla Trachonas* (Westholm 1939a), Angastina *Vounos* (Karageorghis 1964b; Nicolaou 1972), and Akhera *Chiflik Paradisi* (Karageorghis 1965b) seem to have been established in the MCIII or LCI periods. Although settlement sites must certainly have been located nearby, these rural cemeteries would seem to represent a continuing separation between burial grounds and zones of human occupation, contrasting with patterns of intramural burial emerging at many of the urban sites discussed further on.

As in previous periods, the size and physical settings of Late Cypriot extramural cemeteries varied considerably. The cemeteries of Dhenia and Katydhata, for example, seem to have been quite large, while others such as Ayios Iakovos *Melia*, Myrtou *Stephania*, Akhera and Pendayia appear to have comprised small numbers of tombs, frequently ranging from 12–15 identified features, presumably used by small segments of the local population. Possibly several such cemeteries were in use contemporaneously in the vicinity of any given settlement. Some burial grounds were located on plains, some on ridges, while others, such as Ayia Irini *Palaeokastro*, were located near the sea. Tomb orientations within these cemeteries were also highly variable, influenced perhaps by a combination of factors ranging from the peculiarities of local terrain to shifting sociocultural preferences. The disposition of the tombs at Ayios Iakovos *Melia* helps to elucidate this point: most of the tombs at the northern end of the site were oriented south-north, while those at the southern end tended to be oriented west-east (Gjerstad *et al*. 1934: Plan V, 3). This could be interpreted as evidence for social divisions between clans or moieties or other groupings within the community, but the placement of tombs appears to have shifted northward over time, suggesting that chronology could have been a factor as well (Keswani 1989a:260–61). It remains unclear, therefore, as to whether the orientation of tombs was an adjustment to micro-topographical conditions, a deliberately constructed opposition between kin groups, a contradiction of past practices and social predecessors, or the consequence of other unknown variables.

While EC–MC traditions of extramural burial persisted in rural regions, new patterns of association between habitation zones and tombs were emerging elsewhere. The mortuary precinct at Korovia *Nitovikla*, made up of roughly 15 tombs located within an unusual walled area outside the fortress, affords one distinctive example of changing practices (Gjerstad *et al*. 1934:407–15).

Elsewhere the boundaries between spaces for the living and spaces for the dead were seldom clearly demarcated, but often deliberately juxtaposed and, over time, superimposed. The six MCIII–LCI/LCII tombs excavated by Vermeule at *Toumba tou Skourou*, for instance, were located just outside the so-called House B, a well-constructed, perhaps elite residential building that contained several large storage pithoi and was located immediately to the south of probable pottery workshops (Vermeule and Wolsky 1990: Fig. 2). At Enkomi, hundreds of tombs were located within the settlement proper and in the crags along the eastern edge of the site (Murray *et al*. 1900; Courtois 1981: Fig.1; 1986b: Fig. 4). Some of the tombs may originally have been located in streets and open areas around the buildings, while others were located in courtyards and sometimes even beneath contemporaneous buildings. Eventually many tombs were either destroyed or permanently sealed in the context of later construction activities. As in the extra-mural cemeteries of the Late Cypriot period, considerable variation in tomb orientations is evident, again perhaps attributable to localized building constraints, diverse social and ideological symbolism, and diachronic shifts in the expression of that symbolism (Keswani 1989a:343–4, Tables 5.34, 5.36, 5.43, 5.49, 5.52).

Swiny once suggested that the limitations of space in walled towns, along with 'a reluctance towards extra-mural burials', may have resulted in the overlapping of mortuary and habitation zones (1981:79, note 90). Presumably the cause of such 'reluctance' would have been the prevalence of warfare or raiding, which would also have necessitated the fortification of settlements. However, intra-settlement burials seem to have occurred both prior to and in the absence of circumvallation at a number of town sites, and it is therefore possible that more fundamental transformations in community structure and the symbolic relations between the living and the dead were responsible for this change in practice.

Part of the explanation for the placement of tombs within or in very close proximity to habitation areas may lie in the circumstances under which the earliest Late Cypriot towns were established. The earliest settlers of the newly founded towns at *Toumba tou Skourou* and Enkomi presumably moved to these sites from various communities in outlying areas in order to take advantages of the opportunities for trade (Keswani 1996). The convergence of heterogeneous social groups may be reflected in the open areas surrounding the earliest architectural units at Enkomi and the occurrence of several discrete mounds at *Toumba tou Skourou* (Vermeule and Wolsky 1990:15), perhaps representing hamlets occupied by kin groups from different 'ancestral' villages. If this hypothesis is correct, then the founders of these settlements may have lacked the sense of corporate identity often associated with communal, extramural burial grounds. Closely related individuals and/or descent groups would have maintained their own burial grounds near their houses and workshops, apart from those of more distant or unrelated groups within the community. At Enkomi, where the evidence is most detailed, relationships between the living and the dead seem to have been overtly expressed through the location of tombs in courtyards and entranceways that were traversed by the living on a daily basis (Dikaios 1969: Pls 248, 267, 268). The intensity of these associations between particular residential groups and their respective ancestors may have actively reinforced the social boundaries between those groups over time.

However, the practice of intra-settlement burial is represented not only in 'frontier towns' like *Toumba tou Skourou* and Enkomi, but also in centers located in areas where a fairly continuous sequence of prior occupation is evident. At Kition, the very rich Tomb 9 was located near a contemporaneous copper workshop (Karageorghis 1976: Fig. 7). At Maroni and Kalavasos, LCIIC ashlar buildings appear to have been deliberately constructed above and adjacent to earlier tombs of which the builders could scarcely have been unaware (Cadogan 1996; Manning 1998a; Manning and Monks 1998; South 2000). At Kourion *Bamboula* and Alassa *Pano Mandilaris*, the juxtaposition of tombs and houses seems to have been established during the earliest phases of settlement, with tombs located in streets and open areas where the living pursued their daily activities (Hadjisavvas 1989: Figs 3.1, 3.3; Benson 1972: Pl. I; Weinberg 1983:36–7). It is possible that the placement of tombs in close proximity to domestic and workshop areas was linked not only to the heterogeneous origins of urban settlers, but also to an incipient and widespread 'privatization' of the ancestors in the context of increasing inter-familial, as opposed to inter-community, competition in the Late Cypriot period. The proximity of the ancestors would have served as an important symbolic validation of the users' rights of ownership or control over land, houses, and productive facilities, fostering in turn a strong sense of kin group identity as the tombs of the dead were encountered on a daily basis. Even tombs located in ostensibly 'public' areas (streets, squares etc.) may have been closely identified with specific elite groups, testifying to their hereditary legitimacy.

Manning (1998a; Manning and Monks 1998) has recently argued that the sealing and destruction of some earlier LCIIA–B tombs at Maroni *Vournes*, taking place in conjunction with the construction of the ashlar building and associated craft and storage facilities at that site in LCIIC, may be interpreted as a strategic appropriation of ancestral authority by the new ruling power responsible for these building activities. He explains the apparent cessation of elite burials at both *Vournes* and nearby Maroni *Tsaroukkas* as a deliberate suppression by the new ruling authority of traditional modes of prestige display among competing lineages. This is an engaging hypothesis and one that is not without ethnographic parallels, such as the Dahomeyan and Chinese examples discussed in Chapter 2 (Law 1989; Rawski 1988). Assuming that the dearth of LCIIC burials is not a function of changing locational preferences or sampling and preservation factors, however, there is also the possibility that the florescence of the power group established in the ashlar complex at Maroni was relatively short-lived, having been eclipsed or even terminated by the expansion of a rival urban center at Kalavasos. This too could account for the seeming absence of elaborate LCIIC burials at Maroni. Both 'upper' and 'middle' elite chamber tomb burials clearly persisted in the Kalavasos area during the heyday of the *Ayios Dhimitrios* ashlar building, as attested by the finds from disturbed tombs in the West Area of that site and at Kalavasos *Mangia*, discussed later in this chapter.

## Programs of mortuary ritual

The identification of different burial treatments and their relative frequencies in Late Cypriot tombs is often problematic because of the complicating effects of prolonged tomb reuse and the incomplete reporting and analysis of skeletal remains. Nevertheless, many different patterns of tomb use and a mixture of burial programs—primary, secondary, sequential and collective—are apparent at sites throughout the island (Table 5.1). Practices of collective secondary burial that had developed previously in the EC–MC periods seem to have been practiced quite intensively in the Late Cypriot I and early Late Cypriot II periods, when numerous instances of so-called 'mass burials' are evident, particularly in northern Cyprus. Episodes of collective secondary treatment are also evident at Enkomi and Kourion, along with many other patterns of mortuary ritual. Protracted forms of mortuary treatment clearly persisted in the later LCII–LCIII periods, but the appearance of numerous shaft graves suggests that traditional sequences of ritual treatment were to some degree abbreviated and transformed in the context of ongoing urbanization. Below, the evidence for the ritual treatment of the dead is surveyed on a regional basis, with detailed consideration of the especially informative samples from Ayios Iakovos, Enkomi, and Kalavasos, followed by a general discussion of the cultural significance of the practices observed.

## Northwestern sites

One of the most striking occurrences of collective secondary treatment in northwestern Cyprus is associated with Pendayia *Mandres* Tomb 1, one of three MCIII/LCIA tombs investigated by the Department of Antiquities at this Morphou Bay coastal site in 1960. Tomb 1 contained two intact burial strata, the earlier of which yielded 12 skulls and other skeletal material, with only one burial found in its approximate original position (Karageorghis 1965c: Fig.4). The second burial stratum, which does not appear to have been significantly later than the first, consisted of a 30 cm thick layer of loose reddish earth, sand, and gravel with finds and disarticulated bones, including 22 skulls, mixed up throughout the thickness of the stratum (Karageorghis 1965c:17–8). Although the deposits may have been somewhat disturbed by flooding, the large number of skulls as opposed to other body parts reported is suggestive of secondary treatment, and the excavator himself regarded the interments as possible 'mass burials' similar to those observed in LCI tombs at Myrtou *Stephania* and Ayios Iakovos *Melia* (Karageorghis 1965c:54–6).

Pendayia Tombs 2 and 3, which were found devoid of skeletal material, may provide evidence for other complex post-burial activities. Tomb 2 was said to have been looted quite some time prior to the excavation but numerous copper-based artifacts and many sherds were recovered from the reddish fill (Karageorghis 1965c:19). Tomb 3 also seems to have been looted at an unspecified date, but inside the dromos some sherds and a cache of 14 copper-bronze items was found, apparently left behind by the 'looters' (Karageorghis 1965c:19–21). It is curious that so many metal goods were left behind in both of these tombs. The presence of what were presumably the most valuable grave goods raises the question of whether the tombs were truly 'looted', or were deliberately emptied of skeletal remains and less valuable items that were to be reburied elsewhere, with the metal objects being left behind in a more accessible hoard. The same consideration might apply to the hoard of approximately 20 hook-tang weapons and four cast-hilted Near Eastern daggers reportedly found in a tomb dromos at Katydhata (Megaw 1957:44; Catling 1964:111, 128).

Other episodes of secondary treatment and possibly collective burial may be discerned at three sites located along the northern flanks of the Troodos mountains. The first of these is Katydhata, where five apparently 'squatting' burials had been placed around the periphery of Tomb 42 during the second, LCI burial period, with skulls and other bones found in distinct heaps (Gjerstad 1926:82). This configuration was also observed in the Middle Cypriot Ayios Iakovos Tomb 6 (see discussion in Chapter 4) and in three LC tombs at Enkomi (Swedish Tombs 2, 6, and 17) discussed further on. The second site is Akhera *Chiflik Paradisi*, where all of the burials encountered in the three intact LC tombs excavated by the Department of Antiquities in 1960 (Karageorghis 1965b) appear to have been disarticulated to varying degrees. Remains of at least two burials were encountered in Tomb 1, dated to LCIA, and a minimum of fourteen and six individuals were discerned respectively in Tombs 2 and 3, both dated to LCIIC. Although the excavator ascribed the jumbled condition of the burials and the excessive breakage of the pottery to repeated instances of filling, emptying, and refilling in conjunction with successive interments (Karageorghis 1965b:74), it is curious that none of the burials in any of these tombs—not even the latest burial in each chamber—was preserved intact. It is thus possible that some or all of the burial remains in the Akhera tombs were redeposited from primary graves elsewhere. The third occurrence is associated with Politiko *Ayios Iraklidhios* Tomb 6, an intact pit tomb of LCIIB date excavated by the Department of Antiquities (Karageorghis 1965a). In the first burial level of Tomb 6, 12 skulls and other bones were found, heaped up with the goods on the east side of the pit, and two infant burials were encountered, more or less *in situ*, on the west side. In the upper level, seven skulls (including one of an infant) were found piled up with most of the goods on the east side. Near these was an extended skeleton, thought to have been the latest burial in the chamber, but lacking its skull

(Karageorghis 1965a:11–14). Even if the disarticulation of the earlier burials in this tomb might be explained in terms of repeated tomb reuse and disturbance, it is difficult to account for this individual's condition without reference to some form of post-interment ritual.

Moving further north, complex programs of mortuary treatment are also apparent at the important LCI coastal town of Morphou *Toumba tou Skourou*. Six tombs with a total of 12 chambers used mainly in the LCI period were excavated within the settlement area by the Harvard Boston Museum expedition in 1972–74 (Vermeule 1974, 1996; Vermeule and Wolsky 1977, 1978, 1990). Although the burial remains were clearly affected by post-depositional flooding and repeated reuse of the chambers, a number of lines of evidence, including the relative scarcity of articulated, largely complete skeletons, suggest the practice of some form of secondary treatment. In the case of Tomb 1, the disturbed and disarticulated state of the skeletal remains and the discovery of joining sherds in different chambers prompted Vermeule to suggest that the burials were first laid on the floor of the dromos until the flesh had decayed, after which the bones were swept haphazardly into the adjacent chambers. A largely complete adult skeleton was in fact found on the east side of the dromos floor (Vermeule and Wolsky 1990:164). In other cases, most notably Tomb 6, the number of skulls present seems to have exceeded the number of individuals represented by other body parts (Vermeule and Wolsky 1990:309), likewise suggesting that some protracted form of mortuary processing and/or storage had occurred prior to chamber tomb burial. Some unusual features were found above Tombs 1 and 2, and these were interpreted by the excavators as a 'funeral pyre'. The features consisted of a patch of paving rocks and burned areas, a partially consumed wooden plank, a mass of smashed pithos sherds, fine ware sherds, charcoal, a series of stones interpreted as tomb markers, and animal bones that included remains of sheep, possibly deer, and a cow's horn, found in a clay pan beneath one of the marker walls (Vermeule and Wolsky 1990:169, 245–6). The presence of pithos fragments on the floor of the Tomb 2, Chamber 2 suggested to the excavators that 'whatever burial and burning rites went on up above…were also connected with the placing of the bodies in the tomb' (Vermeule and Wolsky 1990:246). Possibly, as in the case of the features associated with Tumulus 1 at Korovia *Paleoskoutella*, this area was used for feasting and funerary sacrifices and/or for defleshing or other ritual processing of corpses.

It also seems likely that infants and children at *Toumba tou Skourou* were recipients of secondary treatment. Vermeule and Wolsky note that the 18 burial niches cut in the dromos of Tomb 1 were generally too small to accommodate primary burials of even small infants (1990: 163). Therefore, the burials found in the niches were most probably defleshed, disarticulated remains at the time of their insertion. Whether these secondary burials took place when the tomb was reopened for new burials of adults or as independent funerals for the children themselves is unknown, but in either case the accordance of such extended treatment to the very young is notable.

Further indications of multi-stage ritual practices may be discerned in the sample of 14 MCIII and LCI–II mortuary features excavated at the inland cemetery of Myrtou *Stephania* by the Sydney University expedition in 1951 (Hennessy 1964). Here, too, the tombs had suffered from repeated episodes of flooding that caused collapse of the walls and ceilings of the chambers and serious disturbance of the tomb deposits, rendering the interpretation of burial practices problematic. However, the excavator argued that a 'mass burial' of LCI date was represented by the 14 skulls and other skeletal material recovered from Tomb 12 (Hennessy 1964:31–3). This may represent another instance of collective secondary burial. The presence of sherds in Tombs 3 (described as a 'niche') that joined with larger pots in Tomb 5 suggest the redeposition of burial deposits from Tomb 3 to Tomb 5, rather than from Tomb 5 to Tomb 3 as Hennessy (1964:3) suggested, and similar transfers of material are evident between the side and main chambers of Tomb 14 (Hennessy 1964:44). It is possible that the 'niche' Tomb 3, Tombs 1 and 11 (described as unfinished tombs or Roman wells, Hennessy 1964:1–2, 30), and Tombs 6 and 8 (thought to have been either unused or emptied in antiquity, Hennessy 1964:1, 18, 25) were in fact temporary mortuary features used for burials that were later redeposited in other chamber tombs. No traces of skeletal remains were found in the otherwise intact MCIII Tombs 10 and 13 (Hennessy 1964:27–30, 33–4), but it is impossible to determine whether the single interments thought to have occurred in these tombs had been deliberately removed or were simply destroyed because of the wet conditions in the chambers.

Finally, additional instances of 'mass burial' may have taken place in one or more of the eight LCIA–LCIB tombs excavated by the Italian Istituto per gli Studi Micenei ed Egeo Anatolici in the predominantly Iron Age coastal cemetery at Ayia Irini *Paleokastro* (Pecorella 1977). All of these tombs had been disturbed by some combination of erosion, flooding, and looting (some associated with Iron Age reuse of tombs) prior to excavation, making it very difficult to determine programs of mortuary ritual and individual treatments. However, 15 and 14 burials were preserved in Tombs 20 and 21 respectively (Pecorella 1977:103–107, 133–40), some relatively complete and articulated in extended dorsal positions, others disarticulated. In Tomb 20, the upper bodies of four skeletons (B–E) were preserved seemingly *in situ,* placed side by side in dorsal position. Pecorella speculated that they were victims of an epidemic, interred hastily and carelessly (Pecorella 1977:107), but in light of the complex mortuary practices noted at other sites, it is possible that one or more episodes of collective secondary treatment, along with various sequential interments, are represented here as well. At least 37 burials including adult males, females, and one child were found in another contemporaneous tomb located a short distance inland from *Paleokastro* (Quilici 1990:145), but the mortuary programs associated with these interments cannot be reconstructed.

## Late Cypriot tombs at Ayios Iakovos *Melia* and other northeastern sites

The cemetery at Ayios Iakovos *Melia*, already introduced in Chapter 4, was in use from MCII–LCII. Ayios Iakovos Tombs 8, 10, and 14 are especially renowned for their deposits of so-called 'mass burials' of LCI–LCII date.

The LCIA burial level in Tomb 8 contained the remains of nine burials, mainly represented by skulls that were found in the central basin of the chamber. The next burial group, dated to LCIB, comprised at least 35 individuals, with groups of skeletons in the central area and the left niche seemingly piled on top of each other simultaneously (Gjerstad *et al*. 1934:325–35). It is unclear as to whether the 18 additional burials which occurred in the final, LCIIA burial period, were sequential or simultaneous. Interestingly, however, it would seem that some of the skulls of the second burial period were removed and placed around the walls of the east niche at this time (Gjerstad *et al*. 1934:325–35).

The lowest burial layer in Tomb 14, dated to LCIB, was spread over both the floor and niches of the chamber and contained the remains of approximately 25 well-preserved and 10 very poorly preserved skeletons, found mainly in extended dorsal or slightly lateral positions, often superimposed (Gjerstad *et al*. 1934:349–54). In the second burial layer (LCIIA), five skeletons were found, all in extended dorsal positions, arranged so that there was one in each part of the chamber (Gjerstad *et al*. 1934:352 and Fig. 133:2). It is possible that simultaneous interments occurred in one or both of these burial periods.

The so-called 'mass burial' in Ayios Iakovos Tomb 10 consisted of approximately ten skeletons, partially superimposed in the central basin of the main chamber (Gjerstad *et al*. 1934: 339). The raised part of the chamber or niche was covered by a mixed layer of 'culture earth,' *havara*, and burial remains seemingly disturbed during the construction of Tomb 14. The side chamber B of Tomb 10 yielded a number of copper-bronze artifacts and three skulls, deposits which the excavators believed to have been sacked, perhaps contemporaneously with latest burials in the main chamber, as the side chamber's doorstone was *in situ* and the dromos filling was intact (Gjerstad *et al.* 1934:338). As noted in the case of the seemingly looted tombs of Pendayia, however, it is curious that so many metal objects were left behind. Possibly some portion of the Chamber B deposits had been removed for reburial elsewhere (in the main chamber, for example), with the metal goods left behind as a 'cache', or perhaps the skulls and bronzes in Chamber B were not disturbed at all but were instead deposited there as secondary burials at the outset.

Although large-scale collective mortuary celebrations comparable in scale to those of Ayios Iakovos have yet to be observed at other northeastern sites during LCI, practices involving secondary treatment may nonetheless have taken place in neighboring communities. Among the much smaller tomb groups encountered in the cemetery associated with the LCI–II fortress at Korovia *Nitovikla*, only two tombs (1 and 2) appeared to be intact. Tomb 1 yielded two skulls and some disintegrated skeletal fragments representing a total of perhaps three burials (Gjerstad *et al.* 1934:407–10). Two burial strata were discerned in Tomb 2, with four interments ascribed by the excavators to the MCIII period and three to LCIA. The skeletal remains from both levels were very disintegrated and poorly preserved, and no identifiable fragments are shown on the published plan (Gjerstad *et al.* 1934:410–14, Fig. 159:8). The condition of the burials in both tombs suggests that they could have been secondary burials. Alternatively, they may represent the remains of primary interments, portions of which were later removed to more elaborate chamber tombs elsewhere, leaving only a few fragmentary skeletal remains, small ceramic vessels, and 'caches' of copper/bronze items behind. *Nitovikla* Tomb 3 was a shallow, rock-cut shaft of rectangular shape that had been disturbed by modern plowing. No skeletal remains were recovered but the presence of five pots, three spindle whorls, and a fragment of a bronze pin suggests that this was indeed some sort of mortuary feature (Gjerstad *et al.* 1934:414–5). It is possible that Tomb 3 was used as a temporary grave for interments that were subsequently relocated.

At the site of Milia *Vikla Trachonas*, the SCE excavated several LCI tombs that had been looted or otherwise disturbed to varying degrees (Westholm 1939a). Tomb 10 was looted overnight in the midst of the excavation, which is particularly unfortunate as it contained one burial of particular interest, an individual placed in a semi-contracted position within a 'sarcophagus' built from stone rubble (Westholm 1939a:1–2). Such a 'sarcophagus' is not paralleled in any other Late Cypriot tombs excavated to date. Tomb 12 was found completely empty, and was thus interpreted as a 'cenotaph' by the excavators (Westholm 1939a:14), although it might alternatively be interpreted as an emptied temporary burial feature. Tomb 13 had been partially disturbed long prior to excavation. Only Tomb 11 was relatively intact, with four LC burials and another of Iron Age date (Westholm 1939a:10). Amongst the LC burials, Skeletons I–III seem to have been missing their skulls and other upper body parts, perhaps an indication of some form of secondary treatment.

Concluding the mortuary sample from northeastern Cyprus, four tombs ranging in date between MCIII/LCIA–LCIIC were discovered in 1962 near Angastina during bulldozing for the Nicosia–Famagusta road and were subsequently excavated by the Department of Antiquities (Karageorghis 1964b; Nicolaou 1972). It is difficult to differentiate the degree to which the disturbed condition of the burial deposits resulted from bulldozing and/or the effects of mortuary practices and repeated episodes of tomb reuse, but the presence of an empty (or emptied) burial feature (Tomb 4; Nicolaou 1972:58, note 2) suggests the possibility of multi-stage rituals in this locality as well.

## Enkomi *Ayios Iakovos*

Approximately 180 tombs and other mortuary features dating between MCIII–LCIII were excavated at Enkomi by the British, Swedish, French, and Cypriot expeditions between 1896 and 1969 (Murray *et al*. 1900; Gjerstad *et al*. 1934; Schaeffer 1936, 1952; Johnstone 1971; Courtois 1981; Lagarce and Lagarce 1985; Dikaios 1969). Rock-cut chamber tombs were the most commonly observed mortuary facilities, but a variety of other mortuary features were discovered as well, including ashlar-built tombs, tholos tombs, shaft graves, pit graves, and infant burials placed in pots. Although protracted sequences of tomb reuse often make it difficult to reconstruct mortuary practices in detail, the presence of what appear to have been 'emptied' mortuary features, the differential representation of body parts, and the distinctive burial configurations observed in various tombs make it possible to reconstruct a range of complex burial programs and patterns of tomb use.

In the vicinity of the Level I fortress, Dikaios excavated three small tombs (Cypriot Tombs 16, 17, and 21) dating to MCIII and LCI and one other in Q4W (Cypriot Tomb 15) which appeared to have been emptied at some time during LCI. All of the burial deposits had been removed, with only isolated ceramic objects being left behind. Noting that all of these tombs were subsequently covered by later construction, Dikaios explained their clearing in terms of the builders' aversion to retaining tombs within the expanded architectural complexes (Dikaios 1969:422–3). However, in other areas of the site many tombs continued in use long after they were partially built over, and some may even have been deliberately constructed within architectural complexes. It is therefore conceivable that these tombs were emptied in the course of a ritual program involving exhumation and secondary burial.

Swedish Tomb 8, an intact rectangular chamber tomb of LCI date, may offer other evidence for the practice of a multi-phase ritual system. Two burial levels separated by a layer of horizontally stratified sediments were discerned inside the chamber, the first consisting of a thin layer of decomposed, organic matter and completely mouldered skeletal remains that covered the floor and the niches. Burials of the second period were represented by two femora and a tibia from a body that seemed to have been placed in a sitting position to the right of the entrance, and another femur and tibia found on the opposite side of the door (Gjerstad *et al*. 1934:502). Although the limited preservation of skeletal material may have been affected to some degree by roof collapse and flooding, the condition of the finds supports the hypothesis that these burials were secondary. Tomb 8 contained one of the earliest accumulations of gold jewelry encountered in the Enkomi tombs, and it is likely that the associated burials were individuals of relatively high social status and wealth (Table 5.9a). However, the pottery associated with these burials is not consistent with the quality of the other finds. A large percentage of the pots recovered were damaged and incomplete. This anomaly only seems explicable if it is postulated that the goods had been exhumed from another context and reinterred in Tomb 8, in which case such damage would be understandable.

The configuration of burial remains reported in the intact Swedish Tomb 2 (Gjerstad *et al*. 1934:470–75), used between LCIB and LCIIA, bears a strong resemblance to that observed in Ayios Iakovos *Melia* Tomb 6 and Katydhata Tomb 42, with complications added by multiple episodes of tomb reuse. Tomb 2 was an approximately circular chamber tomb with a small bay to the left of the entrance to the chamber. A total of 11 skeletons were distinguished in two burial layers. On the floor of the chamber the skeletal remains of six individuals were found, five of whom were apparently seated with their knees drawn up and their backs along the walls of the chamber (VI–IX and XI, see Gjerstad *et al*. 1934: Fig. 188:6). These burials may represent an episode of collective secondary reburial, in the course of which an earlier extended dorsal burial (X) found near the rear of the chamber was partially displaced. In the upper burial layer, two skeletons (III and IV) were found jumbled and incomplete along the north wall of the chamber, perhaps disturbed by the interment of three more skeletons (I, II and V) in extended dorsal positions near the front of the chamber. Whether the incomplete preservation of Skeleton V, represented by skull, legs and feet only according to the plan (Gjerstad *et al.* 1934: Fig.188:7), as well as the absence of the right arm from Skeleton II, should be interpreted in terms of ritual practices or the very damp post-depositional conditions in the chamber remains a matter for conjecture.

A somewhat similar pattern of tomb use may be exemplified in Swedish Tomb 17 (Gjerstad *et al.* 1934:541–6), another intact tomb dated to LCIB–LCIIA. On the floor of the chamber the remains of five individuals were found, at least four of them apparently interred in sitting positions, perhaps collectively. These skeletons are represented primarily by skulls, long bones, and vertebral columns on the plan (Gjerstad *et al.* 1934: Fig. 204:8). The remains of at least six other individuals and finds ascribed to an intermediate burial period were found near the walls of the chamber. These skeletal remains were surmounted by a layer of *havara*, on top of which a single skeleton, possibly a primary burial, was placed in an extended dorsal position, oriented roughly southeast-northwest (Gjerstad *et al.* 1934: Fig. 204: 9). This skeleton, identified as an old male (Fischer 1986:36), was equipped with a unique Mycenaean chariot crater (the so-called Zeus crater, Dikaios 1969:918–25; Knapp 1986b:30–37), a very heavy golden bowl, and a very elaborate gold pin. The finds from Tomb 17 overall were considerably richer than those from Tomb 2 (Table 5.9b), which is interesting in the light of the great similarity of the mortuary programs and the number of burials observed in these two contemporaneous tombs.

Swedish Tomb 6 (Gjerstad *et al.* 1934:491–7), a tomb of low to moderate wealth (Table 5.9c) apparently cut in LCIIC and perhaps reused in LCIIIA, affords a striking illustration of the continuation of collective mortuary practices during the later part of the Bronze Age. The remains of 13–15 individuals, represented mainly by skulls and long bones, were found on the floor of the chamber. Although the burials had been somewhat disturbed by flooding, they all seemed to have been seated along the walls facing the center of the chamber, with the grave goods placed around them. The skeletal remains were found mainly along the walls, the skulls were frequently found with or on top of the legs close to the walls, and there were several instances of crossed femora with cubital bones and humeri nearby (Gjerstad *et al.* 1934: Fig. 194:1). The burials were overlain by a layer of silty clay containing much ash and carbonized material, which the excavators suggested might have resulted from a purification fire. Since none of the finds or the bones within the chamber showed any indications of direct burning (traces of ash on the crania noted by Fischer may have derived from the ash layer itself), it is possible the carbonized material, perhaps produced by the burning of decompositional residues or other ritual practices, was deliberately introduced from outside (see also the discussion of Swedish Tomb 11 below). A small cupboard located in the dromos contained the skulls of at least six individuals and two objects. This feature was interpreted by the excavators as an ossuary used for older burial remains that were cleared from the main chamber. Alternatively, the skulls might have been collected from other burial contexts. In either case, the preferential curation of the skull is noteworthy.

Other tombs such as Enkomi French Tomb 1851 (Lagarce and Lagarce 1985) and Enkomi Cypriot Tomb 19 (Dikaios 1969:404–14) may have been used largely for sequential primary interments. These were tombs of modest wealth whose use may be dated to the LCIB and LCIIA periods respectively. French Tomb 1851 was a bilobate chamber tomb with raised benches on either side of the chamber, on each of which three skeletons were found superimposed in extended dorsal positions, most in alternating west-east, east-west orientations (Lagarce and Lagarce 1985: Figs 1, 3). The excavators believed that there had been a significant time lapse between the burials on each bench because subsequent interments had sometimes apparently displaced the defleshed remains of earlier burials. Curiously, no stratigraphic separations were observed among the grave goods, primarily ceramics, found in the central basin, prompting the suggestion that goods from earlier burials were removed with each successive inhumation and replaced with new items (Lagarce and Lagarce 1985:30). The remains of a newborn found in a Monochrome bowl on the south bench probably represent a secondary interment placed in the tomb along with one of the adult burials. In the rectangular Cypriot Tomb 19, the superimposed remains of three extended dorsal burials were found on a rock-cut bench along the wall opposite the entrance. The lowest burial was oriented west-east, the two above east-west. The lower jaw and one leg bone of another skeleton were also found on the bench. These bones and an isolated skull found on the floor may represent the remains of the first burial or burials in the chamber, disturbed or largely removed in the course of subsequent reuse, or of secondary burials redeposited from another context, possibly at the inception of the tomb's use. Another skeleton, possibly a primary burial, was found on the floor of the chamber next to the bench, the body oriented east-west in extended dorsal position, but with the lower legs drawn together and slightly flexed because of the person's height (Dikaios 1969: Pl. 289).

French Tomb 2 (1949), a rectangular tomb with rock-cut benches along the left and rear walls of the chamber, was used for three very rich, intact LCIIA burials whose mortuary programs are difficult to infer (Schaeffer 1952:111–35 and Pl. 12). The remains of two very scantily preserved individuals, both identified as males by Schaeffer based on their stature and the absence of 'feminine' jewelry, and both interpreted as elderly because of the disintegrated condition of their bones, were found on the south bench to the left of the entrance. The first skeleton, represented only by a powdery shadow and a few fragments of the pelvis, long bones, and spinal column, was oriented east-west in an extended dorsal position. The other individual on the south bench, similarly represented only by powdery remnants of bone and the left femur and foot fragments, was placed in an extended, apparently lateral position, oriented west-east (Schaeffer 1952: Pl. XII). On the west bench opposite the entrance was a single skeleton oriented north-south, of which only powdery traces of bone and the right femur were preserved. The lesser height and relatively wide pelvis of this skeleton, in conjunction with the associated jewelry types and other finds (see Table 5.9b), led Schaeffer to conclude that this was a female, also elderly based on the state of the bones. Whether the extremely fragmentary preservation of the bones in this chamber is indeed to be explained in terms of the great ages of the deceased, as Schaeffer argued, the post-depositional conditions in the chamber (it should be noted that the roof had collapsed on the burials), or ritual programs which affected the preservation of the skeletal remains can no longer be determined. However, the fact that relatively fragile body parts such as vertebrae and foot bones were observed while cranial and arm bones were either absent or had almost completely disappeared could be an indication of secondary treatment. The arrangement and quality of the burial goods, many of which appeared to have been brand new (Schaeffer 1952:129, 134 describes them as a mortuary 'trousseau') are certainly suggestive of carefully planned mortuary festivities, which might in turn have entailed long-term storage of the deceased prior to the final interment.

It is also unclear as to whether the 10–12 burials in Swedish Tomb 18 (Gjerstad *et al.* 1934: 546–59), another extremely rich and intact tomb dating to LCIIC, were primary or secondary interments. This tomb was approximately rectangular in shape, with a large shallow basin in the center of the chamber and a continuous narrow bench along the south, west, and north chamber walls. Two side-chambers were cut at different levels from the main chamber, possibly after the original construction of the tomb (Gjerstad *et al.* 1934: Fig. 209). The earliest burials in Tomb 18 were represented by the disturbed and incomplete remains of three or four individuals found in the central basin of the main chamber and in the side-chamber to the northwest, and perhaps by some isolated skull fragments found on the bench in the main chamber. Prior to the second burial period, a horizontal layer of *havara* was put in the central basin bringing it roughly level with the surrounding benches, and thereafter, seven more bodies were interred, all or most oriented west-east. They were placed side by side, the five in the center in extended dorsal positions. The spinal columns of two of these were slightly curved, perhaps indicating that they were positioned or repositioned after some degree of decomposition had taken place. Skeleton I, lying next to the south wall, had been partly moved by subsequent burials and its position and articulation could not be determined. Skeleton VII, along the north wall, was represented by the left leg and skull only. The careful parallel placement of the other individuals is suggestive of simultaneous interments, but the primary or secondary status of the burials cannot be determined.

In tombs with large numbers of burials, it is probable that a mixture of primary and secondary interments had taken place. Swedish Tomb 11 (Gjerstad *et al.* 1934:510–25; Figs 195, 198) was used for at least 21 moderately rich burials dating from LCIIA/LCIIB–LCIIC, several of which were represented primarily by skulls, either remnants of secondary burials or of primary interments that had been partially removed to make way for others. A final, poorly preserved burial was found in the dromos of Tomb 11, the body placed on its left side in a slightly flexed lateral position. The associated goods were very poor and it was suggested by the excavators that this could have been a servant's burial or a victim of 'sacrifice'. However, it is equally possible that this was a primary burial awaiting secondary treatment or that the individual was left in the dromos simply because the chamber itself was too full. Layers of ash found in the dromos were thought to be evidence of a purification or sacrificial ceremony, carried out after the last burial in the chamber, although no ash was found in the chamber itself.

French Tomb 5 (Schaeffer 1952), a bilobate chamber tomb of moderate wealth (Table 5.9c), was one of many tombs including Enkomi Cypriot Tombs 2 and 10 (Dikaios 1969), French Tomb 1907 (Lagarce and Lagarce 1985), and Swedish Tombs 3, 13, and 19 (Gjerstad *et al.* 1934) that were used over remarkably extended periods of time. Schaeffer dated the inception of French Tomb 5 to the Middle Cypriot period, but the LCIB date proposed by Åström (1972b:46) is more consistent with the ceramic finds. Subsequently, the tomb seems to have continued in use throughout LCII and LCIII, with a minimum of 55 individuals interred over this 300–400 year period. At least 17 burials were associated the first stratigraphic horizon that terminated in LCIIC, most oriented east-west, some in extended dorsal postures and occasionally superimposed. One was apparently interred in an unusual extended ventral position, with the legs crossed at the ankles. Interestingly, only three skulls were recovered from the LCI–II burial stratum, but it is possible that many were displaced to the periphery of the chamber in the course of repeated reuse (Schaeffer 1952:229), perhaps deliberately as in Ayios Iakovos *Melia* Tomb 8. The LCIII burial level, separated from the lower stratum by a layer of sterile soil, was further divided into two layers by a lens of roof collapse (Schaeffer 1952:160–1). At least 20 different skeletons, some in extended dorsal positions, along with a total of 28 skulls and other disarticulated remains, were found in the lower LCIII level. Most of these skeletons seem to have been oriented west-east, exactly the opposite of those in the LCI–II stratum. Schaeffer attributed this reversal to the influx of a new ethnic group (1952:220), but it might alternatively be explained by the use of a new dromos on the east side of the tomb (suggested by the published section, Schaeffer 1952: Pl. 32). In the uppermost LCIII burial layer 10 skeletons and 24 skulls were found, most of the skeletons once again oriented east-west, mainly in extended dorsal positions, sometimes superimposed. The powdery bones of a young infant were found between the legs of one skeleton. Schaeffer noted that the bones of most of the skeletons in this level were in anatomical articulation (1952:156–60). It is likely that there were several instances of deposition in Tomb 5 during the LCIII period, but whether these entailed incremental primary inhumations or the co-interment of more than one well-preserved corpse at a time is uncertain.

While older chamber tombs such as French Tomb 5 continued in use and some new ones such as French Tomb 6 (Schaeffer 1936) were constructed during the LCIII period, burials in shallow pit or shaft graves seem to have become equally if not more common during the last phases of occupation at Enkomi. Most of these graves were used for one to three burials that might include either adults or children and sometimes both, as in Swedish Tomb 14, where two children were placed side by side and oriented perpendicularly to the body of what may have been an adult female, their upper bodies resting on the older person's thighs (Gjerstad *et al*. 1934: Fig. 204:1). Some graves contained either a combination of primary and secondary burials or two or more sequential primary interments. In Swedish Graves 11A and 16, for example, two burials were found in each shaft, one *in situ* in extended dorsal position, the other represented by a heap of disarticulated and incomplete remains (Gjerstad *et al*. 1934:540; Fig. 204:4; 510–25; Fig. 195:7). It is difficult to determine whether these incomplete burials were originally secondary or whether they had merely been disturbed in the course of reuse. However, it is notable that Cypriot Grave 24 contained the disordered bones of a single adult individual, whose mandible was found between the femora (Dikaios 1969:433), and in this case secondary burial seems quite likely. Other shaft graves such as Cypriot Grave 4A, Cypriot Grave 23, and Swedish Graves 10A, 13C, and 15 contained single extended dorsal burials that may have been primary inhumations, although at least two of these seem to have been affected by distinctive pre- or post-interment treatments, as explained below.

In at least three shaft graves it was noted that specific body parts were missing from otherwise complete skeletons. The single skeleton in Swedish Grave 15, identified by the excavators as an adult female (apparently on the basis of associated finds), was missing the right arm (Gjerstad *et al*. 1934:538; Fig. 204:3). The lower left arm was absent from the single adult burial in Cypriot Grave 4A (Dikaios 1969:432). Dikaios was of the opinion that these were both women who lost their arms while defending themselves from attacking invaders (1969:433), but it is more likely that some systematic mortuary practice is represented here instead. In Swedish Grave 7A, the lower of the two burials was placed face down in a slightly contracted position with the legs crossed and both arms missing (Gjerstad *et al*. 1934:498–500; Fig. 194:2), an atypical position that may have indicated something unusual about this person's death or social status. Other examples of missing arms may be observed in much earlier chamber tombs: the single skeleton in Swedish Tomb 12 (of MCIII/LCIA date) was missing the right arm (Gjerstad *et al*. 1934:525–6; Fig. 201:1), and arms may have been removed from other individuals in Swedish Tomb 2 (Gjerstad *et al*. 1934:470–5; Fig. 188) and French Tomb 5 (Schaeffer 1952: Pl. 36) as well. Possibly certain body parts were deliberately removed from some individuals for reasons upon which we can only speculate.

Since the shaft graves excavated to date contained a total of only 28 individuals, it is difficult to be certain that this form of burial was the normative mortuary treatment in the LCIII period. It is possible that we are missing a considerable segment of the larger mortuary population, which may have been interred in chamber tombs in other locations, possibly outside the settlement proper. However, the occurrence of the shaft graves themselves still remains to be explained. Inasmuch as shaft graves required less energy to construct than chamber tombs (cf. the floor areas of shaft graves and chamber tombs in Table 5.5), it is tempting to regard their use as a 'low status' method of disposal, or as evidence for catastrophic circumstances (epidemics, invasions) requiring burial in haste (Niklasson-Sönnerby 1987:225). However, Tombs 13, 15, and 16 excavated by Schaeffer in 1934 (Schaeffer 1936) all contained items of gold jewelry, which was comparatively scarce in this period, and some of the very rich LCIII burials excavated by the British Museum expedition (e.g. Tomb 75, described in Williamson and Christian n.d.) may have been found in shaft graves, suggesting that this type of burial facility was used by individuals and kin groups of both greater and lesser wealth in circumstances that did not inhibit all forms of display. To a considerable extent, the use of this relatively low cost form of interment may reflect the decreasing importance of mortuary rites as a means of legitimizing and enhancing kin-group status, resulting in a truncation or shortening of traditional sequences involving exhumation and reburial in familial chamber tombs. Several of the shaft graves excavated by the Swedish Cyprus Expedition were located off the dromoi of or otherwise in close proximity with older chamber tombs (e.g. shaft graves 7A, 10A, 11A, 13C, and 19A), suggesting perhaps an intention, never fulfilled, of one day reinterring the shaft grave burial in the associated chamber tomb.

## Southeastern sites

Although numerous tombs of Late Cypriot date have been excavated in the vicinities of Pyla, Aradhippou and Klavdhia in southeastern Cyprus (known mainly by their Mycenaean pottery, see Karageorghis and Vermeule 1982), as well as at Larnaca *Laxia tou Riou* (Myres 1897), Kalokhorio (Karageorghis 1980:766), Livadhia *Kokotes* (Åström 1974), Dromolaxia *Trypes* (Admiraal 1982), and Arpera *Mosphilos* (Merrillees 1974), information about the ritual treatment of the dead can be elicited only from some of the more recent excavations at Hala Sultan Tekke and Kition. In 1968, the Department of Antiquities excavated two large pit tombs (Tombs 1 and 2) east of Hala Sultan Tekke *Vizaja* near the banks of the salt lake (Karageorghis 1968, 1972b). None of the skeletal remains were found in articulation, a circumstance which the excavator attributed to the effects of removing the fill in the course of repeated burials (Karageorghis 1972b:71). Inasmuch as the tombs had been damaged by recent construction activity and partly looted, the question of whether any of the interments were in fact redeposited secondary burials is problematic. A Late Cypriot III shaft grave containing the intact burial of an adult male, oriented east-west in an extended dorsal position and presumably representing a primary inter-ment, was found inside the corner of a building within the *Vizaja* settlement (Niklasson 1983; Schulte-Campbell 1983).

At the nearby town site of Kition, three skeletons that had been placed in extended dorsal positions were found in Tomb 3, but few conclusions can be drawn about the circumstances of their interment (Karageorghis 1960b:515–20). A pit tomb in the same vicinity contained several burials, including four of children, with the skeletal remains dispersed through a thick layer 0.46 m deep along with extremely fragmentary pottery finds (Karageorghis 1960b:515). Based on this description, it might be inferred that these burial remains had been redeposited from another context, but as the stratification of the tomb was reportedly disturbed by later reuse of the area, it is impossible to be certain of this. At least 48 individuals were buried in the large and extremely rich Kition Tomb 9 in two burial periods dating to LCIIC and the onset of LCIIIA (Karageorghis 1974; Kling 1989). Seven burials were associated with the first burial level, confined to the eas-tern portion of the chamber, and 41 were associated with the second. Amongst all of these, only one skeleton was found *in situ*, the head and torso in an oblique position as if the body had been pushed in over the goods (Karageorghis 1974:42–3). The scarcity of articulated remains is sug-gestive of secondary burial practices.

## Kalavasos

In the Kalavasos area, tombs with Late Cypriot burials are known from Kalavasos Village, the outlying locality *Mangia*, and the nearby town site at *Ayios Dhimitrios*. Kalavasos Village Tomb 51 contained a single, intact LCIA burial of a late adolescent under 25 years of age, possibly male in light of the accompanying copper-based sword and other finds (Pearlman 1985). Among the eight LCII tombs found at the site of Kalavasos *Mangia* (McClellan *et al.* 1988), intact burial remains were encountered only in Tomb 6, where three articulated skeletons were found in extended positions. Skeleton 1 belonged to an adult male 24–35 years of age; Skeleton 2 was that of a young to middle-aged female; and Skeleton 3 belonged to a child of 7–8 years. According to the excavators, Skeletons 1 and 2, which were placed parallel to one another, may have been interred at the same time (McClellan *et al.* 1988:207–8, Fig. 4). If these burials were indeed simultaneous, it is possible that at least one of them, perhaps Skeleton 2, whose hands and arms were positioned in such a way as to suggest that she had been wrapped in some type of shroud, was secondary. Remains of two other adult individuals found in a cluster of disarticulated bones may represent earlier burials, disturbed in the course of reusing the tomb.

Although the majority of the *Ayios Dhimitrios* tombs discovered to date were found looted or otherwise disturbed, the careful excavation and detailed reporting of burial remains at this site (South 1997; 2000; South *et al.* 1989) nevertheless make it possible to discern a variety of mortuary programs including both primary and secondary treatment of the dead. Some of the smallest tombs (e.g. Tombs 2, 3, 7, 10, 16) seemed to have been emptied, and it might be speculated that some of these were temporary burial facilities from which the burial remains were later exhumed. Other tombs contained the remains of incomplete and/or disarticulated burials whose condition cannot be explained solely in terms of tomb reuse or other post-depositional processes.

Some of the most detailed evidence for mortuary practices comes from Tomb 11 (South 1997: 161; South 2000:349–53; Goring 1989: Fig. 13:1), an intact, bilobate chamber tomb dated to LCIIA: 2, located beneath the street to the west of the LCIIC ashlar Building X in the Northeast Area and parallel to a wall from an early phase of construction in the vicinity. This tomb contained the burials of three young adult females who were very richly equipped with gold jewelry (including signet rings) and other higher order valuables such as Mycenaean pictorial craters. The articulated and largely complete Skeleton I, belonging to a female aged 19–20 years, was found in extended dorsal position on the larger rock-cut bench on the west side of the chamber. At the north end of the narrower east bench were the remains of Skeleton II, a female aged 21–24 years, represented only the skull, patellae fragments, the right tibia, fibula, and scapula, bones of the hands and feet, and a few vertebrae. A second burial on the east bench, found with the skull to the south, was identified as a female aged *c.* 17 years. This skeleton was relatively complete but highly fragmented and several of the long bones had been placed in anatomically impossible positions. It is possible that Skeleton II was a secondary burial, perhaps exhumed from another mortuary facility. Skeleton III could have been a primary burial, originally laid out on the larger west bench, but moved in partially decomposed condition when Skeleton I was interred as a primary burial at a later date.

On the floor of the chamber, the disarticulated remains of a three-year-old child (Skeleton IV) were found near the north end of the east bench, and the commingled bones of three infants (Skeletons V–VII) that must have been interred in a macerated state were found in a localized deposit 5 cm above the chamber floor. These burials were almost certainly secondary, presumably exhumed from some other context and reburied in Tomb 11. Other secondary remains were found in a small niche in the east wall of the dromos, from which very fragmentary remains of a two-year-old and an adult individual aged 17–25 years were recovered. Tomb 9, a cupboard located at the south end of the dromos directly opposite Tomb 11, contained some fragmentary LCII fine wares, bits of ivory, and the bones of a newborn infant.

Other evidence for complex ritual practices was observed in Tomb 14 (South 1997:165, South 2000:353–4) of LCIIB date, another a bilobate chamber tomb located to the south of Tomb 11. A rich array of grave goods was found scattered upon the chamber floor, and the disarticulated and incomplete remains of at least two adults (one definitely and one probably male) were commingled with a dense cluster of pottery inside the entrance to the chamber. The burials may have been swept aside from the large rock-cut bench on the left side of the chamber, on which only a few small fragments of bone and one or two objects remained. Inasmuch as the chamber door-stone was found in place and there were no indications of looting, the excavators have speculated that the bench was cleared in preparation for a burial which never occurred, or in a Late Bronze

Age (rather than a more recent) episode of looting. Nevertheless, many valuable goods including gold diadems, beads, and finger rings, three bronze daggers, objects of ivory, faience, and glass and a considerable quantity of pottery (including a Mycenaean pictorial crater and three Red Lustrous Wheelmade arm-shaped vessels) were left behind. It might be speculated that some of the skeletal remains were removed from this tomb to inaugurate the use of another chamber or for some other ritual purpose.

Less is known about the original disposition and condition of the burials in the other Northeast Area tombs excavated so far, but the evidence, while limited, attests to several instances of mortuary obsequies for infants and children, and other instances of contemporary looting or deliberate removal of burial remains. Tomb 15, a small niche carved in the dromos wall of Tomb 14, contained only a White Slip bowl (South 1997:167); it is conceivable that an infant burial had been removed from this feature. Tomb 16, a small chamber tomb possibly dating to the LCIIB period and located to the north of Tomb 11, was found looted; fragmentary remains of an adult female and an infant were found there, and a small dromos niche (Tomb 16A) also contained the remains of an infant (South 1997:167; South 2000:349). The very small Tomb 17 contained bones of a child aged 8–12 years (South 1997:167; South 2000:355–6). Tomb 13, a large circular chamber tomb of LCIIB–C date located immediately to the south of Tomb 14, was ostensibly looted during the Late Bronze Age, but numerous pots and luxury items of alabaster, faience, and ivory remained, along with the scattered bones of several individuals on the chamber floor and the skeleton of a middle-aged woman found on a stone bench to the left of the entrance. Tomb 12, the subsidiary chamber or dromos cupboard of Tomb 13, contained the bones of three infants and a child (South 1997:163–5; South 2000:354–5). Tomb 8, the other tomb discovered in the Northeast Area, was largely eroded and empty (South 1997:159).

Mortuary practices are more difficult to reconstruct in other parts of the site. In the West Area, looting and prolonged reuse, spanning the LCIIB–C periods and the Geometric period or later, obscured the original disposition of the burials in Tomb 18, and the burials in Tomb 19, located on the opposite side of the same dromos, were thought to have been disturbed in the Iron Age or Classical period. Tomb 20, the dromos niche associated with these tombs, contained only a skull fragment and a few scattered sherds (South 1997:167–70; South 2000:356–61). In the Central Area, remains of at least nine burials were recovered from Tomb 1 (LCIIB–C), two from Tomb 4 (LCIB–LCIIA), and eleven from Tomb 5 (LCIIA–C), but inasmuch as all of these tombs had been looted prior to excavation, little can be concluded about the mortuary programs represented (South *et al.* 1989:42–53). Tomb 6, an intact chamber clearly stratified beneath walls of LCIIC date in the Southeast Area, contained the very fragmentary skeletal remains of one adult individual aged 25–30 years and a child aged 3–8 years old whose burial programs cannot be determined (South *et al.* 1989:54–5). Nevertheless, the 'looting' or emptying of Tombs 2, 3, 7 and 10 in the Central and Southeast Areas of the site (South *et al.* 1989:47–8, 55–7) may have occurred in conjunction with programs of exhumation and reburial for some individuals.

## Kourion and other southwestern sites

In southwestern Cyprus, the best documented sample of mortuary data comes from the town site of Kourion *Bamboula* (Benson 1972). Although the burial deposits of the Kourion tombs, as in so many other cases, had been extensively damaged as a result of looting, flooding, and disturbance in the course of reuse, a diverse range of burial practices are evident. Most of the burials occurred in chamber tombs, but one shaft grave with an individual burial (Tomb 14) and a trio of circular to ovoid cist graves with unidentified skeletal remains (Tomb 37) were also found (Benson 1972: 18, 31). The disarticulated and incomplete representation of skeletal remains in the intact Tomb

6, dated to LCIIB or LCIIC, may attest to practices involving secondary treatment. A large pile of bones including two skulls was found on the east side of the chamber, perhaps swept aside in preparation for later burials which did not take place, according to Daniel, or disturbed by flooding, according to Benson. No objects were found *in situ* with the bones (Benson 1972:13–4). The skeletal remains were studied by Angel, who distinguished the presence of at least five individuals. Among these he identified two adult males (one definite, one probable) represented by cranial material, another adult male represented by leg, forearm, and clavicle bones, an adult female, represented by the tibia and forearm, and another adult female represented by an unspecified arm bone. The individuals represented by postcranial bones only were distinguished from one another and from the two individuals represented by crania on the basis of estimated stature (Angel 1972:160). It is curious that different individuals were represented by different body parts. Assuming that Angel's identifications are correct and that all of the osteological material present in the tomb had been saved, it is possible that these disarticulated remains represent secondary burials, the bones of which were incompletely exhumed or incompletely removed from the primary packaging.

Other indications of complex ritual processing come from the intact Tomb 21 of LCIIA date. In this tomb, according to Benson, 'Masses of bones and skulls were found along the south wall in a careful arrangement which appears intentional' (1972:24). The tomb plan (Benson 1972: Pl. 11) shows three piles of disarticulated skeletal remains which indeed appear to have been carefully stacked, and it is possible that these remains represent simultaneous secondary burials, rather than an episode of housekeeping as suggested by Benson. The bones were studied by Angel, who distinguished five adult individuals, including two adult males and three adult females (one definite, two probable). Further instances of collective secondary burials might conceivably be represented in Tombs 19 (LCIIA–LCIIIB) and 40 (LCIIA, LCIIIA), with 52 and 20 or more burials respectively. However, there were multiple episodes of use in both of these chambers, and it is very difficult to reconstruct the associated mortuary programs with certainty.

Although many tombs have been excavated in the Kouklia area, the problems of post-depositional disturbances and the incomplete publication of the burial remains makes it difficult to reconstruct the forms of mortuary ritual practiced or to ascertain the demographic composition of the tomb groups. At *Teratsoudhia*, where two intact and seven variably disturbed chambers were investigated, skeletal remains from earlier burials often appear to have been swept into the characteristic central cist of the tomb chamber to make way for succeeding burials that were placed on the surrounding benches (Karageorghis 1990a:19). Sequential primary interments may have occurred in some of these tombs. The extremely fragmentary and incomplete representation of body parts in the intact chambers B and N of Tomb 104 could indicate that secondary treatment was also practiced in some cases, but the evidence as reported is inconclusive (Karageorghis 1990a:6–7, 11–2). At *Eliomylia,* the discovery of skull fragments and leg bones of an adult burial in a dromos niche (Karageorghis 1990a:77–8) is strongly suggestive of secondary treatment.

## Ritual practices and ideology

For much of this period, continuities with Middle Cypriot mortuary traditions involving complex physical treatments of the dead and occasional collective secondary treatment are evident at sites throughout the island. The disposition of the burials in the LCI stratum of Katydhata Tomb 42, where approximately five individuals seem to have been arranged in squatting positions around the walls of the chamber, bears a strong resemblance to the arrangement observed in Ayios Iakovos Tomb 6 of Middle Cypriot date. The same pattern of multiple, carefully distributed burials in squatting positions is also attested in Enkomi Swedish Tombs 2, 6, and 17. The stacks of bones encountered along the walls of Kourion *Bamboula* Tomb 21 may also have been arranged in the course of a similar ritual involving the exhumation and reburial of several persons at the same time, or they may have been rearranged in the course of other post-interment rituals.

Although the LCIIC date of Enkomi Swedish Tomb 6 attests to the long duration of these practices, the performance of large-scale collective mortuary celebrations seems to have been most frequent during the LCI period and, interestingly, is at present more often attested at rural sites than in town centers. The most striking instances of 'mass burials' come from localities such as Pendayia *Mandres*, Myrtou *Stephania*, Ayia Irini *Palaeokastro*, and Ayios Iakovos *Melia*, sites which, on the basis of current evidence, do not appear to have been associated with large towns. The unusually large numbers of seemingly simultaneous interments in some of the tombs at these sites have sometimes been explained in terms of a catastrophic phenomenon such as an epidemic or intensive warfare (e.g. Åström 1972b:764), although Fischer's (1986) restudy of the skeletal remains from Ayios Iakovos yielded no support for this theory. However, in light of the evidence for collective burial events in the Middle Cypriot period, it seems more likely that an intensification of the earlier tradition of large-scale mortuary celebrations is represented. It is possible that these events were a manifestation of the great importance placed on descent group identity and membership as differential access to trade goods and copper contributed to increasing social inequality in the more remote 'rural' areas as well as in the coastal towns. Another, not unrelated possibility is that some of the individuals interred were members of local kin groups who had migrated to towns such as Enkomi and *Toumba tou Skourou*, and were buried in the towns temporarily upon death. Subsequently their bodies may have been returned to their ancestral villages for periodic, collective mortuary celebrations.

In addition to various instances of multiple simultaneous burials, there are many other indications of complex mortuary processing in pre- and post-interment rituals during the LC period. The complete or nearly complete absence of articulated skeletons in a number of intact tombs at Myrtou *Stephania*, Akhera *Chiflik Paradisi*, Morphou *Toumba tou Skourou*, Pendayia *Mandres*, and Kourion *Bamboula* suggests that at least some of the burials in these tombs were secondary reinterments exhumed from temporary graves or other storage facilities. Small 'emptied' or at least 'empty' mortuary features have been observed at Myrtou *Stephania*, Enkomi, Kalavasos *Ayios Dhimitrios*, Korovia *Nitovikla*, Milia *Vikla Trachonas*, and Angastina *Vounos*. These comprise a variety of pits and small chamber tombs that could have been used as temporary graves. Cross-joins between ceramics from *Stephania* Tombs 3 and 5 (also between the side and main chambers of Tomb 14) indicate the redeposition of material from smaller into larger tomb chambers. Fragmentary secondary burial remains were also sometimes deposited in dromos niches, as at Kalavasos *Ayios Dhimitrios* (dromos of Tomb 11) and Kouklia *Eliomylia*. In other cases such as *Ayios Dhimitrios* Tomb 14, episodes of Late Bronze Age disturbance characterized as 'housecleaning' or looting (with many valuables left behind nonetheless) may have been associated with further post-interment rituals, sometimes involving the partial transference of skeletal remains to other tombs, dromos niches etc.

A continuing emphasis upon the special treatment of the skull is evident at a number of LC sites. At Ayios Iakovos *Melia*, the nine burials in the LCIA burial stratum of Tomb 8 were represented predominantly by skulls. During the final (LCII) period of use in the same tomb, skulls from the preceding levels were carefully arranged around the periphery of the east niche. The side chamber of Tomb 10 contained three skulls, along with a number of bronze objects. Six skulls were also found in the side chamber of Enkomi Swedish Tomb 6. At Morphou *Toumba tou Skourou*, cranial and mandibular fragments appear to have been predominant in the skeletal remains from Tomb 1 (chambers 1 and 2), Tomb 5 (chamber 2), and Tomb 6. Skulls also appear to have been preponderant in the skeletal material from Korovia *Nitovikla* Tomb 1, Pendayia *Mandres* Tomb 1, Politiko *Ayios Iraklidhios* Tomb 6, Ayia Irini *Palaeokastro* Tombs 20 and Tomb 21 (upper burial), and Kourion *Bamboula* Tombs 12, 18, and 19, although in some of these cases their prominence may well have been exaggerated by reporting biases. But it is also noteworthy that the latest and the only more or less *in situ* burial in Politiko Tomb 6 consisted of an extended skeleton from which the skull had been separated; whether it was included among the other skulls found in the chamber or whether it had been altogether removed from the tomb is unknown.

There are also indications that certain body parts were selectively gathered for secondary burial, while others were ignored, removed, or perhaps distributed amongst different tomb groups. For example, in Kalavasos *Ayios Dhimitrios* Tomb 11, Skeleton II, in contrast to the more or less complete Skeletons I and III, was represented only by the skull, patellae, right tibia, fibula, scapula, bones of the hands and feet, and a few vertebrae. A burial niche in the dromos of Tomb 11 contained only a few teeth and fragmentary bones from a child and an adult individual; similarly, a dromos niche at Kouklia *Eliomylia* contained only skull fragments and leg bones. Various individuals in Kourion *Bamboula* Tomb 33, Enkomi Swedish Tomb 8, and Milia *Vikla Trachonas* Tomb 11 were represented primarily by long bones and, in some cases, other portions of the lower postcranial skeleton. Also, several burials from Enkomi ranging in date from MCIII–LCIII appear to have had portions of the right or left arm deliberately removed at the time of burial or perhaps shortly thereafter (e.g. Swedish Tombs 12 and 2, French Tomb 5 (1949), and shaft graves Swedish Tomb 15 and Cypriot Tomb 4A). These practices would seem to illustrate a rich complex of symbolism associated with human bone, in which the physical remains of the deceased were perhaps used to express ongoing social relations among the living and the dead (e.g. Weiner 1976).

While examples of disarticulated, either secondary or otherwise, disturbed burials are numerous in tombs of Late Cypriot date, there are also many instances of apparently articulated burials, usually in extended dorsal positions. The appearance of extended burials, which were relatively uncommon in the EC–MC period, implies at the very least a change in the customs for positioning the deceased at the time of death. Whether this was related in turn to a change in methods for storing the deceased prior to reburial or to an increasing frequency of single stage primary interments cannot be determined from the evidence presently available. Interestingly, however, extended and 'crouching' burials were sometimes found in the same tombs at Enkomi (e.g. Swedish Tombs 2 and 17). In these chambers the initial burials appear to have been placed in seated or squatting positions around the walls of tomb, but subsequently both crouching and extended burials were interred. It is possible that the extended burials in these tombs represent individual primary burials. Conceivably, in conjunction with the prolonged reuse of particular tombs, single stage funerals may have succeeded single or collective secondary interments, and the laying-out of the deceased in an extended position may have been deemed a more prestigious or suitable posture for these occasions than traditional flexed positions. It may also have afforded a better opportunity for displaying the clothing and jewelry of the deceased.

As in the EC–MC periods, the treatment of the dead in the Late Cypriot period probably encompassed a variety of alternative practices, including 'grand' and not-so-grand primary funerals that served as the principal mortuary observance for some people, along with more protracted sequences of exhumation and reburial for others—both individuals and groups. However, as Cypriot communities grew more complex and socially heterogeneous, it is likely that their mortuary practices and associated beliefs about death also became more diverse and subject to change. The completion of a full ritual program, involving primary burial and secondary treatment (exhumation and reburial), may have been a 'cultural ideal' for some communities or kin groups but not for others. And even for those who subscribed to this ideal, its accomplishment may have been contingent on the social and economic circumstances of the moment. Over time, as social competition and social stratification increased, the ideological basis for earlier traditions of secondary treatment and collective burial may have eroded. As suggested earlier, the appearance of numerous shaft grave burials in LCIII, many of which appear to represent primary inhumations, might be explained in part as a truncation of older systems of dual obsequies, such that the completion of the ritual sequence, entailing exhumation and reburial in a family chamber tomb, never took place for many people. It may also reflect the presence of numerous individuals who were detached from their original communities inside or outside of Cyprus, and thus did not belong to any local tomb groups (cf. Liverani 1975 on the contemporary social milieu in Syria.)

## The grouping of the dead

As in the preceding EC–MC periods, LC chamber tombs were generally used for multiple burials. There is slightly more information concerning the representation of ages and sexes at least at a few sites than is available for the previous era and these data suggest that infants and children, while underrepresented in the mortuary samples at many sites, may nevertheless have been accorded chamber tomb burial more often than in the past. Furthermore, the mortuary samples from three sites indicate that adult females were much less likely to receive chamber tomb burial than adult males, although other evidence shows that there may have been important exceptions to this pattern among high status groups. Mean tomb group size increased significantly relative to earlier periods, in part due to the frequency of 'mass burials' early in the Late Cypriot period, but more importantly because of the greatly extended duration of tomb use throughout this era. Towards the end of the period, it appears that the reuse of older tombs gave way at some sites to the use of shaft or pit graves for individuals or very small burial groups, a phenomenon that may have been related to the changing importance of kin group identity and mortuary symbolism in the urbanized communities of this period. The evidence for burial group composition from the best-documented samples is examined below, after which patterns of diachronic change, along with their social and ideological significance, are considered.

### A regional overview of burial group composition

Approximately 97–100 individuals were buried in the six tombs (a total of twelve chambers) excavated at Morphou *Toumba tou Skourou* (Vermeule and Wolsky 1990:325). The number of burials per chamber varied greatly, but the overall mean was relatively high (7.9; see Table 5.2). The ratio of males to females is unknown, but infants and children seem to have comprised roughly one-quarter to one-third of the entire burial population, a high proportion relative to the figures reported at other Cypriot Bronze Age sites. It is possible that the practice of special mortuary rites for infants and young children was most common among higher status social groups, as suggested by the unusually large number of burial niches in Tomb 1, the richest of all the *Toumba tou Skourou* tombs (see also Lapithos Tomb 804 with 18 burial niches; Herscher 1978).

At Ayios Iakovos *Melia* on the opposite side of the island, approximately 152 individuals were observed in the seven tombs of MC–LC date (Gjerstad *et al.* 1934). A total of 37 crania were published by Fürst, who identified 25 adult males, 8 adult females, and 4 children overall in the samples from Tombs 8, 13, and 14 (Fürst 1933:52). This surprising sex imbalance was recently verified by Fischer (1986:7, 16–17), who examined 41 crania from the same tombs, identifying 25 individuals as definitely adult male, 3 as possibly male, 7 as definitely adult female, and 1 as possibly female. Four children and one individual of uncertain age and sex were also present (Table 5.3). Such a disparity in the ratio of the sexes rather suggests that males and females were frequently subject to different forms of mortuary treatment, with many females being disposed of in other features such as pit graves or perhaps by abandonment and exposure. The very small number of sub-adults recovered from the Ayios Iakovos tombs, less than 10% of the sample analyzed by Fischer, suggests that many infants and children were also excluded from burial within the chamber tombs.

Approximately 350 individuals were reported in the chamber tombs and shaft graves excavated by the Swedish, French, and Cypriot expeditions at Enkomi *Ayios Iakovos* (Keswani 1989a: Tables 5.24–5.27). Burial group sizes ranged from 1–55 burials per tomb, with a mean of 10.7 individuals per chamber (Table 5.2). The 55 burials in French Tomb 5 (1949) constitute an exceptionally large group. Group size seems to have been correlated, in a general way, with the total span of time over which the tombs were used (Fig. 5.1). Many of the tombs were repeatedly reutilized over several centuries, a practice perhaps necessitated in part by the lack of space for new constructions within an urban context. However, the deliberate construction of tombs in close proximity to houses, public buildings, workshops etc. may also have served as an expression of ties to the past and of affiliation with particular social groups. It is interesting that extended periods of use are evident in the tombs of both extremely rich groups (e.g. Swedish Tomb 3 and British Tomb 12) and of groups of lesser wealth (e.g. Cypriot Tomb 10, Swedish Tomb 11, French Tomb 5 [1949]), indicating that this phenomenon was not merely restricted to the uppermost social groups.

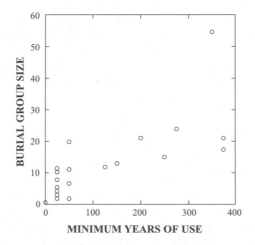

**Figure 5.1**     Enkomi burial group sizes scatterplotted against minimum years of use.

Infants and children are highly underrepresented in the Enkomi mortuary sample. To some extent their presence may have been underestimated due to the imperfect preservation and recovery of softer and smaller bone material. However, it is also apparent that there were separate disposal facilities for many sub-adults, who were sometimes placed in pit graves (e.g. Cypriot Tomb 20; Dikaios 1969:414–5), in fragmentary jars (e.g. French Tomb 325; Courtois 1981:291–2), or in the dromos cupboards cut adjacent to many chamber tombs. No information is available concerning the form of the grave known as French Tomb 134, the so-called 'tombe de l'enfant', but this feature yielded a large collection of small juglets dating to LCI (Courtois 1981:125–30), and six crania of infants or young children bearing this provenience label have been found in the Cyprus Museum in Nicosia (personal observation). Other mortuary features containing collections of small juglets (e.g. French Tomb 240, French Tomb 3 (1947); Courtois 1981:11–30 and 113–24) may also represent single or collective burials of juveniles, but unfortunately no osteological data are available to establish this with certainty. The neonates observed in the main chambers of French Tombs 1851 and 1907 (Lagarce and Lagarce 1985) were placed in ceramic vessels, suggesting that their skeletons were disarticulated at the time of interment. It is possible that many infants and children were initially interred in dromos cupboards or other features and then exhumed when a chamber tomb was being reopened for one or more adult burials.

As also observed at Ayios Iakovos *Melia*, there seems to have been a pronounced tendency for adult females to be excluded from chamber tomb burial at Enkomi. Osteological remains from some of the Swedish and French tombs were originally studied by Fürst (1933) and Hjörtsjö (1946–47) and were recently reanalyzed by Fischer (1986), who obtained essentially the same results as Fürst and Hjörtsjö, identifying 36 adult males (25 definite, 11 probable) and 15 adult females (10 definite, 5 probable), along with 5 children of indeterminate sex aged between 6–8 years (Table 5.3). Males thus outnumbered females in the Enkomi sample by a ratio of more than 2:1. Since a significant number of sex attributions made by both Fischer and Fürst are attended by question-marks, indicating that the sex suggested is probable but not definite, it is possible that the number of males has been somewhat inflated by biases on behalf of each analyst. Nevertheless, it is difficult to escape the conclusion that males were present overall in larger numbers than females, whatever the actual proportions may have been.

It is important, however, to note that in several tomb groups at Enkomi and other sites, female burials may have predominated. In Swedish Tomb 11, for example, four adult females and three adult males were identified amongst a minimum of 21 burials overall. Other Enkomi tombs from which the skeletal material has been lost may also have contained relatively larger proportions of females; some of the richest tombs excavated by the BM expedition contained high concentrations of jewelry types such as earrings that may indicate a strong female presence, e.g. British Tomb 19 (31 earrings) and British Tomb 93 (37 earrings). Even assuming one woman might wear as many as 3 pairs of earrings simultaneously (Goring 1989:103–4), these numbers imply that several females were present in each of these tombs. Seven of the nine sexed skulls (among a total of 48 individuals present) in the very rich Kition Tomb 9 were identified as female (Schwartz 1974), suggesting a strong representation of women in this tomb as well. Also relevant are the cases of Kalavasos *Ayios Dhimitrios* Tomb 11, where three young women were buried with considerable quantities of wealth, and Tomb 13, which also contained a rich female burial (South 1997, 2000). Once again it appears that within high status social groups, some women were entitled to burial treatments involving very high levels of material expenditure.

Burial groups excavated at *Ayios Dhimitrios* to date range from 1–11 individuals in size (statistical calculations are omitted as the final publication of the skeletal material is pending). As at *Toumba tou Skourou* in the north, a substantial proportion of the burial sample was made up of infants and children (approximately 38% of the individuals represented in Tombs 1–11; see Keswani 1989a: Table 5.50), contrasting with the underrepresentation of the very young characteristic of the EC–MC periods and of other LC sites such as Enkomi. The construction of special niches, dromos cupboards, and, in the case of Tomb 17, independent chamber tombs, for children and infants is noteworthy.

The remains of at least 132 individuals were preserved in the LC tombs at Kourion *Bamboula*. The range of burials recorded (3–52) is comparable to the range observed at Enkomi, and the mean number of burials per chamber is nearly identical: 10.75 at Kourion, 10.73 at Enkomi (cf. Table 5.2). Some of the tombs were used over periods of 200 years or more. With regard to adult burials, the Kourion *Bamboula* mortuary sample displays the same 2:1 ratio of males to females noted at Enkomi. Angel (1972:148) identified 40 adult males (34 definite, 6 probable) and 21 adult females (12 definite, 9 probable), along with 20 non-adults (Table 5.3). Whether the bodies of women were placed in pit graves which were not preserved, or whether they were exposed or disposed of in some other way is unknown. Like that of Kalavasos, the Kourion *Bamboula* mortuary sample contained a rather high percentage of infants, children, and sub-adults relative to Enkomi and other LC sites. Within the sample examined by Angel, approximately 20% of the individuals present were infants or children up to 12 years of age. Most of these individuals were found in Tombs 19 and 40. At least six skulls and other skeletal remains said to represent 'many babies' were reported in Tomb 19, which had a mininum of 52 burials overall. In Tomb 40, infants and children accounted for 60% of the sample of 20 burials (Angel 1972:148). Angel suggested that some of the deaths in this tomb might have been caused by thalassemia, an inherited anemia attested in skeletal remains by porotic hyperostosis (Angel 1972:154–6).

## Tomb groups and social groups in long-term perspective

Considerable increases in mean tomb group size took place between the EC, MC, and LC periods. The average group size doubled from two to four burials per chamber between the Early and Middle Cypriot periods, and nearly tripled in the Late Cypriot, when the mean for all sites rose to approximately eleven. The high LC mean cannot be attributed solely to 'inflation' by the inclusion of sites such as Ayios Iakovos and Pendayia with their unusually large groups of 'mass burials'. The mean burial group sizes at Enkomi and Kourion, where very large numbers of simultaneous interments seem to have been relatively uncommon, are nearly identical to the mean for all LC sites.

At many sites, the overall increase in burial group size would seem to have been related not only to collective interment practices but also, or even more so, to the increasing timespans—sometimes extending over two to three centuries—through which the tombs were reused. In the major towns, the protracted reuse of old tombs (as opposed to the frequent construction of new ones) may sometimes have been a necessity imposed by the constraints of urban living space. Yet the long duration of tomb use evident in rural contexts—Dhali *Kafkallia* Tomb G, for example, was used from MCIII–LCIII (Overbeck and Swiny 1972), similarly Nicosia *Ayia Paraskevi* Tomb 6 (Kromholz 1982) was used from the Middle Cypriot period through LCIII—suggests that other social and symbolic considerations were involved. Whether a continuous chain of biological descent linked the individuals buried in a particular chamber remains an open question. However, it might be suggested that putative continuity in descent may have been of greater importance to those who repeatedly reused ancient tombs, or reappropriated them after long hiatuses in use. The long-term reuse of LC tombs could have served as an important affirmation of kin group identity and social status and as a means of legitimizing claims to surrounding residential and productive complexes.

Considering the long periods over which so many tombs were in use, it is somewhat surprising that burial group sizes were not actually much larger than those observed. Most LC burial groups, with the possible exception of those from Ayios Iakovos, were nowhere near as large as would be expected if large lineal groups like the Bara of Madagascar (Huntington 1973; Metcalf and Huntington 1991) were represented. Assuming a simple 'geometric' accumulation of the descendants of a single individual over six generations or roughly 150 years (less than half the time that tombs such as Enkomi Cypriot Tomb 10 and Enkomi French Tomb 5 were in use), a minimum of 63 burials $(1+2+4+8+16+32)$ per tomb might be expected. Yet as Figure 5.1 illustrates, relatively few burial groups even approached such a figure. As Cassimatis has observed, 'Les tombes sont réutilisées longtemps, mais chaque genération ou chaque époque peut n'avoir que deux morts' (1973:124).

Cassimatis attempts to explain the anomalously small size of LC burial groups by suggesting that multiple tombs were used by a single group in order to avoid opening any one tomb too soon after an earlier death, that is, before decomposition was complete, which would in turn explain the absence of evidence for fumigation. The association of groups of more or less contemporaneous tombs with particular buildings in Dikaios' Areas I and III at Enkomi lends some support to this hypothesis, although in light of the evidence for occasional secondary treatment, one might question the extent of the practitioners' squeamishness with regard to decaying flesh.

Another part of the explanation for the paradoxically 'small' size of LC burial groups may be found in the exclusion of some individuals from chamber tomb burial on the basis of age. Infants and children, as in previous periods, are greatly underrepresented in chamber tombs at sites such as Enkomi and Ayios Iakovos. Factors of preservation, recovery, and analysis may have exaggerated this bias to some degree, but overall it seems that chamber tomb burial was a prerogative infrequently extended to the young. Many juveniles may have been interred instead in jars, pit graves, and dromos cupboards, occasionally to be transferred to the principal tomb chambers when adults were buried, but not invariably. The remains of children and infants were relatively more numerous in chamber tombs at sites such as *Toumba tou Skourou*, Kalavasos, and Kourion, however, possibly marking a transformation in EC–MC customs regarding the treatment of the young in some communities.

Still another factor which could have limited burial group size is the exclusion of some family members from chamber tomb burial on the basis of sex. One of the most startling features of the Late Cypriot mortuary sample is the pronounced underrepresentation of adult females. This trend is apparent from the work of three different osteologists working on the material from three different sites, with Fischer essentially confirming the earlier studies of Fürst at Ayios Iakovos and Enkomi, and Angel obtaining very similar results at Kourion. In the sample from Ayios Iakovos, a four-to-one ratio of males to females was observed, while at Enkomi and Kourion, the ratio was two-to-one (Fürst 1933; Fischer 1986). Such a bias is suggestive of a male-focused, male-dominated society. Nevertheless, some high status burial groups at Kition, Kalavasos, and possibly Enkomi may have had predominantly female interments, suggesting that at the apex of the social hierarchy, considerations of descent in the female line were politically significant. Among groups of lesser status, it is possible that the cohort of females excluded was predominantly young (Table 5.3), and that successful child-bearing and child-rearing were the *de facto* prerequisites for full mortuary honors.

A final factor limiting burial group size may have derived from political and economic competition both within and between households and descent groups during the Late Cypriot period. This could have resulted in the periodic fissioning or fragmentation of kin groups, such that long-term attachments to or identifications with particular tomb groups were not invariably sustained. The widespread appearance of shaft graves in LCIII may represent the intensification of this tendency and at least a partial eclipse in the importance of kin group identity or membership as the basis of social status. It has already been suggested that the shaft grave burials might represent the truncation of older systems of mortuary ritual, with the final incorporation of the deceased in the collective tomb of his or her real or putative ancestors being omitted. This in turn may reflect a waning emphasis amongst the survivors on the affirmation of social status through ties to the recently deceased and other more distant forebears. As wealth and prestige might now be obtained through entrepreneurial success in various productive activities (e.g. copper working) other than agriculture, in local or long-distance exchange transactions, and through political appointments and maneuvering within emergent court and temple institutions—rather than exclusively within the matrix of competitive kin groups prevalent in earlier periods—the shift from collective to individual burials may have followed accordingly. Furthermore, as towns became more and more heterogeneous in their social composition, with many non-indigenous residents from other parts of Cyprus and probably other parts of the Mediterranean world at large, there may have been a substantial population of individuals who claimed no ties to local kin groups or their ancestral tombs, and whose status—both high and low—depended instead upon wealth and/or position within diverse institutional contexts.

# Variability and change in Late Cypriot tomb architecture

As in the EC–MC periods, rock-cut chamber tombs continued to be the most common type of burial facility in the LC period (see Figs. 5.2–5.4 for examples by region). And, as in previous periods, distinctive regional peculiarities of tomb construction are evident, along with widely shared traditions of tomb form. In comparison to chamber tombs of EC–MC date, however, LC tombs were generally smaller and less elaborate in form, and multi-chambered tomb complexes were less common. The sociocultural factors that may have contributed to this decline in ostentation are considered following a brief regional overview of architectural traditions and variability.

## Northern sites

In northwestern Cyprus, the most elaborate LC tombs that have been found to date are those of Morphou *Toumba tou Skourou*. In general, the *Toumba tou Skourou* tombs were constructed with one to four chambers opening from a central dromos, with chambers that were either roughly ovoid or bilobate in form, with projecting, quasi-rectangular buttresses cut opposite the entrance to the chamber (see Vermeule and Wolsky 1990: Figs 30, 32, 34, 36, 37, 40, 43). Tomb 1 is the largest and most complex of the tombs excavated at *Toumba tou Skourou*, with three chambers, placed on the east, west, and south sides of an unusual, circular, chimney shaft type dromos, and fourteen niches cut for infant burials in the sides of the dromos shaft (Fig. 5.2a). The general design of this tomb is similar to contemporaneous tombs at Maroni *Kapsaloudhia* in the south (Herscher 1984; Cadogan 1984; Fig. 5.4c here). Although the range of chamber sizes at *Toumba tou Skourou* varied considerably from slightly over one square meter in floor area (Tomb 5, Chamber 1) to nearly 10 sq m for Tomb 1, Chamber 1, the mean chamber floor area for all of the tombs, calculated at 4.38 sq m (Table 5.4), was relatively low compared to the MC tombs of Lapithos and contemporaneous LC tombs at Ayios Iakovos.

Elsewhere in the northwest, tomb plans seem to have been fairly simple. The LC tombs excavated at Ayia Irini *Palaeokastro* (Pecorella 1977) and Myrtou *Stephania* (Hennessy 1964) had single ovoid or rectangular chambers and oblong dromoi, with relatively few architectural elaborations (Fig. 5.2 b, 5.2c). The Ayia Irini tombs had chamber floor areas ranging between 2.7–11.7 sq m, with a mean floor area of 7.9 sq m, while the Myrtou *Stephania* tombs were somewhat smaller in size, with chamber floor areas ranging from 2.5–7 sq m, averaging 4 sq m (Table 5.4). A dual-chambered tomb complex with rounded chamber plans and relatively large floor areas (approximately 9.5 sq m and 10.5 sq m) was found at the locality Kazaphani *Ayios Andronikos* (Nicolaou and Nicolaou 1989: Figs 1, 3). At Lapithos *Ayia Anastasia*, two tombs that were apparently cut in LCII and reused in both LCIII and the Iron Age had differing shapes; Tomb 501, with a floor area of roughly 6.3 sq m, was square in plan with a shallow oblong pit in the center of the chamber floor, a niche off the rear left (southwest) corner, and what appears to have been a faintly defined pillar in the center of the rear wall opposite the entrance (Gjerstad *et al*. 1934: Fig. 61:2–4), while Tomb 502, with chamber dimensions of 2.68 × 3.42 m, was reportedly round with an oval basin sunk in the floor (Gjerstad *et al*. 1934:164). There is little other information regarding tomb architecture along this part of the north coast.

Small, rounded, single-chambered rock-cut and pit tombs seem to have been the norm along the northern foothills of the Troodos (Figs 5.2d, 5.2e). The Pendayia tombs were all approximately ovoid in plan with small, pit-like dromoi entering high up in the chamber wall, nearly level with the roof, in the fashion of the so called 'chimney tombs,' and they were small in size, with

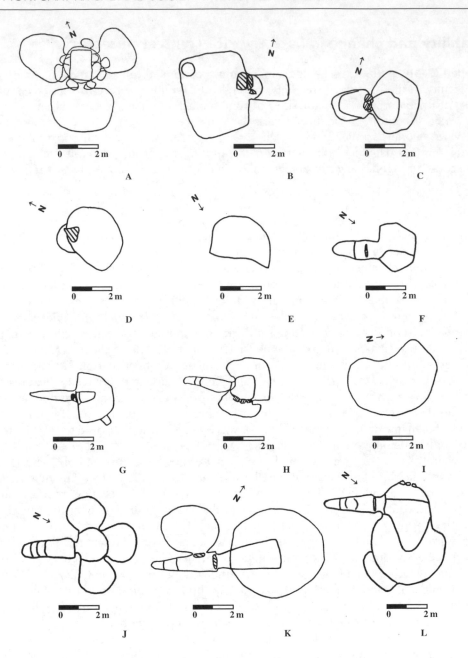

**Figure 5.2**     Plans of tombs cut in MCIII/LCIA and LCI at sites in northern Cyprus: (a) Morphou *Toumba tou Skourou* Tomb 1 (after Vermeule and Wolsky 1990: Fig. 30), (b) Ayia Irini *Palaeokastro* Tomb 20 (after Pecorella 1977: Fig. 251), (c) Myrtou *Stephania* Tomb 5 (after Hennessy 1964: Pl. 31), (d) Pendayia *Mandres* Tomb 1 (after Karageorghis 1965c: Fig. 4), (e) Akhera *Chiflik Paradisi* Tomb 1 (after Karageorghis 1965b: Fig. 20), (f) Korovia *Nitovikla* Tomb 1 (after Gjerstad *et al.* 1934: Fig. 159:1), (g) Milia *Vikla Trachonas* Tomb 11 (after Westholm 1939a: Fig. 1), (h) Milia *Vikla Trachonas* Tomb 13 (after Westholm 1939a: Fig. 4), (i) Angastina *Vounos* Tomb 2 (after Nicolaou 1972: Fig. 9),( j) Ayios Iakovos *Melia* Tomb 8 (after Gjerstad *et al.* 1934: Fig. 126:1), (k) Ayios Iakovos *Melia* Tomb 10 (after Gjerstad *et al.* 1934: Fig. 126:9), (l) Ayios Iakovos *Melia* Tomb 14 (after Gjerstad *et al.* 1934: Fig. 130:10).

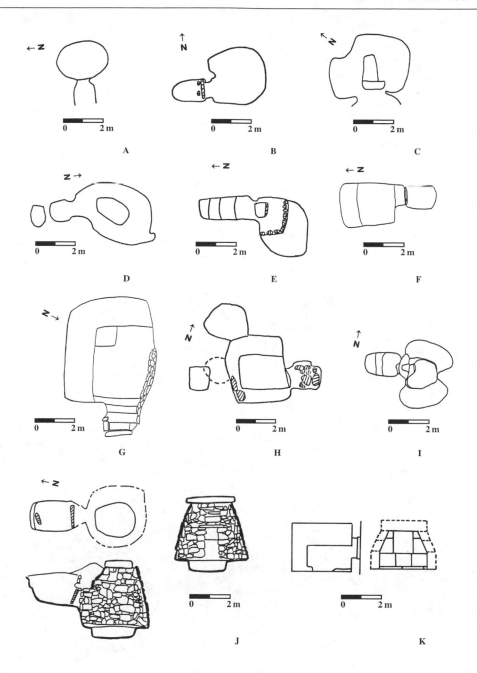

**Figure 5.3**    Diverse Late Cypriot tomb plans from Enkomi: (a) Cypriot Tomb 21 (after Dikaios 1969: Pl. 289:6), (b) Swedish Tomb 2 (after Gjerstad *et al.* 1934: Fig. 188:3), (c) French Tomb 10 (1934) (after Schaeffer 1936: Fig. 29), (d) French Tomb 12 (1934) (after Schaeffer 1936: Fig. 35), (e) Swedish Tomb 8 (after Gjerstad *et al.* 1934 : Fig. 194:7), (f) Cypriot Tomb 19 (after Dikaios 1969: Pl. 289:8), (g) French Tomb 2 (after Schaeffer 1952: Pl. 12), (h) Swedish Tomb 18 (after Gjerstad *et al.* 1934 : Fig. 209:1), (i) French Tomb 1851 (after Lagarce and Lagarce 1985: Fig. 3), (j) Swedish tholos Tomb 21 (after Gjerstad *et al.* 1934 : Fig. 213:8–10), (k) British ashlar Tomb 66 (after Murray *et al.* 1900: Fig. 5).

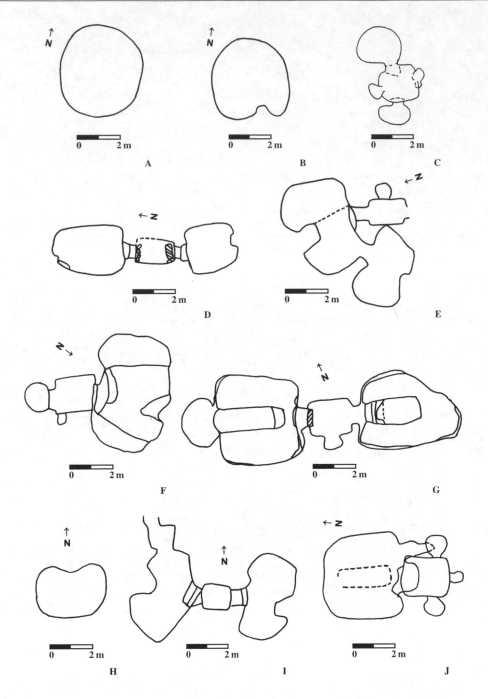

**Figure 5.4**    Tomb plans from Late Cypriot sites in southern Cyprus: (a) and (b) Hala Sultan Tekke Tombs 1 and 2 (after Karageorghis 1972: Pl. 74 ), (c) Maroni *Kapsaloudhia* Tomb 2 (after Cadogan 1984: Fig. 4), (d) Kition Tombs 4+5 complex (after Karageorghis 1974: Pl. 118), (e) Kition Tomb 9 (after Karageorghis 1974:134), (f) Kalavasos *Ayios Dhimitrios* Tombs 11+9 complex (after South 1997: Fig. 5), (g) Kalavasos *Ayios Dhimitrios* Tombs 18+19 complex (after South 1997: Fig. 10), h) Kourion *Bamboula* Tomb 21 (after Benson 1972: Pl. 11), (i) Kourion *Bamboula* Tombs 17+17A complex (after Benson 1972: Pl. 9), (j) Kourion *Bamboula* Tomb 32 (after Benson 1972: Pl. 11). Note that the scales associated with Kourion *Bamboula* Tombs 21 and 32 may be inaccurate (see Benson 1972: Pl. 11 notes).

floor areas between 3.3–5.4 sq m (Karageorghis 1965c: Figs 4–6). LC tombs in the vicinity of Katydhata, like those of earlier periods, are generally described as 'beehive' tombs with oval to circular chambers of small to medium size, for example, 1.55 × 1.5 m (Tomb 90) to 2.9 × 2.35 m (Tomb 85) (Gjerstad 1926; Åström 1989). The tombs of Akhera *Chiflik Paradisi* were small pits (4–5.2 sq m in floor area, see Karageorghis 1965b: Figs 20–22) rather than typical chamber tombs; they were semi-circular or ovoid in shape, with no dromos or clearly marked entrance, a type of construction possibly necessitated by the nature of the local subsoil (Karageorghis 1965b: 72). Similarly, Politiko *Ayios Iraklidhios* Tomb 6 was a small pit tomb with a floor area of *c.* 5 sq m (Karageorghis 1965a: Figs 2–3).

The LCI–II tombs of Ayios Iakovos *Melia* in northeastern Cyprus continued the MC architectural tradition of large niched chambers with long tapering dromoi (Fig. 5.2j–l). They also attest to unusually high levels of energy expenditure in tomb construction, with floor areas ranging from 8.3–24.8 sq m. Tomb 8 was particularly elaborate in form, with three raised, quasi-circular lobes or wings, all of approximately the same size, opening off from a circular central well (Gjerstad *et al.* 1934: Fig. 126:1). Tomb 10 had a long rectangular trough cut in the center of the chamber (Gjerstad *et al.* 1934: Fig. 126: 9–10), a feature paralleled in the contemporaneous Tombs 10 and 11 at Milia *Vikla Trachonas* located several kilometers to the south. The Milia tombs, with their semicircular and/or 'lobed' chambers and their long narrow dromoi, share a certain resemblance to the chamber tombs of Ayios Iakovos and *Paleoskoutella*, but they were considerably smaller in size, with chamber floor areas of only 3–4.6 sq m (Westholm 1939a: Figs 1–5; Fig. 5.2g and 5.2h here). The LCI Tombs 1 and 2 at Korovia *Nitovikla* (Gjerstad *et al.* 1934: Fig. 159; Fig. 5.2f here), with their irregular ovoid chambers accessed through short oblong dromoi, were also small and unimpressive (floor areas of approximately 4.8–5.8 sq m) in comparison to the elaborate and quasi-contemporaneous chamber tombs of nearby *Paleoskoutella* and Ayios Iakovos. The tombs of Angastina *Vounos* had irregularly ovoid to kidney-shaped plans and somewhat larger floor areas ranging from 9–13 sq m (Nicolaou 1972: Figs 3, 9, 11; Karageorghis 1964b: Figs 1a–b; Fig. 5.2i here).

## Enkomi *Ayios Iakovos*

As noted earlier, many different types of mortuary features have been found at Enkomi (Fig. 5.3). Chamber tombs, which were by far the most common, were constructed throughout LCI and LCII; relatively few seem to have been cut in the LCIII period, but it is possible that more existed in parts of the site that were never investigated, or beyond the periphery of the settlement (Dikaios 1969:431). Most of the chamber tombs were cut with a single chamber entered via a small, shaft-like dromos. Occasionally two chambers were cut at opposite ends of a common dromos (e.g. Cypriot Tombs 1/10 and 7/11, Dikaios 1969: Pls 285 and 287; see also French Tomb 1907, Lagarce and Lagarce 1985: Figs 1–3). There are several examples of chamber tombs having one or more small dromos cupboards, which may have been intended primarily for infant burials, but remains of older individuals may have been placed in the 'ossuary' of Swedish Tomb 6 (Gjerstad *et al.* 1934:491–7, Fig. 191:15–16).

Enkomi chamber tomb plans were diverse (see Fig. 5.3 and Keswani 1989a: Tables 5.28–5.32). The most common tomb form appears to have been the simple rounded chamber tomb, entered via a small shaft-like dromos either through the roof of the chamber in 'chimney' fashion or relatively high up in one of the chamber walls (Fig. 5.3a). A few tombs were cut in asymmetrical shapes, with special niches or extensions on the left side of the chamber (e.g. Swedish Tombs 2, 12, and 17, Cypriot Tomb 13, French Tombs 10 [1934] and 1907E; Fig. 5.3b and 5.3c), somewhat reminiscent of Tombs 6 and 7 at Ayios Iakovos. Two examples of asymmetrical tombs with extensions or niches on the right side of the chamber (French Tomb 12 (1934 ) and Swedish Tomb 8; Fig. 5.3d and 5.3e) were also observed. Other chamber tombs were roughly rectangular in shape, and were sometimes equipped with rock-cut benches to accommodate extended burials (e.g. Cypriot Tomb 19, French Tomb 2 (1949), Swedish Tomb 11, and Swedish Tomb 18; Fig. 5.3f–h). There are also a few examples of bilobate chamber tombs with or without rock-cut benches on opposite sides of the chamber (e.g. French Tomb 1851 and French Tomb 11 (1949) with benches; French Tomb 5(1949), Swedish Tomb 3, and Cypriot Tombs 2 and 10 without benches; Fig. 5.3i).

It is difficult to link differences in tomb shape with variables such as chronology, status, or wealth. It might be speculated that the rectangular chamber tombs, for which earlier prototypes seem to be lacking, were constructed in emulation of the rectangular ashlar built tombs, but they were not invariably associated with extremely rich tomb assemblages (Cypriot Tomb 19, for example, was notably modest in wealth; see Table 5.9b). Since tombs of different shape sometimes shared a common dromos (e.g. the rectangular Cypriot Tomb 7 was adjacent to Tomb 11, a simple rounded chamber, and the rectangular Cypriot Tomb 1 lay opposite Cypriot Tomb 10, a bilobate chamber tomb), it is also problematic to assert that the variability observed is linked to the peculiar customs of different residence or 'ethnic' groups. Moreover, most tomb forms were found in several different residential 'quartiers' or blocks of the site, and most blocks had multiple tomb forms. Some of the architectural diversity may be explicable, however, in terms of function, as some of the smallest and simplest tombs such as Cypriot Tombs 15, 16, 17, and 21 were found to have been emptied (Dikaios 1969:400–406, 415–6), and may have been intended only for temporary use. Some of the variability in the form of more permanent burial chambers could have been related to the nature of the mortuary treatment and positioning intended for the deceased, with tombs of approximately circular shape constructed to accommodate seated or crouching burials, and those with bilobate or rectangular chambers and benches cut to accommodate extended burials. Over extended periods of reuse, however, diverse burial postures are evident in some chambers (e.g. Swedish Tombs 11 and 17; Gjerstad *et al.* 1934).

In comparison to the Middle Cypriot tombs of Lapithos *Vrysi tou Barba*, Ayios Iakovos *Melia*, and Korovia *Paleoskoutella*, the Enkomi tombs were small, with floor areas ranging between 1–10 sq m, averaging just under 5 sq m (Table 5.5). Floor area was rather weakly correlated with other variables such as the number of burials per chamber and various wealth measures such as total pots, bronzes, and gold weight per chamber (Table 5.6); only French Tomb 2 (Schaeffer 1952: Pl. 12), with an estimated floor area of 6.4 sq m, stands out in terms of the exceptional area allotted for each of its three very richly equipped burials. It is probable that the situation of the tombs adjacent to settlement architecture placed significant limitations on the potential sizes of both chambers and dromoi, and it may also account for the irregular configuration of some chambers and dromoi (e.g. Swedish Tombs 13 and 19 (Gjerstad *et al.* 1934: Figs 201:6, 213:1; Cypriot Tomb 2, Dikaios 1969: Pl. 283). However, changes in social structure and in the scale of mortuary observances may also have been factors that affected energy expenditure in tomb construction, as discussed at greater length in the conclusion of this chapter.

Given the spatial constraints affecting chamber tomb construction, the use of built tombs, rather than large and ornate chamber tombs, may have presented a more feasible means of architectural elaboration and display for those kin groups with access to the services of skilled architects, masons and laborers. During LCI and LCII, at least four and possibly as many as five tholos tombs with partially corbelled superstructures of stone and/or mudbrick were built at Enkomi. All of the tholoi had been looted prior to excavation, but fragmentary gold finds from Swedish Tomb 21 (Gjerstad *et al.* 1934:571) and British Tomb 71 (Williamson and Christian n.d.) suggest that they were used by groups of considerable wealth. Extrapolation between the site maps of the various excavators suggests that Swedish Tomb 21 (Fig. 5.3j) may have lain in the vicinity of Q6W; British Tomb 48 (if this was indeed a tholos tomb; the description in Williamson and Christian n.d. is ambiguous) may also have been located in Q5W or Q6W; British Tomb 71 probably lay somewhere in the western part of the site; French Tombs 1336 and 1432 were definitely located in Q5E (Courtois 1986b; Johnstone 1971). Thus, although not all of the tombs can be precisely located, they seem to have occurred in more than one part of the site, suggesting that they were not simply the idiosyncratic constructions of a particular residence group.

Superficially, these tombs bear some resemblance to the great tholoi of Mycenaean Greece, but as Pelon (1973) has observed, the Enkomi tombs are much smaller and more irregular in construction than their counterparts in Greece, and their general plans are more similar to the local rock-cut chamber tombs than to the Aegean tholoi. The location of the tholoi within the bounds of the settlement, sometimes within the courtyards of contemporaneous houses, is also inconsistent with Aegean tradition (Pelon 1973:252–3). Rather than being the work of Aegean builders, it is probable that the Enkomi tholos tombs represent a distinctively Cypriot adaptation of either Aegean or even earlier Near Eastern prototypes known from Megiddo, Tell Dan, Chagar Bazar, and other sites (Ilan 1995:138; Gilmour 1995:163–4; Sjöqvist 1940:18, 149–50; Mallowan 1936:55–6). Alternatively, they may have been built by Levantine immigrants, but if so, those settlers would seem to have been well assimilated within the community, given the central locations of most of the tombs and the long periods over which some were used (e.g. French Tomb 1336; Johnstone 1971).

At least five rectangular, partially corbelled tombs built of ashlar masonry were constructed in the neighborhoods of Q4E (British Tomb 1/French Tomb 1409?, British Tomb 12, British Tomb 66/French Tomb 1322, and French Tomb 1394) and Q3E (British Tomb 11). These tombs, with finds dating mainly to LCII, were probably inspired by the ashlar tombs built beneath the houses of the Ras Shamra elite (Westholm 1939b; Schaeffer 1939; Courtois 1969; Salles 1995), although in their construction they were less sophisticated than the contemporaneous, fully corbelled tombs of the Syrian site. Of all of the ashlar tombs discovered at Enkomi, only British Tomb 66 (French Tomb 1322), illustrated in Figure 5.3k, was found intact, yielding an exceptionally rich array of gold, bronze, faience, and other exotic goods (Table 5.9c). However, fragmentary finds from the other ashlar tombs suggest that they too were used by groups of exceptional wealth. The nature of the tomb architecture, the finds, and the clustering of the tombs in association with substantial, well-built houses (Courtois 1986b: Fig. 4) are suggestive of the establishment of important elite groups in this area. But it is problematic to characterize the ashlar tombs as the paramount 'royal' tombs of Enkomi, as rock-cut tombs of comparable and even greater wealth are found in several other areas of the site, among them French Tomb 2 (Schaeffer 1952), British Tombs 19, 67, and 93 (Murray *et al.* 1900), and Swedish Tomb 18 (Gjerstad *et al.* 1934). As with the tholoi, it is very difficult to determine whether these tombs were introduced by Syrian immigrants or by Cypriots who were impressed by the burial practices of foreign elites.

At the opposite extreme in terms of energy expenditure were the LCIII shaft graves. These features were fairly shallow in depth and had average floor areas of under 2 sq m (Table 5.5). Despite their overall simplicity, there were some variations in the construction of the shaft graves; some, such as Swedish Tombs 5 and 14 (Gjerstad *et al.* 1934: Figs 191:6–11, 204:1–2), French Tombs 15(1934) (Schaeffer 1936: Fig. 85) and 1(1947) (Schaeffer 1952: Fig. 87), were lined with stones, while others were not (Keswani 1989a: Table 5.36). Whether the stone linings represent deliberate attempts at 'architectural elaboration' or the need to stabilize the sides of the graves in some locations is unclear.

## Southern sites

The limited evidence for tomb architecture at Hala Sultan Tekke *Vizaja* suggests that at this important town site, as at Enkomi, there was considerable variation in LC mortuary features. Although few details concerning tomb architecture were recorded for any of the tombs excavated by Walters and Crowfoot on behalf of the British Museum in 1897–98, a sketch plan of Crowfoot's Tombs 11 and 12 depicts two rectangular chambers opening from opposite ends of a square dromos shaft, each rendered bilobate by a long projection or buttress opposite the door (Bailey 1972: Pl. I:c). The two tombs (Tombs 1 and 2) excavated by the Department of Antiquities in 1968 were both large pit tombs with floor areas of approximately 9.9 sq m and 8.5 sq m respectively (Karageorghis 1968; 1972b: Pl. 74; Fig. 5.4a and 5.4b here). Looted chamber tombs investigated more recently by the Swedish expedition (Åström 1983: 145–54, Fig. 380b) may have been comparable in size, with variably bilobate heart-shaped (Tomb 20) and irregularly ovoid plans (Tombs 20–21). Tomb 22 had a short rectangular dromos and a small rectangular niche inside the chamber immediately to the left of the entrance. Shaft grave 23, which had been cut in the corner of an earlier building, was an unlined feature measuring 2.25 × 1.05 × 0.7 m, possibly marked by a large rectangular worked stone block found above it (Niklasson 1983:170).

The Kition tombs also display a wide range of variability in size and form. Chamber Tombs 1 and 3 were fairly simple in their construction, with ovoid and rounded rectangular plans respectively (Karageorghis 1960b: Figs 9–12). Tomb 1 was equipped with a rock-cut bench around the inner periphery of the chamber. Tombs 4 and 5, which shared a common dromos, also had rounded rectangular plans. Tomb 4 had a small rectangular projection in the center of the south wall immediately opposite the entrance, while Tomb 5 had a small, irregular projection in the left rear (northwestern corner) of the chamber (Karageorghis 1974: Pls 118–9; Fig. 5.4d here). Tomb 9 was by far the most complex of the Kition tombs in plan, having a tripartite chamber that may originally have been bilobate, with a third lobe possibly added to the southwest at a later date (Karageorghis 1974: Pl. 134; Fig. 5.4e here). The tomb was entered via a short rectangular dromos with a small round cupboard opening off the center of the east wall. The floor area of the chamber, estimated at *c.* 12 sq m, was quite large in comparison to the other Kition tombs, which ranged from 2.4 to 6.9 sq m in area. At the opposite extreme in energy expenditure was a small (2.4 sq m) pit tomb, elliptical in shape and lined with two courses of mudbricks, used for the burial of several individuals including four children (Karageorghis 1960b: 515).

Virtually no information pertaining to tomb architecture was recorded in the account of the British Museum excavations at Maroni, although in one rare instance, Walters described *Tsaroukkas* Tomb 10 as a 'triple tomb' (Johnson 1980:9). This could either mean that it was tripartite in form like Kition Tomb 9 or that it had three chambers opening off from a common dromos, similar to two tombs of MCIII/LCIA date found at Maroni *Kapsaloudhia* which had multiple chambers entered from a chimney-like dromos (Cadogan 1984: Fig. 4; Fig. 5.4c here). Recent attempts to recover more architectural data from these tombs have been hampered by the destructive character of the BM investigations (Manning and Monks 1998). However, one tomb was reported to have had two benches with a 'chamber' in between (possibly similar to Kalavasos *Ayios Dhimitrios* Tombs 18, 19 and 21 as well as others discussed below), and another tomb was also provided with two benches (Manning and De Mita 1997:131).

A number of tomb plans are evident in the Kalavasos area. The LCIA Tomb 51, discovered near the mosque in Kalavasos Village, was roughly circular in plan, measuring 2.45 × 2.35 × 1.32 m (Pearlman 1985: Fig. 1). The tombs found at the locality *Mangia* were variably circular, ovoid, or bilobate in form (McClellan *et al.* 1988: Figs 3–6). Tombs 5 and 6 were of medium size, with floor areas that may be estimated at 6.7 sq m and 4.9 sq m respectively. The *Ayios Dhimitrios* chamber tombs varied considerably in form, some having relatively simple, roughly circular to ovoid floor plans, others having more elaborate, bilobate chambers divided by central buttresses and provided with benches (either rock-cut or stone-built) along the sides of the chamber (South *et al.* 1989: Figs 37–41; South 2000: Figs 2–3; Fig. 5.4f here). Tomb 14 was divided by a rectangular rock-cut buttress reinforced by a stone-built pillar. Tombs 18 and 19 in the West Area (Fig. 5.4g) and Tomb 21 in the Northeast Area had rectangular plans and benches formed on either side of the chamber by a long central depression or cist (South 2000:362 and Fig. 4), a form of tomb architecture that seems to have been quite common in the southwestern part of the island. There was considerable variation in chamber floor area, perhaps reflecting a combination of factors relating to the age of the deceased, the elaboration of the associated mortuary rituals, and whether or not the tomb was intended as a permanent burial facility. Tombs 2, 3, 7, and 10, which were found empty and may have served as temporary burial facilities, had floor areas of less than 4 sq m; Tomb 16, likewise emptied, was also quite small. Tombs 1, 4, 5, and 6, all of which presumably served as permanent collective burial facilities for groups of low to medium wealth, were very uniform in size, with floor areas ranging between 5.7–6.9 sq m. The bilobate chamber of the very rich Tomb 11 was considerably larger, having a floor area of approximately 10 sq m; the energy expenditure represented by the construction of this tomb is particularly high relative to the number of adult burials present. Tombs 13, 14, 18 and 19, which also seem to have had impressive assemblages of grave goods, had comparably large floor areas ranging from 8.6–11.5 sq m (South 2000:361).

Most of the mortuary features encountered at the town site of Kourion *Bamboula* were rock cut tombs with single chambers of irregular form, variably ovoid, kidney-shaped, or rectangular in plan, with few distinctive structural features (e.g. Fig. 5.4h). The dromoi, when preserved, tended to be shallow, pan-shaped pits. Tombs 9 and 32, both cut in LCI, had long, relatively shallow rectangular troughs running lengthwise in the center of the chamber floors (Benson 1972:15, 27 and Pl. 11; Fig. 5.4j here), features similar to those observed in Ayios Iakovos Tomb 10, Milia *Vikla Trachonas* Tombs 10 and 11, and Enkomi French Tomb 10 (1934). Occasionally two chambers opened off a common dromos, as in the case of the dual-chambered Tomb 33 (Benson 1972: 28) and the complex of Tombs 17 and 17A (= BM Tombs 53 and 102; see Fig. 5.4i here). The very rich Tombs 17 and 17A (Benson 1972: Pl. 9) were notably large and elaborate, with bilobate chambers divided by projecting rectangular buttresses into two asymmetrical portions, of which the left side was the larger in both cases (as in Kalavasos *Ayios Dhimitrios* Tombs 11 and 14). An approximately square side-chamber was cut back from the right rear corner of Tomb 17A (cf. Enkomi Swedish Tomb 18). Chamber floor areas varied between 3–10 sq m, averaging 5.4 sq m (Table 5.7), comparable to the range of tomb sizes observed at Enkomi and Kalavasos. Daniel suggested that Kourion *Bamboula* Tomb 2 may have had an interior facing of rubble masonry, and that Tombs 9 and 16 had superstructures of mudbrick covered with earthen mounds, but Benson has discounted the presence of built tombs, interpreting the presence of stone or mudbrick fragments as remnants of various filling incidents unrelated to the original construction of the tombs (1972:7).

Other mortuary features discovered at Kourion included one pit or shaft grave of LCIII date (Tomb 14, Benson 1972:7, 18 and Pl. 8)) and another unusual complex, Tomb 37, consisting of three circular to ovoid cist graves covered with flat stones and containing some unidentified skeletal remains and sherds (Benson 1972:7, 31 and Pl. 12). Very different, 'Mycenaean'-type rectangular tombs with long, sometimes stepped dromoi appeared nearby at Kourion *Kaloriziki* in LCIIIB (Daniel 1937; Benson 1973), along with another unique shaft grave or chamber tomb containing a cremation burial (McFadden 1954).

Further to the west, a tremendous range of variation in tomb form may be noted in the many tombs of LCI–CGI date excavated in the Kouklia area since the 1950s. The *Teratsoudhia* and *Eliomylia* tombs, as well as others at *Asproyi* and *Evreti*, display the distinctive southwestern chamber tomb plan with one or two oval or square chambers opening off small, square, pit-like dromoi, very low, slightly vaulted ceilings, rectangular cists in the center of the chamber floor and benches along the side and rear walls of the chamber (Catling 1968; Karageorghis 1990a: Figs 2–4). The tombs at *Teratsoudhia* and *Eliomylia* were uniformly small, with floor areas ranging between 4.2–5.9 sq m. Other instances of this form have already been noted at Kalavasos *Ayios Dhimitrios*, and further examples are known from Alassa *Pano Mandilaris* (Hadjisavvas 1989:39) and Yeroskipou *Asproyia* (Nicolaou 1983). Subsequent variations in mortuary practice are attested by the twenty pit graves dating to LCIIIB found at the locality Kouklia *Kaminia* (Maier and Wartburg 1985:146, 151–2); from the published information it is unclear as to whether these features represent 'poor' burials or partially emptied primary graves. Meanwhile, at other Kouklia localities such as *Xerolimni* and *Lakkos tou Skarnou*, still other new types of tombs came into use in LCIIIB. These were rectangular chamber tombs with short, steep dromoi (Maier and Wartburg 1985:151; Karageorghis 1967). The LCIIIB–CGI tombs of Palaeapaphos *Skales* (Karageorghis 1983) were of the particularly elaborate 'Mycenaean' type with elongated dromoi and square chambers, an innovation in form that has often been ascribed to the intrusion of new ethnic groups (Maier 1973; Maier and Wartburg 1985; Karageorghis 1983; see also Steel 1993; 1995:199).

## The social dimensions of diminishing architectural expenditure

Mean chamber floor area for all sites decreased from *c.* 8.3 sq m in the MC period to 6 sq m in the LC period (Fig. 5.5), marking a substantial decline in energy expenditure in tomb construction, especially in light of the concurrent increases in burial group size. Only the LCI tombs of Ayios Iakovos seem to have continued the MC tradition of architectural grandeur. At Enkomi, the only site with sufficient data to permit such calculations, there appears to have been a rather low correlation among the variables of tomb size, wealth, and number of burials present (Table 5.6), suggesting that the rich as well as the poor were in some way constrained from building large facilities for their dead. At other sites, the richest tombs such as Kition *Chrysopolitissa* Tomb 9, Kalavasos *Ayios Dhimitrios* Tomb 11, and Kourion *Bamboula* Tomb 17/17A seem to have been somewhat larger than tombs of lesser wealth, but their chamber floor areas seldom exceeded 10–11 sq m, making them far smaller than the grand tombs of MC Lapithos, Dhenia, *Paleoskoutella*, and Ayios Iakovos, which sometimes attained chamber floor areas of 20 sq m or more.

The smaller chamber floor areas observed in the Late Bronze Age may be partially attributable to the limitations on tomb construction within the context of urban space, a factor which also promoted long-term reuse of many tombs. The innovation of ashlar tombs and tholoi at Enkomi, tomb types that were probably inspired by Near Eastern prototypes, may have presented a more feasible avenue for tomb elaboration and assertions of social status than the construction of large chamber tombs. Yet in addition to the practical constraints on tomb construction, the decreasing scale of kin and communal participation in mortuary ritual and the associated 'privatization' of tombs in residential space could also have contributed to the decline in the scale and ostentation of chamber tomb architecture. It has already been suggested that the frequency of shaft grave burials in LCIII, representing a further diminution of the energy expended in tomb construction, was also related to a trajectory of social and ritual transformations in which traditional programs of mortuary treatment were simplified or shortened. As social stratification, urbanization, and socioeconomic competition between groups increased, the number of individuals involved in active sponsorship, labor, and material contributions to particular funerals may have declined relative to earlier periods.

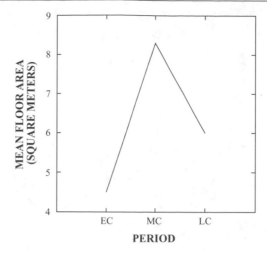

**Figure 5.5**    Changes in mean chamber floor area by period throughout the Cypriot Bronze Age.

## Changing patterns of mortuary expenditure and status symbolism in grave goods

Although some elements of the EC–MC prestige complex persisted in LCI tomb assemblages, significant changes in both the complements of status symbolism deployed in mortuary rituals and their distributional patterns took place during the Late Cypriot period. Along with a changing array of locally produced ceramic and metal goods, Aegean and Levantine ceramics, local and imported glyptics, balance weights, and exotic items made of gold, silver, faience, glass, ivory, ostrich egg, and semi-precious stones were now included in some tombs. Per capita consumption of pottery and copper-based artifacts (hereafter referred to as bronzes, see Swiny 1982:77) actually decreased relative to the Middle Cypriot period (Fig. 5.6), and large faunal remains seem to have been scarcer, while the new types of luxury goods became increasingly common. The most dramatic illustration of material expenditure in LC mortuary ritual and of the differentials in wealth between groups may be found in the disposal of gold, a material rarely encountered in earlier periods. Although scarce in the LCI period, gold objects were frequently deposited in urban tombs of LCII date. However, the distribution of gold types indicates that the richest groups not only had access to more of this precious metal but also had exclusive access to the heaviest, most intricately worked, and most informationally rich items such as signet rings with engraved designs, hieroglyphs, and Cypro-Minoan inscriptions. The deposition of other exotic materials followed approximately the same pattern as that of gold, with the largest and most diversified arrays of materials, along with the most ornate and iconographically rich items (e.g. Mycenaean pictorial craters, carved ivories with pictorial representations), concentrated in the richest tombs overall. Thus in contrast to the EC–MC period, when some of the most valuable prestige goods such as copper hook-tang weapons were not only widely available but subject to broadly based competitive elaboration, the highest order prestige goods of the LC period evince an unmistakably hierarchical distribution, suggesting that they were the exclusive perquisites of elite groups. The concentration of the highest order valuables in certain tombs that were reused over several generations further suggests that access to such items, and to the elite status which they represented, were to some extent hereditary.

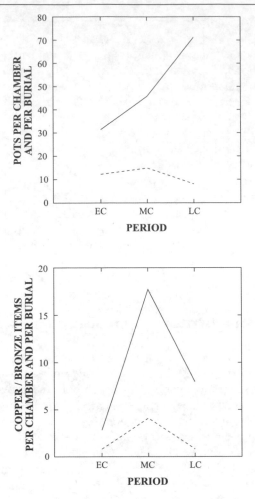

**Figure 5.6**    Changes in mean pots and mean copper or bronze items per chamber (solid line) and per burial (dashed line) by period throughout the Cypriot Bronze Age.

Much of the interpretation of Late Bronze Age mortuary symbolism and social structure rests upon the evidence from the east coast town of Enkomi, at present the only site with an extensive and well-documented mortuary sample spanning the entire Late Cypriot period (Keswani 1989a: 335–42; 1989b). However, many of the patterns observed at Enkomi can be discerned in the as yet less comprehensive samples from several of the south coast centers, where it appears that elite groups employing similar complements of prestige goods had also emerged by the beginning of the LCII period if not sooner. These data support the proposition that the south coast centers were for the most part independent 'peer' polities not under the domination of Enkomi, for otherwise we might expect to see a more hierarchical distribution of valuable types, with the most valuable items being retained in the paramount center. Meanwhile, it appears that the relations between coastal and inland sites had become to some extent hierarchical, with the assemblages of prestige goods from inland sites being comprised mainly of the lower order valuables common in tombs of low to average wealth at Enkomi. Thus it might be suggested that while the overall affinities in the prestige complements found in coastal and inland areas are strong (indicating that their communities participated in a common system of prestige symbolism), access to the most valuable goods may have been controlled by the coastal centers.

The following discussion begins with a brief review of LCI and early LCII mortuary assemblages known primarily from sites in the northern part of the island. Patterns of mortuary expenditure and symbolism spanning the LCI–LCIII periods at Enkomi are then considered, followed by a comparative analysis of the evidence from LCII–LCIII sites in other coastal and inland regions. Finally, I attempt to explicate the meanings of the various transformations in mortuary consumption and the symbolism of prestige that are evident over the course of the entire LC period.

## Burial complements of the early LCI–LCII periods

The burial assemblages from the MCIII/LCIA tombs of Pendayia *Mandres* (Karageorghis 1965c) and the LCIA Tomb 1 at Akhera *Chiflik Paradisi* (Karageorghis 1965b) bear strong affinities to the collections of grave goods from tombs of the preceding Middle Cypriot period. The intact Pendayia Tomb 1 yielded 77 pots and a total of 46 bronzes, over half of which were comparatively heavy implements or items of weaponry such as knives, hook-tang weapons, and razors, with small ornaments and implements making up most of the remaining bronze finds. The looted Tombs 2 and 3 contained 11 and 14 objects respectively, and 10 of the 14 objects from Tomb 3 were knives or hook-tang weapons. Exotic goods were limited to a group of faience beads from Tomb 1 and single silver-lead rings from Tombs 1 and 2. Akhera Tomb 1 was rich in both ceramic and metal finds, with a total of 80 ceramic pots and 32 bronzes, including 3 knives, 6 hook-tang weapons, and other typical ornaments and implements. This represents a substantial array of goods relative to the small number of burials observed (two), and the quantity of metal weaponry is notable. As at Pendayia, however, the range of exotic finds was limited, encompassing a steatite scarab, some faience beads, and a silver ring. In contrast to the finds from other coastal settlements in the northwest, the material from Pendayia and Akhera suggests that while the sources of metal wealth were readily accessible to these communities, their external trade contacts were more restricted (Table 5.8).

The tombs of Morphou *Toumba tou Skourou* (Vermeule 1974; Vermeule and Wolsky 1990) attest not only to the development of wealth and status differentials among spatially proximate and perhaps closely related tomb groups, but also to the early emergence of an elite prestige complex of which exotic goods were an important component. Tomb 1 contained by far the richest array of goods of all those excavated at the site, yielding at least 516 pots, mostly of local manufacture but exceptional in their quality and diversity. The presence of possible horse bones and the skull of a small dog or fox on the dromos floor (Vermeule and Wolsky 1990:164) is also notable. Tomb 1 contained more than 90 bronze objects, far outnumbering the metal finds from the adjacent tombs. However, it should be noted that the types present were mainly small ornaments and implements such as pins, needles, spiral beads and hair rings, rather than the heavier weapons and other objects that dominated the assemblage at nearby Pendayia (Table 5.8). The array of imported valuables from Tomb 1 was more impressive, comprising several examples of imported Late Minoan IB pottery, over a dozen small silver objects, several beads of faience, glass, carnelian, and amethyst, and a possible knife handle made of ivory with two gold studs. Numerous other small objects of worked bone were also found, including gaming pieces and fragments of a gaming table incised with a Hathor head, a figure in a spotted robe and another in a pleated kilt, described by Barnett as 'Syro-Egyptian offbeat work of the late Hyksos period' (Vermeule and Wolsky 1977:85). Other important symbolic or 'sociotechnic' objects included two hematite balance weights and three cylinder seals, one of Syrian and two of probable Cypriot manufacture. A similar but somewhat smaller array of exotic goods including two ostrich eggs was recovered from Tomb 2. Interestingly, Tomb 2 may have been remembered as a locus of elite burials over an extended time period, as it was reused in LCIIB for a rich, possibly female burial (found in Chamber 4) provided with two Mycenaean flasks, a stirrup jar and a pyxis, numerous small ivory fittings and container fragments, a few gold beads, a gold capped lapis lazuli cylinder seal, and fragments of two glass bottles. In contrast to Tombs 1 and 2, the contemporaneous assemblages from Tombs 3–6 contained a much less diverse range of goods, mostly local ceramics which, in the case of Tomb 4 particularly, were described as 'relatively unsophisticated and inexpensive pottery' (Vermeule and Wolsky 1990:273).

Several of the LCI tombs found at Ayia Irini *Palaeokastro* (Pecorella 1977) yielded arrays of bronze artifacts and other local and imported valuables roughly comparable to finds from *Toumba tou Skourou* Tomb 1 (Table 5.8). The largest and most diverse collection of metal finds (29 objects) came from Tomb 21, with ornaments such as pins, spirals, beads, and other small finds predominating in the assemblage. The paraphernalia of metalworking and the metals trade also figures prominently in the burial goods from Ayia Irini, with a pair of tongs and an ingot fragment having been recovered from Tomb 21 and fragments of bronze scale or balance pans found in both Tombs 20 and 21 (Pecorella 1977:252). Sets of hematite balance weights of Syrian type occurred in Tombs 3 and 21, and weights made of local stone were recovered from Tombs 11 and 20. Other items likely to have had status connotations included a total of ten cylinder seals, one of probable Mitannian origin (Pecorella 1977:268; Figs 190, 210:67), and seven stone maceheads. Among the exotic goods present in various tombs were faience beads, semi-precious stones, an ostrich egg, small ornaments of gold, silver, and lead, and a few ivory objects. Two Late Helladic II cups were found in Tomb 3, and another was recovered from Tomb 20. Another very rich albeit partially disturbed LCI tomb, possibly associated with a different segment of the Ayia Irini community, was found a short distance inland from *Palaeokastro* (Quilici 1985; 1990). This tomb contained a rich array of local pottery and two Late Helladic II cups, diverse bronzes including numerous small ornaments (rings, spirals, pins etc.), a few hook-tang weapons and knives, a razor of Aegean type, and other metal types seldom seen in LCI tombs such as a bowl, a jug, and a mirror. Lumps of iron slag and pyrite were also found in the chamber. Non-ceramic exotic finds included two faience vases, various beads of faience and glass, small items of lead and silver, an imported hematite cylinder seal of the Mitannian Elaborate Style, and several balance weights made of hematite and other stone, variously associated with the Syrian and Microasiatic metrological systems. The users of this tomb, like those who buried their dead in the nearby *Palaeokastro* cemetery, would seem to have had significant long-distance exchange contacts, presumably linked to their participation in the copper trade.

Finds from LCI–II tombs at Kazaphani *Ayios Andronikos* on the north coast near Kyrenia attest to elements of both MC and LC prestige complexes. The presence of a bronze warrior belt in one of the two tombs excavated by the Cyprus Survey (Tombs 5 and 6, as yet unpublished) in 1971 (Karageorghis 1972a:1011, Fig. 13; Philip 1991:84), suggests that communities in the Kazaphani region shared in the same general complex of symbolism observed in late Middle Cypriot and early LCI tombs at Ayios Iakovos *Melia*, Dhali *Kafkallia*, Klavdhia *Tremithios*, and other sites where similar belts and/or shafthole axes have been found (Buchholz 1979; Philip 1991). The Kazaphani tombs contained hundreds of grave goods, predominantly locally made LCI–LCII pottery in the cases of Tombs 2A and 2B published by Nicolaou and Nicolaou (1989). These also contained a few imported ceramics including objects of Mycenaean pottery (mainly small containers), a LMIIIB jar, and several examples of Painted Red Slip Wheelmade ware, which Hennessy (1964:46) suggests had Anatolian affinities. Also noteworthy is the unusually large collection of 51 Red Lustrous Wheel-made spindle bottles, flasks, and bowls from Tomb 2B, paralleled to date only in the LCIIA–B Tomb 11 at Kalavasos *Ayios Dhimitrios* (South 1997; 2000; Goring 1989) and to a lesser extent in Enkomi French Tomb 2 (Schaeffer 1952). A Pastoral Style crater from Chamber 2A attests to the long time period over which this tomb was in use. Chamber 2B yielded an interesting Plain White Handmade ship model with parallels at Maroni *Tsaroukkas* (Tomb 1, no. 15, Tomb 7, no. 60, Johnson 1980: Pls IX and XVI), a rectangular sandstone weight, a locally made cylinder seal, and a locally made stamp seal. The number of copper-bronze objects recovered from Tombs 2A and 2B was quite small, comprising roughly a dozen small ornaments and fragmentary knives and daggers from each chamber. Items of exotic origin included faience beads, small silver ornaments, two alabaster jugs, fragments of an ostrich egg, and small finds of ivory, carnelian and glass. It is difficult to determine the extent to which the quantities of bronze and other valuables preserved in these chambers had been affected by looting.

Burial assemblages from the inland cemetery of Myrtou *Stephania* (Hennessy 1964), located at the southwestern end of the Kyrenia mountain range, reveal a marked difference in the consumption, and presumably also in the availability, of both metal and exotic goods relative to the sites just discussed above. The accumulations of ceramics were small (Table 5.8), and the most notable concentration of metal finds (a total of six items) came from Tomb 10 of MCIII date, from which a copper axe, a hook-tang weapon, a knife, a pin, a razor, and tweezers were recovered (Hennessy 1964: Pl. 48). Probable imports from the cemetery included only a few silver or lead rings, a small gold ring, a Black Lustrous Wheelmade juglet, and some Painted Red Slip wares (Hennessy 1964:46, 53). These items were broadly distributed among the various tomb groups, and it is difficult to single out any particular tomb as being richer than the others. It is possible, however, that there may have been greater differentiation in the size and resources of the groups making use of the Myrtou *Stephania* tombs than is readily apparent from the finds published by Hennessy. The looted Tomb 2, described only briefly in Hennessy's report (1964: 2), yielded approximately 200 pots mostly of LCI date, including a Bichrome tankard, a worn Minoan stirrup jar, large quantities of White Slip wares, and a very fragmentary Mycenaean IIIB:1 chariot crater with grooms, one bearing a 'Naue II' type sword (Karageorghis and Vermeule 1982:39, 201; see Catling 1956 on the sword type). Although these finds are not overly impressive in comparison to those from contemporaneous coastal sites, they are still suggestive of wealth and status differentials within the *Stephania* community, especially given the possibility that non-ceramic valuables may have been removed by looters.

Collections of grave goods from the LC tombs of Ayios Iakovos *Melia* in northeastern Cyprus reflect a pattern of austerity (Table 5.8) that was also observed at this site in the preceding MC period. The largest accumulation of pots and metal items occurred in Tomb 8, with 66 ceramic finds and 18 copper-bronze items in association with 62 burials of LCI–LCII date (Gjerstad *et al.* 1934:325–35). A few items made of exotic or imported materials, primarily Mycenaean containers and small finds of gold, faience, and ivory, were found in the LCII burial deposits of Tombs 8, 12, 13, and 14. Bronze objects were more plentiful at the cemetery adjacent to the fortress at Korovia *Nitovikla* along the south coast of the Karpass peninsula, with 19 metal objects having been recovered from Tomb 1 and 24 from Tomb 2. These assemblages included several bronze knives, scrapers, and various types of ornaments, along with several stone maceheads (Table 5.8). The quantities of metal present may have been affected by the relationship of the deceased to the political authority associated with the fortress complex. Interestingly, however, exotic goods were extremely scarce, seemingly limited to some faience beads in Tomb 2 and perhaps some scarabs and cylinder seals from other looted tombs (Gjerstad *et al.* 1934:407–15).

In contrast to the burial assemblages of the Karpass region, tomb finds from some of the inland sites in eastern Cyprus attest to significant external exchange contacts as well as considerable access to bronze goods during the LCI period. One of the many tombs opened by Ohnefalsch-Richter at Nicosia *Ayia Paraskevi* in 1884–85 (Tomb 1, 1884; Merrillees 1986; see also Ohnefalsch-Richter 1893) contained ceramics of probable LCI date along with stone and terracotta whorls, a whetstone, and several copper-based artifacts including two daggers, two awls, two pairs of tweezers, and a flat axe, types characteristic of earlier EC–MC tomb assemblages. Exotic finds included a paste cylinder seal or bead incised with a Levantine 'Egyptianizing' design (Merrillees 1986:137), along with a gold-mounted hematite cylinder seal, originally cut with an Old Babylonian presentation scene featuring two deities and a king with a mace, to which North Syrian or Cypriot style figures of a bull or griffin type monster and a goat-like creature were later added (Merrillees 1986:120–29).

The LCI tombs of Milia *Vikla Trachonas* (Westholm 1939a) were also notably rich in pottery, metal objects and exotic goods. The burials in Tomb 10 were equipped with 94 pots, including a Bichrome Wheelmade jug decorated with a frieze of birds, four Tell el-Yahudiyeh juglets, and five Mycenaean vessels (two kylikes, a pyxis, and two stirrup jars). Twenty copper-based objects, a lead ring, four faience beads and a faience statuette were also recovered from this tomb. The four burials in Tomb 11 were provided with 19 copper-based artifacts and a few small ornaments of faience and gold. Tomb 13, which had been robbed, nevertheless yielded over half a dozen metal items, some faience objects and five Mycenaean bowls (Table 5.8). It is possible that the residents of this community derived some of their wealth from connections with the nearby coastal town of Enkomi.

LCI tomb groups in southeastern Cyprus are scantily documented, but Gjerstad's (1926:82–3) brief description of LCI tombs at Arpera *Mosphilos* indicates that two burials in Tomb 204 and one in Tomb 205a were each provided with daggers and axes. Tombs at Arpera *Mosphilos* (Merrillees 1974) and other cemeteries such as Livadhia *Kokotes* (Åström 1974), Dromolaxia *Trypes* (Admiraal 1982), and Klavdhia *Tremithios* (Kenna 1971:22) have also yielded rare finds of imported Syrian and Egyptian pottery and/or Near Eastern cylinder seals dated to the MCIII–LCI transitional period, attesting to the development of long-distance trade contacts at this time. The looted Tomb 1 at Dromolaxia *Trypes* yielded a Bichrome Wheelmade tankard depicting a warrior brandishing a lance in his left hand and a sword in the right, exemplifying a form of Near Eastern iconography which is paralleled on contemporaneous cylinder seals (Courtois 1986a:84; Admiraal 1982:49, note 6; Karageorghis 1979).

Even less is known about LCI burial groups in south central and southwestern Cyprus. In Kalavasos, the intact Tomb 51 contained the remains of a single LCIA burial with two stone beads positioned on either side of the skull (possibly to fasten a garment at the shoulders), a copper or bronze sword found near the waist, an imported Canaanite jar and eleven other ceramic vessels placed around the upper body and feet. The deceased was apparently a late adolescent under 25 years of age, possibly male in light of the finds (Pearlman 1985). This tomb is of considerable interest inasmuch as it illustrates a complement of grave goods associated with an individual burial. However, it is unfortunately as yet impossible to assess the extent of local variations in wealth or overseas trade contacts in the Kalavasos area or elsewhere along the south coast during the LCI period.

## Enkomi *Ayios Iakovos*

Amongst all of the Late Cypriot mortuary complexes excavated until the present, it is only the evidence from the east coast town of Enkomi that yields a relatively comprehensive picture of variability in tomb assemblages dating to different phases of the Late Cypriot period, and of the changing patterns in mortuary consumption and symbolism that were underway between the beginning and the end of this era. For the purposes of this analysis, tombs with intact or substantially preserved collections of goods have been grouped into four chronological categories: MCIII–LCI, LCI–LCIIB, LCI–LCIIC, and LCIIC–LCIII (Tables 5.9a–5.9d). The overlap between these groups reflects the extended periods of time over which many of the tombs were used. The separation of the second and third groups is particularly important because of the greatly increased availability of bronze during the LCIIC period, which in turn affected the quantities of metal deposited in tombs with LCIIC components of use.

In contrast to the mortuary complexes of the Early and Middle Cypriot periods, the wealth differentials among tomb groups at the Late Cypriot site of Enkomi were not merely quantitative but qualitative and distinctly hierarchical or discrete. This pattern is evident among tomb groups dating from the earliest periods of the site's occupation and all of those which followed. Although the sample of intact and/or well-documented tombs dating to MCIII–LCI (Table 5.9a) is small, there is significant variability among the tomb assemblages recorded. The artifact assemblage of Enkomi French Tomb 32 (Courtois 1981:39–81) stands out among the others included in this group on the basis of both the local and non-local goods present, with 24 or more copper/bronze rings and spirals, a copper/bronze axe, 2 stone maceheads, 4 hematite balance weights, 3 faience beads, a faience scarab, and numerous Tell el-Yahudiyeh and Syrian Burnished juglets. The presence of imported Syrian and Egyptian pottery, maceheads, and axes may represent the persistence of certain elements of the Levantine complex of prestige symbolism observed elsewhere in the MC and MCIII/LCI periods. The balance weights suggest the importance of involvement in the metals trade, an emphasis which is also evident in the more modest burial assemblage of French Tomb 1851 (Lagarce and Lagarce 1985), which contained a pair of balance pans and a single balance weight, and in French Tomb 126 (Courtois 1981:83–111), which similarly yielded a single (perhaps token?) balance weight. French Tomb 126 is also noteworthy for the presence of a very early Mycenaean pot and a small quantity of gold jewelry, items which may have been associated with an early LCIIA component of use. Swedish Tomb 8 seems to have been the earliest of the excavated Enkomi tombs to yield a substantial assemblage of gold jewelry; it was also distinguished by the presence of copper-based weaponry and early imports of faience (not merely beads, but a vase as well) and semi-precious stones (Gjerstad *et al.* 1934:500–504). The most impressive valuables contained in these tombs do not occur even sporadically in the 'poor' tombs listed at the beginning of Table 5.9a.

In tombs with LCIIA–B components, there is continuing and increasing evidence for wealth and prestige differentials between groups. It should be noted at the outset that 'poor' to middling groups are likely to be underrepresented in Table 5.9b, in part because many tombs groups containing only local pottery were inadequately reported and curated by the British Museum expedition. Also, some long-used tombs with fairly unremarkable LCI and early LCII assemblages are presented in Table 5.9c, for example, Swedish Tomb 13, 19, and 11, Cypriot Tomb 10, and French Tombs 110 and 5. Nevertheless, striking contrasts between 'rich' and 'poor' tomb groups of comparable size are quite apparent. The intact French Tomb 3(1934) (Schaeffer 1936:135–6), Swedish Tomb 2 (Gjerstad *et al.* 1934:470–5), and Cypriot Tomb 19 (Dikaios 1969:404–14) were devoid of copper and most categories of imported valuables other than Mycenaean pottery. French Tomb 11 (1949) (Schaeffer 1952:135–56) contained minor valuables of copper, gold, and other exotic materials, along with another single hematite weight, but lacked the heavy and elaborately worked items of gold jewelry and/or gold and silver bowls variably observed in the contemporaneous French Tomb 2, Swedish Tomb 17, and British Tombs 92, 67, and 19. Several of these very rich tombs were also distinguished by the presence of large sets of balance weights, copper or bronze knives, ceremonial vessels or 'rhyta' and Mycenaean pictorial craters not found in the 'poor' tombs. The depiction of Mycenaean conical rhyta as filling ornaments in the chariot procession on the 'Parasol' crater fragment from British Tomb 67 (which also contained a Mycenaean bull's head rhyton) suggests that these and other unusual vessels such as Red Lustrous arm vessels, faience zoomorphic rhyta and cups modelled in the form of female heads, may have been utilized and/or displayed by members of the elite on ceremonial occasions. The iconography of the so-called 'Zeus crater' from Swedish Tomb 17, portraying a chariot scene, an elaborately robed individual holding a pair of scale pans, and a lower status individual apparently bearing a copper oxhide ingot would seem to indicate that the symbols of the copper trade were significant components of elite status paraphernalia as well, as is also suggested by the sets of balance weights found in the richest tombs (Keswani 1989b:57–62).

Although objects of gold and silver were more widely distributed and plentiful in tombs with LCIIC components (Table 5.9c), a remarkably hierarchical or exclusive patterning nevertheless continues to be apparent in the occurrence of specific types that may be classified as 'higher-order valuables'. With respect to the category of gold and silver objects, tombs with relatively low gold content tended to contain only the simpler, lighter weight types such as gold foil diadems and mouthpieces, small groups of beads, earrings, and hair spirals. Heavier and more ornate items—toggle pins, elaborate necklaces, and finger rings with intricate filigree work or bezels inscribed with hieroglyphs, Cypro-Minoan signs, and other emblems with high iconographic or informational content—were concentrated in the tombs with the greatest gold wealth overall. Gold and silver vessels, the rarest and most precious objects by far, were found only in Swedish Tombs 3 and 18 and British Tombs 66 and 93. The gross disparities in wealth between tomb groups are striking; Swedish Tomb 18 and the ashlar British Tomb 66, which contained impressive accumulations of over 300 g of gold jewelry were even further surpassed by British Tomb 93 (a rock-cut chamber tomb), with its total of more than 800 grams of gold, including extraordinary items of jewelry, a smashed gold cup, and other miscellaneous, often quite heavy pieces of gold (Keswani 1989b:62–3).

Similar patterning is evident in the distribution of other exotic goods such as faience, ivory, and Mycenaean pottery, as well as in the distribution of locally produced copper or bronze items, with the most elaborate arrays and the most ornate objects concentrated in the richest tombs overall. For example, faience and glass beads were frequently deposited in tombs of both greater and lesser wealth, but the ashlar-built British Tomb 66 contained an extraordinary array of faience bowls, a faience pail, a faience mortar and pestle, two glazed bottles and five glass bottles. Likewise, a number of tombs contained ivory pyxis lids, rods, buttons, and other ivory small finds, but the very rich Swedish Tomb 18 yielded a large collection of ivory box fragments and an unusual comb decorated with an engraved roebuck. Mycenaean containers and cups or bowls were present in almost all tombs of LCII date, but pictorial craters occurred mainly in tombs that were rich in gold jewelry and other valuables. Small items of bronze jewelry, along with mirror discs and simple bronze bowls, occurred in a number of tombs, but items of weaponry and vessels of more intricate workmanship were concentrated in the wealthiest tombs overall. The ashlar British Tomb 66 contained two spearheads, five daggers, several whole or fragmentary bronze bowls, and a bronze mirror. Swedish Tomb 18 (a rock-cut chamber tomb) contained 12 bronze bowls, a bronze jug, bronze greaves, a Naue II type sword (possibly imported, Catling 1956), as well as other weapons, jewelry and mirrors. French Tomb 1394 (another ashlar tomb, Courtois 1981:279–84), although looted, nevertheless yielded the most extraordinary array of bronze vessels found in the Enkomi tombs, including fragments of several bronze jugs and bowls, a strainer, and other objects (Keswani 1989b:63–6).

The evidence from the tombs summarized in Tables 5.9a–5.9c suggests the presence of elite groups who made use of a distinctive complement of prestige symbolism incorporating lavish forms of self-ornamentation, the display of exotic luxury goods and the deployment of ceremonial vessels and other objects that were rich in symbolism and iconography. Jewelry items such as stamped diadems, mouthpieces, and frontlets display a rich complex of natural and supernatural symbolism in motifs such as rosettes, palmettes, rams, bucrania, lions, and sphinxes. The iconography of the Mycenaean pictorial craters suggests that special forms of dress, especially spotted robes, and processions of individuals riding in horse-drawn chariots were also important elements of elite symbolism. The redundant accumulations of certain types of prestige goods, especially jewelry and Mycenaean pictorial craters, repeatedly deposited in some tombs over several generations, suggest that the status differences which are represented were by now at least partly hereditary. However, this 'hierarchical' social order was not necessarily pyramidal nor permanently fixed in structure. The lack of correlation among the dimensions of tomb location, architecture, and assemblage quality (i.e. 'rich' tombs were widely distributed throughout the settlement, and chamber tombs might surpass built tombs in wealth; see Keswani 1989a: Tables 6.9–10) may indicate a considerable degree of competition, diversity, and fluctuation in the ascendancy of elite groups at this site, consistent with the evidence for periodic destructions, rebuildings, reorganizations, and relocations of administrative and elite residential complexes throughout the occupation of the settlement.

While the massive accumulations of gold and other exotics seen in LCII tombs are not observed in the shaft graves and chamber tombs of LCIII, significant differences in wealth between groups and hierarchical distributions of valuables are still evident (Table 5.9d). The richer burial groups often contained impressive arrays of bronze and sometimes iron weaponry, along with elaborately wrought bronze vessels and tripods or offering stands of particularly complex workmanship (Catling 1964; 1984; Lagarce and Lagarce 1986:84–100). Their assemblages were also frequently distinguished by the presence of artistically impressive carved ivory mirror standards and boxes of various forms, in some instances decorated with scenes of warriors in triumphant combat with supernatural beings such as griffins, and in one case (the gaming box from British Tomb 58, Murray *et al.* 1900: Pl. 1) with a hunting scene that has obvious links to Near Eastern royal iconography. Some shaft graves were also provided with rich complements of gold jewelry. 'Middle' status burials were often accompanied by ceramic vessels, various bronzes, and other small objects of stone, ivory and other exotic materials (e.g. the assemblages from Cypriot Tomb 23 through French Tomb 13 [1934] in Table 5.9d ), while other, presumably lower status burials had only pottery or minor items of personal adornment (e.g. Swedish Tomb 13C, Cypriot Tombs 4A, 4, 24). Thus a significant level of social stratification seems to have persisted, but the types of symbolism employed and the scale of mortuary consumption were clearly changing (Keswani 1989a:66–8).

Looking at gross consumption patterns from LCI–LCIII, pottery was by far the most ubiquitous category of grave goods at Enkomi. Most assemblages were dominated by locally made wares such as Base Ring, White Slip, Monochrome, Plain, White Shaved, and Red Lustrous Wheelmade (Eriksson 1991), with small percentages of imported Syrian and Levanto-Egyptian pottery (e.g. Tell el-Yahudiyeh ware) in LCI, along with a few Minoan imports and more substantial quantities of Mycenaean or locally imitated Mycenaean wares in LCII. Mycenaean pottery, especially containers (e.g. piriform jars, stirrup jars, pyxides, and flasks) and pictorial craters, often appears to have been heavily used prior to deposition in the mortuary context, with abundant signs of wear such as chipping and abrasion along the rims, bases, and handles (personal observation), damages which are not readily attributable to post-depositional factors. French Tomb 2 is one important exception to this, with newly made pottery that Schaeffer referred to as a mortuary 'trousseau' (Schaeffer 1952:134). In contrast to the majority of Aegean imports, however, much of the locally made pottery appears to have been undamaged by previous use, suggesting that some pots at least were newly made 'gifts' rather than long-held personal possessions of the deceased. In some cases 'wasters' or less than perfect specimens may have been acceptable for this purpose; Schaeffer observed a number of misfired and misshapen pots in some of the poorer tombs that he excavated (Schaeffer 1936:75–80; 1952:156). Keeping the possibility of specific funerary production in mind, it is interesting to note that the disposal of pottery, as measured by the mean per capita value of 'pots per burial' calculated for each chamber, rose dramatically between LCI and LCII, and then plummeted sharply in LCIII (Table 5.10). The decline in 'ceramic expenditure' in LCIIC and LCIII may have been related to an overall decrease in mortuary consumption, with a lower level of group participation (fewer funeral guests, thus fewer 'givers' and fewer pots). The increasing availability of and preference for bronze vessels may also have been a factor, particularly among the elite groups that could best afford them.

References to faunal remains in any of the Enkomi tombs are remarkably lacking, perhaps because very little attention was devoted to analyzing any of the bones recovered, human or otherwise. Schaeffer (1936; 1952) refers to the presence of animal and bird bones in French Tomb 16 (1934) and animal bones in French Tomb 5(1949), and other tombs excavated by the French and other archaeological investigators might perhaps have contained fauna that went unreported. However, the dearth of references to any types of animal bones in the account of the Swedish expedition may be significant, as the participants did identify faunal remains at the EC–MC cemetery of Lapithos *Vrysi tou Barba*. Presumably if large bones from cattle or equid species and/or substantial remains of smaller animals had been encountered, these would have been noted in the excavation reports. As they were not, it may be tentatively suggested that major animal sacrifices were a less significant part of ritual practice at Late Bronze Age Enkomi than they were at Early–Middle Bronze Age *Vounous* and Lapithos.

The general scarcity of copper-based artifacts in tombs of LCI–LCIIB date (Tables 5.9a, 5.9b, 5.10) is a perplexing anomaly, considering that the quantities of imported valuables, presumably acquired through the copper trade, increased at this time, and the copper industry indeed may have been Enkomi's '*raison d'être*'. The impressive arrays of weaponry found in the MC tombs of Lapithos and other sites are simply nowhere in evidence prior to LCIIC. Most tombs contained no copper at all, or at most a few small ornaments; even the comparatively rich French Tomb 32 of MCIII/LCIA–LCIIA date (Courtois 1981:39–81) contained mainly rings and spirals, along with a single fragmentary axe. The occurrence of heavier objects such as knives and daggers seems to have been limited to the tombs which were the richest overall in terms of other criteria such as gold, including British Tombs 67 and 92, and Swedish Tombs 8 and 17 (Tables 5.9a, 5.9b). It is possible that the importance of copper as an export (for which other, more exotic valuables were obtained) led to elite or official restrictions on its distribution for general use. This situation apparently changed in LCIIC (Table 5.9c), when evidence from settlement contexts attests to an expansion of copper-working facilities and an increasing number of metal objects in various contexts of discard (Fig. 5.7; see also Courtois 1982, 1984a; Knapp 1988b; Knapp *et al.* 1988). By LCIII, even some of the shaft grave burials that were devoid of other valuables were equipped with bronze bowls (Table 5.9d). It is possible that copper or bronze had now become an element of local 'wealth finance' (D'Altroy and Earle 1985; Keswani 1993) through which various institutional personnel were remunerated for their services.

For much of the Late Cypriot period, objects fashioned from gold and other exotic materials such as silver, faience, glass, ivory, alabaster (calcite), and semi-precious stones (carnelian, agate) would seem to have served as the pre-eminent symbols of prestige and conspicuous consumption. Most of these materials were scarce to non-existent in tombs of MCIII–LCI date, abundant although not completely ubiquitous in tombs of LCII date, and much less plentiful once again in LCIII. Presumably the remarkable displays of valuables in LCII were made possible by increasing levels of overseas trade, fuelling what was by now a well-developed social tradition of conspicuous consumption. However, it does not necessarily follow that a decrease in trade and in the availability of such materials was primarily responsible for their declining frequency in LCIII mortuary contexts, inasmuch as the quantities of exotics recovered from settlement deposits reached an all-time high in LCIII (Fig. 5.8). The quantities of some types of materials may have fluctuated to a certain extent (see Keswani 1989b:66; Vercoutter 1959 on gold in particular) but given the prevalence of 'low energy' shaft grave burials at this time, along with the concurrent decrease in the deposition of locally produced pottery, it seems probable that the falling consumption of gold and other exotics was part of a more general trend towards decreasing mortuary expenditure. This was by no means synonymous with a decline in social stratification between groups, as the contrasts between burial complements in Table 5.9d illustrate. The general decline in mortuary expenditure at this time, when viewed in conjunction with persistent evidence for qualitative differences in burial complements, is probably more indicative of the decentering or diminishing importance of mortuary ritual in the construction of social status than of any diminution in social hierarchy.

## LCII–LCIII assemblages from the south coast

The sample of tomb assemblages from the south coast of Cyprus does not permit the detailed assessment of synchronic variation among tomb groups or of long-term changes in mortuary symbolism that is possible at Enkomi. However, accounts of some of the best-documented tombs, along with the overall palimpsest of finds from sites where substantial numbers of tombs have been opened, attest to the prevalence of an elite prestige complex similar to that of Enkomi (albeit with some distinctive regional features) within the major towns.

Considered collectively, the tombs excavated by the British Museum Expedition (Bailey 1972) and the Department of Antiquities (Karageorghis 1968; 1972b) at Hala Sultan Tekke contained a range of valuables comparable to those found in some of the richest tombs at Enkomi (Table 5.11; cf. Tables 5.9b, 5.9c). Although the great accumulations of gold, bronzes, and exotic materials observed in individual tombs at Enkomi have no parallels at Hala Sultan Tekke, this may be attributable in part to the extensive looting of the site. The presence of objects characterized as higher order valuables at Enkomi—gold and silver signet rings, Mycenaean pictorial craters, faience and calcite/alabaster vases, non-local hematite cylinder seals etc.—were nevertheless present in various tombs, attesting to the presence of very rich elite groups in this coastal town. Moreover, the intact LCIIIA shaft grave (Tomb 23) burial excavated by the Swedish Cyprus Expedition in 1979 was comparable to or surpassing any of the contemporaneous burials known from Enkomi (cf. Table 5.9d), lending additional support to the argument that shaft grave burial was not exclusively a 'low-status' form of mortuary treatment. The adult male buried in this grave was provided with several gold ornaments, a silver finger ring with a pictorial design on its bezel, 15 bronzes (among them four arrowheads, a dagger, an unusual trident, a bowl, a platter, and a jug), an ivory gaming box, faience gaming pieces, and numerous semi-precious stone beads (e.g. agate, carnelian, lapis lazuli, turquoise), several of which had gold mountings (Niklasson 1983).

At nearby Kition, Tombs 1 and 9 (Table 5.11) are clearly identifiable as elite tomb groups. The assemblage of Tomb 1 (Karageorghis 1960b) was small but contained some important prestige items such as a Mycenaean chariot crater, faience vases, and several objects of gold, silver and bronze. The expenditure of material wealth is much more pronounced in Tomb 9, which contained over 40 gold objects, 66 bronzes, an impressive array of alabaster, faience, glass, glazed bottles and ivory objects, and a ceramic assemblage comprised predominantly of Mycenaean and Cypro-Mycenaean vessels (Karageorghis 1974:57, 84–5). The overall composition of this tomb collection is very similar to that of the extremely rich Enkomi Swedish Tomb 18; in both tombs the highest order goods included a fragmentary silver bowl, gold rings with engraved bezels, other elaborate items of gold jewelry, ivory objects with pictorial decoration, diverse bronze types including weapons, bowls, mirrors, and, in each case, single bronze jugs of intricate, albeit differing workmanship (Keswani 1989a:610–11). The looting of Kition Tombs 4 and 5 makes the evaluation of their combined assemblage somewhat problematic, but the high percentages of Mycenaean pottery from these tombs (Karageorghis 1974:34–5), along with the recovery of an extraordinary faience rhyton nearby (Peltenburg 1974), suggest that their original contents were also quite rich.

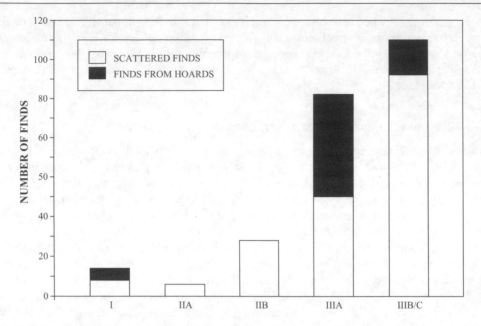

**Figure 5.7** Counts of copper/bronze finds from settlement contexts at Enkomi by level. Note that Level I = LCI, Level IIA = LCIIA–B, Level IIB = LCIIC, Level IIIA = LCIIIA, Levels IIIB–C = LCIIIB.

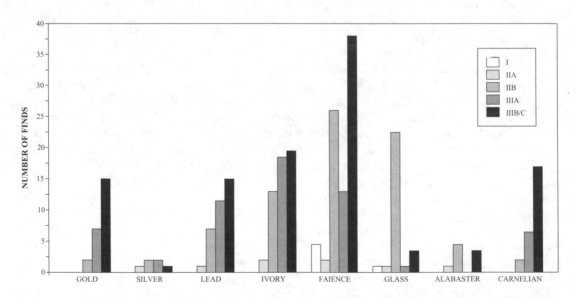

**Figure 5.8** Counts of exotic goods (gold, silver, lead, ivory, faience, glass, alabaster, carnelian) from settlement contexts at Enkomi by level. Note that Level I = LCI, Level IIA = LCIIA–B, Level IIB = LCIIC, Level IIIA = LCIIIA, Levels IIIB–C = LCIIIB.

Elsewhere in the Larnaca region, rich arrays of Mycenaean pictorial pottery and other valuables of LCII date have been recovered from tombs in the vicinity of Pyla (Megaw 1957; Dikaios 1969), Dromolaxia (Witzel 1979; Admiraal 1982), Aradhippou (Pottier 1907), and Klavdhia (Walters 1912). Tombs of lesser wealth are known from Larnaca *Laxia tou Riou* (Myres 1897) and Kalokhorio (Karageorghis 1980:766), and it is conceivable that these communities occupied a lower or less favorable position within the local hierarchy or network of settlement.

Further west, the tomb finds from Maroni *Tsaroukkas* and *Vournes* included a diverse collection of gold and silver jewelry types, glass, faience, ivory and alabaster objects, glyptics, bronze daggers and swords, bronze scale or balance pans, leaded bronze zoomorphic and circular balance weights, as well as some of the earliest Mycenaean containers found on the island. The Maroni tombs also yielded several Mycenaean pictorial craters depicting chariot scenes, a rare boxing scene, dolphins, octopus, birds, and other motifs, along with an exceptional array of ceremonial vessels: a Mycenaean hedgehog rhyton, a Mycenaean conical rhyton, a kernos, a Red Lustrous Wheelmade arm vessel, and numerous Base Ring bull rhyta. Still other interesting finds included terracotta figurines, boat models paralleled at Kazaphani in the north, a Mycenaean boat vase, numerous bull figurines, female figurines, and a Mycenaean *psi* figurine (Johnson 1980). The evidence of the recorded tomb finds thus suggests that some of the Maroni tombs were also used by elite groups employing an array of prestige goods and symbols similar to that observed at other important LC centers. The presence of 'non-elite' or lesser elite tombs may be inferred from Walters' references to 'unproductive' tombs containing only locally made pottery and other fragmentary material of no interest to the British Museum (Manning and Monks 1998:348).

In the Kalavasos area, differences in material expenditure and access to various categories of valuables between groups are quite evident despite the disturbance of many tombs. At the town site of *Ayios Dhimitrios*, high status burials seem to have been concentrated in (although not exclusively restricted to) the northeast area adjacent to the large ashlar Building X. By far the most outstanding burial assemblage encountered so far comes from the LCIIA:2 or LCIIA: 2/LCIIB Tomb 11, which yielded an exceptional 432 g of gold jewelry, an amount surpassed elsewhere only in Enkomi British Tomb 93. Among the finds were multiple pairs of gold earrings and hair spirals, three or four gold necklaces (one a probable Aegean import), three gold bracelets (one with ivory inlays unparalleled in LBA Cyprus), two gold signet rings bearing Cypro-Minoan inscriptions (Masson 1986), and another gold signet ring with an Egyptian inscription. A two-sided stamp seal depicting winged beings on one side and bucrania on the other was also found. The ceramic finds included one Mycenaean crater decorated with dolphins, another decorated with lilies, along with three Mycenaean piriform jars, a flask, a kylix, an unusually large array of Red Lustrous Wheelmade bottles and flasks, two Base Ring bull rhyta, and other Base Ring and White Slip ceramic vessels. Among the exotic luxury goods present were a fragmentary alabaster vessel, two Egyptian glass bottles, a metal-lined, cylindrical glass pyxis decorated with gold and bronze strips (also unparalleled in Cyprus to date), and a sizeable array of small ivory objects (South *et al*. n.d.; South 1997; 2000; Goring 1989). Substantial remains of sheep and goat, along with rock dove and fish bones, were also present (Goring 1989:103; Croft n.d.). The jewelry and other luxury items associated with the individuals in this tomb suggest that they were comparable in wealth and status to some of the highest ranking individuals at Enkomi and other centers along the south coast. The presence of artistically unique objects such as the openwork bracelet, the glass and metal pyxis, and the necklace of ivy relief beads, as well as the quality of the Mycenaean pottery from this tomb, suggest that the Kalavasos elite was by no means dependent on Enkomi or any other Cypriot site for its access to such valuables.

Other tombs in the northeast area were also very rich. Tomb 13, dated to LCIIB, is notable for the presence of three Mycenaean pictorial craters, two with octopus decoration and another with a unique scene depicting a chariot group, horses, and fish flanking a possible shrine with a woman standing inside. The 'shrine' and its antechamber were topped by altogether five pairs of horns of consecration (Steel 1994). Approximately half of the ceramic vessels from this tomb were of Mycenaean fabrics, an unusually high proportion. Other exotic finds included faience and alabaster vases and numerous ivories, including one disc with incised pictorial decoration (South 1997: 161–5; 2000:354–5). Tomb 12, sharing the same dromos as Tomb 13 contained the LCIIC burials of one child and three infants who were provided with a total of seven pots (some 'miniatures' perhaps specially intended for children and/or for mortuary use), some knucklebones, a gold earring, and a silver Hittite figurine (South 1997:163; 2000:355). These were unusually rich goods for a group of children. Higher order valuables were also found in the LCIIA–LCIIC Tomb 14, which yielded several items of gold jewelry including two diadems, a Mycenaean crater decorated with large birds, several other Mycenaean vessels, three Red Lustrous Wheelmade arm-shaped vessels as well as Red Lustrous Wheelmade spindle bottles and flasks, eight pedestalled bowls that were probably Levantine imports, a White Slip crater with unusual bird motifs (Steel 1997), three bronze daggers and other unidentified bronze objects, various ivories, faience gaming pieces and glass plaques presumably deriving from a gaming box (South 1997:165–7; 2000:353–4). Finally, the recently excavated Tomb 21 contained an unusually large Mycenaean crater with anti-thetic chariots and another uniquely decorated with women only, along with fragments of local ceramics, ivories, and five gold diadems or mouthpieces (South 2000:362). Inasmuch as these tombs appear to have been disturbed either in the Late Bronze Age or subsequently, their original assemblages must have been richer still. Again, the similarity of content between the Kalavasos tombs and some of the very rich, possibly 'sacerdotal' burial groups at Enkomi (cf. the ceramics from Kalavasos Tomb 14 and the Red Lustrous Wheelmade finds from Enkomi French Tomb 2 and Enkomi British Tomb 69, as well as the unusual White Slip and Cypro-Mycenaean craters from the latter tomb; Schaeffer 1952: Fig. 42; Murray *et al.* 1900: Fig. 68) attests to comparable access to the ceremonial and other prestige goods at Kalavasos, but the distinctive character of some of the finds, particularly the Mycenaean pictorial vessels portraying women, suggests an independent access to external exchange networks and possibly too an independent local ideology of gender relations and gendered status roles.

In contrast to the tombs of the northeast area, Tombs 1, 4, 5, and 6 located in the central and southeast areas seem to have had relatively unimpressive collections of metal and exotic goods, and lacked Mycenaean pictorial craters. However, the quality and diversity of the local Base Ring and White Slip ceramics present in Tombs 1, 4, and 5 (South *et al.* 1989: Figs 42–59) suggests that these groups may have had considerable access to locally produced goods, and all contained some faunal remains, including bones of more than one caprine in each tomb. Fragmentary remains of pig (Tomb 1), cattle or equid (Tomb 5), rock dove, and fish (Tomb 5) were also noted (Croft 1989:70–2). The looted Tomb 18 in the west area, with finds dating to LCIIB–LCIIC as well as the Geometric and later periods, contained a broader array of exotic goods, including fragmentary remains of gold, silver, ivory, and glass items along with a Mycenaean pictorial sherd, a Pastoral style sherd, and fragments of a Minoan stirrup jar in addition to local ceramics (South 1997:169–70; 2000:356, 359). Tomb 19, approximately contemporaneous and located at the opposite end of the same dromos, had been used for the burial of four adults and two children, accompanied by several Mycenaean containers, a cup and a shallow bowl, locally made ceramics, 13 gold beads and one gold earring, a silver or bronze pin, ivory pyxis fragments, and a stone bead or spindle whorl (South 2000:359). The finds from Tombs 18 and 19 suggest that other elite or 'upper middle-class' residential and burial groups existed in this part of the settlement (South 1997:170).

LCIIA–LCIIC burials have also been excavated at the site of Kalavasos *Mangia* (McClellan *et al.* 1988) a short distance away from *Ayios Dhimitrios*. It is interesting to note that the finds from *Mangia* Tombs 5 and 6, and all of the others excavated at *Mangia* to date, comprise a number of bronze objects and a diverse array of exotic goods, but none of the higher order valuables such as Mycenaean pictorial craters or elaborate gold jewelry found in some of the tombs at *Ayios Dhimitrios*. While persons of elite status may have resided at various localities in the Kalavasos area, it presently appears that the highest status groups were buried at *Ayios Dhimitrios*.

Looking further to the west, the quantity and diversity of valuables found at Kourion *Bamboula* (Benson 1972) appear quite sparse in comparison to other south coast sites (Tables 5.11, 5.12), in part because of the extensive looting of the tombs prior to excavation. This may represent a continuation of the tendency, already observed at EC–MC Episkopi *Phaneromeni*, towards relatively low level of mortuary expenditure in this region of Cyprus. But at least a few tombs may have contained a range of valuables comparable to that observed in elite tombs at other sites. The complex of finds from the adjoining Kourion Tombs 17/17A (identified with BM Tombs 53 and 102), which had been looted before the BM excavations, reportedly included several gold earrings, beads, and other ornaments, silver jewelry, glyptics, a bronze dagger, fragments of one or more glass vessels, and perhaps a Red Lustrous arm-shaped vessel (Benson 1972:20–21; Murray *et al.* 1900). Another important find from this tomb was the famous Mycenaean 'Window Crater' featuring a chariot scene, overlooked through the windows of a building by females of Minoan appearance (Karageorghis 1957). Elite groups at Kourion apparently made use of a complement of prestige symbolism similar to, but perhaps less rich than that observed at other coastal centers. The appearance of women in the Mycenaean iconography from this site suggests that, as at Kalavasos, females may have held significant status positions in this region.

At the important center of Alassa located to the north of Kourion, only a few tombs have been excavated to date. The tombs from the settlement at *Pano Mandilaris* do not seem to have been richly endowed with imported valuables; Tomb 3, the richest, yielded a total of 93 objects, including 5 of gold, 9 of bronze, and a hematite cylinder seal (Hadjisavvas 1989:40; 1991:174, Table 17.1). However, it is possible that much wealthier tombs were located at the nearby locality of *Paliotaverna*, where three ashlar buildings and official storage facilities have been found (Hadjisavvas 1996). The *Pano Mandilaris* tombs may represent only a small segment of the Alassa community.

Little is known about tomb assemblages of the earlier Late Cypriot period in the Kouklia area, but the LCIIC–LCIIIA tombs at *Evreti*, *Teratsoudhia*, and *Eliomylia* attest to significant expenditure in material wealth overall and considerable variation between groups. The complement of prestige goods recovered from Kouklia *Evreti* Tomb 8 (Catling 1968), dated to the LCIIIA period, might be compared with any of several of the richest contemporaneous tomb assemblages from Enkomi, such as British Tombs 58, 75, 16, and 24 (cf. Table 5.9d). However, in terms of the absolute value and diversity of its contents it far exceeds virtually all of them. Among the most outstanding finds of precious metal were two fragmentary silver bowls, a heavy (18.377 g) gold finger ring set with an engraved stone depicting two recumbent bulls, six massive gold finger rings with broad hoops decorated with an impressed flame pattern and gold cloison bezels set with colored enamel (cf. the cloissonné scepter recovered from Kourion *Kaloriziki* Tomb 40; McFadden 1954), a fragmentary gold frontlet with embossed rosettes, two heavy gold ribbed bracelets, four gold toe rings, ten leech-shaped earrings, and four bull-pendant earrings. Other finds included two iron knives set with bronze rivets, two small iron spatulae (one equipped with an elaborate ivory handle decorated with carved relief ornaments of linked spirals and tiny paste beads), one bronze mirror disc, fragments of four bronze hemispherical bowls, and two bronze daggers. A number of ivory objects were also recovered, including a large cylindrical pyxis with relief carving of a warrior in combat with a griffin, and a large ivory mirror handle carved with a warrior thrusting a sword at a rearing lion (Catling 1968:162–9; Iliffe and Mitford 1953). The iconography of the ivories is closely paralleled at Enkomi (cf. Murray *et al.* 1900: Pl. II nos. 872 and 883), and even the two recumbent bulls shown on the seal ring closely resemble those carved on one of the narrow ends of the ivory gaming box from Enkomi British Tomb 58 (Murray *et al.* 1900: Pl. I). The early appearance of iron objects is also paralleled in Enkomi British Tomb 58. The gold earrings are similar to those found in Enkomi British Tombs 24 and 61, but the gold bracelets and cloissonné rings are to date unique.

Other LCIIC/LCIIIA tombs at *Teratsoudhia* and *Eliomylia* were provided with considerably less gold, but were rich in bronzes, ivories, and other items of exotic origin. The intact *Teratsoudhia* Tomb 104 Chamber N, for example, contained only seven small gold ornaments, but it yielded an extraordinary array of bronzes including a laver, an amphoroid crater, two jugs, a mirror, fourteen arrowheads, a spearhead and several other unidentified objects of bronze, along with numerous ivories, fragments of Egyptian alabaster jars, faience gaming pieces, a glass bead, and ostrich egg fragments (Karageorghis 1990a:32–6). The *Eliomylia* tomb contained three small gold ornaments, eleven bronze bowls, a bronze mirror, a spearhead, a mattock, a chisel and an unidentified tool, as well as an iron blade, small ivory objects, a fragmentary Egyptian alabaster jug, a silver ornament and other items (Karageorghis 1990a:79–85). In comparison to the tombs just described, the intact *Teratsoudhia* Chamber 104B appears relatively poor, with only a single bronze needle, three small ivories, a silver ring and a bone plaque in addition to pottery (Karageorghis 1990a:25–6). The numbers, ages, and sexes of the burials associated with each of these chambers will of course have affected the composition of the assemblages to some extent, but the contrasts between these tombs and *Evreti* Tomb 8 suggest considerable social differentiation even among elite groups in the LCIIC/LCIIIA period.

Tombs dating to the LCIIC/LCIIIA period have also been excavated at Yeroskipou *Asproyia* a few kilometers west of Kouklia (Nicolaou 1983). Despite the disturbances of construction and looting at *Asproyia*, an impressive collection of valuables was recovered from Tomb 1, including at least 20 bronze bowls, 2 spearheads, a mirror, 3 gold rings and a gold band, along with some pottery, a steatite scarab, and local stone mortars and pestles. A few small gold and ivory items were also recovered from Tombs 2A and 2B. While these assemblages lack the 'higher order' valuables found in Kouklia *Evreti* Tomb 8, they nonetheless attest to the presence of elite groups with considerable access to bronze in this region.

## Inland sites

LCII–III burial assemblages from the inland regions of the island—the northern Troodos foothills and the Mesaoria Plain—are for the most part few in number and, with rare exceptions, poorly documented. Viewed collectively, however, they show interesting contrasts with the tomb collections from contemporaneous coastal sites, notably in the scarcity of higher order valuables made of gold and other exotic materials and in the near absence of Mycenaean pictorial pottery other than vases of the relatively late, locally made 'Pastoral Style'. Instead, the inland tombs would seem to contain the lower tier of valuables found at Enkomi and other coastal sites, suggesting that local communities participated in the same general prestige complex but had a more limited access to the status goods displayed by high ranking personages in the coastal centers.

Among the best-documented tombs in the inland regions are those excavated by the Department of Antiquities at Angastina (Karageorghis 1964b; Nicolaou 1972), Akhera (Karageorghis 1965b), and Politiko *Ayios Iraklidhios* (Karageorghis 1965a). Angastina *Vounos* Tombs 1 and 5 contained sizeable quantities of pottery relative to the number of burials observed, including some rather unusual shapes of local ceramic wares such as Base Ring flasks, bottles, and pyxis shapes made to imitate Mycenaean and Red Lustrous Wheelmade forms. Tombs 1, 3, and 5 contained sizeable accumulations of Mycenaean pottery, predominantly containers, cups, and bowls, along with seven Pastoral style craters of Cypriot manufacture from Tomb 1, but no Mycenaean IIIA or

IIIB pictorial craters were found. The largest collection of metal objects (20 items altogether) came from Tomb 2, the earliest of the group. Other locally produced valuables including cylinder seals (Kenna in Nicolaou 1972:106–108; Kenna in Karageorgis 1964b:23) and terracotta female figurines were found in various tombs, but the range of non-ceramic exotic goods was small, comprising a few small ivory objects, two gold beads from Tomb 3, and various beads or ornaments of faience and silver (Table 5.13).

Similar observations pertain to the LCIIC Tombs 2 and 3 excavated at Akhera *Chiflik Paradisi* (Table 5.13), both of which contained a high percentage of Mycenaean vessels, but no pictorial craters. Bronzes were scarce in these tombs compared to the metal rich Akhera Tomb 1 of LCIA date, and exotic goods were limited to a few small objects of ivory, faience, alabaster, silver, and gold, along with two cylinder seals, one having Mitannian affinities (Porada in Karageorghis 1965b:151–3). Politiko *Ayios Iraklidhios* Tomb 6 was fairly rich in bronzes, but its Mycenaean assemblage was devoid of pictorial craters, and exotic finds comprised only a few small ivory objects, a bead and a lid of faience or paste (Table 5.13). Nevertheless, stray finds such as the gold Hittite seal from Lambertis hill (Buchholz and Untiedt 1996:71, Fig. 14a) and a fragment from a large Mycenaean IIIB crater from Pera *Kryphtides* (Åström 1972b:317), not far from Politiko, raise the possibility that other richer tombs may have been located in the Politiko area.

Although numerous tombs have been excavated in the vicinity of Nicosia *Ayia Paraskevi*, most are incompletely recorded. Modest arrays of valuables were recovered from Tombs 6 and 10 (Kromholz 1982), as illustrated in Table 5.13, and while many tombs opened in this region may have been looted, gold objects and other higher order valuables are remarkably scarce (Kromholz 1982:282). Various tombs investigated by Cesnola reportedly contained terracotta female figurines, serpentine cylinders, scarabs, and animal-shaped vases (probably Base Ring bull rhyta), objects which can be generally characterized as lower tier valuables, but one tomb also yielded a Mycenaean crater depicting a chariot scene juxtaposed with a woman and floral motifs (Cesnola 1877:246-7; Karageorghis and Vermeule 1982:200). Another tomb excavated by Ohnefalsch-Richter similarly contained a Base Ring II bull rhyton and a terracotta female figurine, along with two higher order valuables: a Mycenaean IIIB pictorial crater decorated with bulls (Karageorghis and Vermeule 1982:53–4, 205) and a large terracotta bull's head of Anatolian origin, the largest Hittite object found in Cyprus to date (Karageorghis 1999). It is conceivable that *Ayia Paraskevi* was a particularly important inland secondary center or quasi-independent peer of the coastal polities, with more extensive access to valuables than most other inland sites.

Dhali *Kafkallia* Tomb G, from which a bronze shafthole axe and 'warrior' belt attributed to MCIII were recovered, yielded a more typical assemblage of 'inland' finds from the subsequent phases of its use, which extended throughout the LC period (Overbeck and Swiny 1972). The goods from this tomb consisted mainly of Cypriot ceramics, along with 17 Mycenaean containers and two cups, two Pastoral Style craters, a fragment of an alabaster chalice, a few small objects of ivory, glass, frit, shell, stone, and terracotta, and a stone cylinder seal of probable local origin (Table 5.13). The lack of gold objects from Tomb G could be attributable to looting, but the absence of Mycenaean IIIA or IIIB pictorial craters in an otherwise large Mycenaean assemblage is striking, and may be indicative once again of the highly concentrated, if not exclusive, distribution of these items among the coastal elites.

At Katydhata, LCII–III tomb goods were comprised predominantly of local ceramics, along with a few Mycenaean containers. Bronze artifacts such as ornaments and weapons were sometimes plentiful, with up to 15 in a single chamber (e.g. Tomb 11, Åström 1989:15). Locally made cylinder seals and an interesting array of Cypriot terracotta figurines, including animals as well as the more common anthropomorphic female types, were also present, along with small, unexceptional ornaments of gold, silver, and faience found in a few of the tombs (Table 5.13; see also

Åström 1989). The paucity of higher order valuables in what is, for this region at least, a fairly sizeable sample of tombs, further exemplifies the pattern observed at other inland sites.

A partially looted LCIIB tomb (Tomb 8) rescued at Dhenia *Mali* yielded five ceramic vessels of local Cypriot wares, a Mycenaean IIIB flask, sherds of a Mycenaean IIIB stirrup jar, a fragmentary Mycenaean bull figurine, an ivory disc, a bronze dagger, fragments of a silver bracelet, and a gold diadem (Hadjisavvas 1985). While these items may be generally categorized as lower order valuables relative to the hierarchy of goods observed at Enkomi, the gold diadem is of particular interest. As Hadjisavvas notes, the decoration of this item, which was embossed with a pattern of 16 figure-of-eight shields, is paralleled at Enkomi, specifically on diadems from Enkomi Swedish Tomb 3 (Gjerstad *et al.* 1934: Pl. 78:3/81) and Enkomi Cypriot Tomb 10 (Dikaios 1969: T. 10 No. 176, Pl. 208/26), and in the necklace made with figure-of-eight-shaped beads from Enkomi Swedish Tomb 18 (Gjerstad *et al.* 1934: Pl.88:2/20, Pl. 147:8) and Enkomi British Tomb 93 (Murray *et al.* 1900: Pl. 7:604). It is impossible to say whether this object was the possession of a coastal official residing in Dhenia or the possession of a member of the local elite with strong social and/or exchange ties to the coast. However, its presence at Dhenia attests to the pervasiveness and relative consistency of the prestige complex shared by communities throughout the island, regardless of prevailing disparities in wealth and access to exotic goods among those communities. Other stamped gold frontlets and crescent-shaped earrings alleged to come from looted tombs at Dhenia may be a further illustration of the links between Dhenia and coastal centers (Åström and Wright 1962:229 and note 5).

## Foreign valuables and the symbolism of prestige in the Late Cypriot period

At the conclusion of Chapter 4 it was argued that the influx of foreign goods and prestige symbolism in Cyprus during the latter part of the Middle Cypriot period promoted important transformations in local ideologies of prestige and the development of new social hierarchies based on the control of copper production and trade. Among the early imports were ornaments of precious metal and faience, Syrian and Babylonian cylinder seals, ceramics of Syro-Palestinian and Egyptian origin, and possibly horses (Courtois 1986a). Other novel goods such as shafthole axes and bronze warrior belts may have been locally manufactured, but seem to have been inspired by Near Eastern prototypes (Philip 1991; 1995). Some of the same types of valuables recurred in LCI contexts, along with an even wider array of exotic items including balance weights, ostrich eggs, worked bone and ivory, gold jewelry and various ornaments of glass and semi-precious stones. The transformative role of these objects lay not merely in the status which the possession of valuable rarities conveyed upon their owners, but also in their symbolic connotations and associations with foreign elites (Flannery 1968; Wheatley 1975; Brumfiel and Earle 1987).

Imported cylinder seals may have served as particularly important emblems of prestige for emergent Cypriot elites. The envaluation of glyptics in general is attested by the rapid establishment of seal cutting workshops in Cyprus, with locally made seals occurring in LCI contexts at Morphou *Toumba tou Skourou* and Ayia Irini *Palaeokastro*. Cypriot seals display a mixture of local, Near Eastern, and Aegean influences in style and in content (Kenna 1968, 1972; Porada 1948, 1986; Courtois and Webb 1987; Webb 2002). The interest which the subject matter of the imported examples held for a Cypriot audience is further reflected in the transposing of compositions from Near Eastern cylinder seals to the decoration of LCI and LCII ceramics. A Bichrome Wheelmade tankard from Enkomi British Tomb 80 depicts four personnages, one of whom is bearded, attired in a short kilt and a belt, holding a mace in his raised right hand and carrying a sword in the left hand. One of three long-robed figures faces him, almost touching the pommel of the sword. Another Bichrome Wheelmade tankard from Dromolaxia *Trypes* Tomb depicts a warrior brandishing a lance in his left hand and a sword in his right, as well as another human figure which is now poorly preserved (Karageorghis 1979; Courtois 1986a). These may have been inspired by glyptics depicting humans paying homage to gods attired as warriors. A somewhat later borrowing of themes from Near Eastern seal iconography, mediated by Mycenaean vase painters, may be seen in a pictorial crater from Aradhippou depicting elite personnages wearing spotted robes in attendance upon an enthroned goddess, similarly attired (Pottier 1907; Karageorghis 1959).

Imported glyptics thus conveyed information about the relationships between foreign elites and gods and about the status regalia of those elites and gods. A number of the appurtenances of personnages depicted in glyptic art—maces, belts, and various forms of weaponry, are also found in tombs of Middle–Late Cypriot I date. Some of these status goods may have been adopted through the inspiration of foreign iconography; others that were already part of the local, traditional prestige complement, particularly items of weaponry, may have been reinterpreted and re-envalued in light of the divine and regal associations suggested in that iconography. It seems somewhat paradoxical that the appearance of local artistic renditions of the Near Eastern complex of 'warrior symbolism' (cf. Philip 1991; 1995) was actually accompanied by a declining frequency of copper weaponry in mortuary contexts. However, the regal and divine associations of metal weaponry conveyed in pictorial representations may have helped to legitimize the more restricted distribution of a once widely consumed commodity.

Concurrently, other status paraphernalia related to the control of metal working and the metals trade began to assume a new importance. Foremost among these items were the balance weights, found singly and in sets in various LCI–II tombs at Morphou *Toumba tou Skourou*, Ayia Irini *Palaeokastro*, Enkomi, and Maroni. Most of these were hematite elliptical or sphenoid weights of Syrian type, but imitations in local stone were found in Tomb 11 at Ayia Irini (Pecorella 1977), and other types of stone and bronze/lead weights were listed among the BM finds in Maroni Tombs 1 and 3 (Johnson 1980). The weights belong to several different systems of measurement—Babylonian, Syrian, Anatolian (Courtois 1983; 1984; 1986a; Petruso 1984). Bronze scale pans were also recovered from Enkomi French Tomb 1851, Ayia Irini Tomb 20, and Maroni Tombs 1 and 10. The association of metal weighing paraphernalia with copper ingots and members of the elite is manifest in the LCII Mycenaean 'Zeus' crater from Enkomi Swedish Tomb 17. It is also interesting to note that at Enkomi, complete or at least multi-piece sets of balance weights were found in some of the richest tombs overall (e.g. French Tomb 32, British Tombs 92, 67, and 19), while single, possibly 'token' weights occurred in tombs of lesser wealth (e.g. French Tomb 1851, French Tomb 11 [1949]);see (Tables 5.9a–5.9b), suggesting not only differential access to the 'tools of the trade' but also a broadly shared envaluation of such paraphernalia. Most of the weights found in mortuary contexts were quite small, ranging in weight from fractions of a shekel to one *deben* or 92 g, and they may have been used to weigh fairly small amounts of precious metal rather than bulk quantities of copper (Courtois 1986a:85–6). However, in ceremonial contexts the display of such items may have symbolized access to both.

Another important component of elite prestige symbolism in Cyprus may have been the display of exotic libation vessels and other ceremonial equipment in ritual contexts (Keswani 1989a:547–53). In addition to the fairly common Base Ring bull rhyta, a diverse array of rare and/or exotic items such as Red Lustrous Wheelmade arm-shaped vessels, faience zoomorphic rhyta, faience cups in the form of a female goddess's face, and Mycenaean conical and zoomorphic rhyta occur in Cypriot tombs. Both contextually and iconographically, these objects seem to be associated with elite groups. Zaccagnini (1987:58) notes textual references for gifts of gold and silver rhyta passing between the Egyptian and Hittite courts, and the king of Hatti also requested rhyta from the king of *Alashiya* (Knapp 1980). Egyptian tomb paintings from the time of Tuthmosis IV portray Syrians carrying bird's head rhyta, and Ugaritic poetry also refers to 'gorgeous bowls shaped like small beasts like those of Amurru' (Peltenburg 1972:135–6). A cylinder seal from Cyprus depicts a ritual scene with Aegeans offering libations from zoomorphic vessels (Karageorghis 1965d:224), and Mycenaean conical rhyta are shown as filling ornaments on the Mycenaean 'Parasol' crater fragment from the very rich Enkomi British Tomb 67. A Mycenaean bull's head rhyton was found in the same tomb (Murray *et al.* 1900: Fig. 65).

Mycenaean pictorial craters also seem to have comprised elite showpieces in ceramic ensembles interpreted as 'drinking sets' (Steel 1998). While it is unlikely that the consumption of alcoholic or other special beverages in general was an exclusively elite activity, given the occurrence of large bowls and craters of locally made White Slip, Base Ring, and Plain White Wheelmade wares in tombs of lower or middle class status groups (e.g. the White Slip and Base Ring finds from Kalavasos *Ayios Dhimitrios* Tombs 1, 4, and 5, the Plain White Wheelmade craters from Enkomi Swedish Tombs 11, 13, and 19), the imagery depicted on the Mycenaean vessels found in high status tombs must undoubtedly have enhanced the prestige of those who displayed them on ceremonial and hospitable occasions. Pictorial craters were decorated with a variety of subjects ranging from the most common chariot scenes and naturalistic motifs (bulls, goats, birds, fish and flowers) to rare or unique portrayals of boxers, women, sphinxes, homage, the Kalavasos shrine, etc. (see Karageorghis and Vermeule 1982 for extensive discussion and illustrations of many of these themes). The frequency and degree of use wear on the rims, bases, and handles as well as on the bodies of these vessels suggests that they were utilized extensively by the living before being incorporated in burial assemblages, as does the occurrence of a chariot crater that had been mended with a lead clamp, found in a settlement context at Pyla *Kokkinokremos* (Keswani 1989a:562; Karageorghis and Demas 1984:82; cf. Manning and De Mita 1997:132).

One of the most overt means of elite image construction may have entailed the use of distinctive modes of personal ornamentation and dress. The proliferation of Mycenaean and other ceramic vessels that were probably used for scented oils and unguents (Leonard 1981; Steel 1998: 294–6), ivory, glass and faience toiletry containers, stone mortars, pestles and bronze spatulae (items that may have been used for preparing cosmetics), along with bronze mirrors and ivory mirror standards, suggests that perfume and make-up were important elements of prestige in Cyprus as elsewhere in the Mediterranean world. It is also evident from the lavish arrays of gold earrings, hair rings, finger rings, necklaces, toggle pins, diadems, mouthpieces, and frontlets recovered from LC tombs that Cypriot elites were richly bejewelled. A few of these items, such as signet rings with hieroglyphic inscriptions, were clearly imports; others show a mixture of Near Eastern, Egyptian, and Mycenaean influences that may represent either direct exchanges or local imitations of foreign styles and motifs (Maxwell-Hyslop 1971:107, 112–31; Lagarce and Lagarce 1986: 109–17; Poldrugo 2002; Catling 1964:45). Although some jewelry may have been produced exclusively for funerary use (see especially Lagarce and Lagarce 1986:117–22), many items show signs of considerable attrition and use, suggesting that they were indeed worn during daily life, if primarily on festive or ceremonial occasions (personal observation, see also Goring 1989:103).

Maxwell-Hyslop (1971) has observed that jewelry played a significant role in Near Eastern religious practice during the second millennium. The Hurrian gods and goddesses depicted in the stone reliefs at Yazilikaya wear earrings as an essential part of their equipment. Hurrian temples were seemingly rich in gold and silver ornaments which were used for the adornment of statues and votive offerings; the Qatna inventories, for example, list gifts of jewelry dedicated to Ningal, the Sumerian moon goddess (Maxwell-Hyslop 1971:110–11). Jewelry was also characteristic paraphernalia for other contemporaneous female deities such as Hathor, Ishtar, and Astarte as well (Lagarce and Lagarce 1986:122). That jewelry was also considered important equipment for female divinities in Cyprus is attested by the Cypriot terracotta 'fertility goddess' figurines whose ears were frequently provided with multiple sets of earrings (J. Karageorghis 1977), and by the discovery of an impressive array of gold objects from the LCII sanctuary at Ayios Iakovos in a deposit which also contained six incomplete Red Lustrous Wheel-made 'bras-encensoirs' or libation/anointment vessels (Gjerstad *et al*. 1934:357–8). The display of elaborate jewelry by members of the elite may therefore have signalled not only the wealth and social importance of the wearers, but also their ritual links to important divinities.

Although items of apparel have not been preserved in Cypriot tombs, it seems likely that elite clothing was also influenced by observations of foreigners and artistic depictions on cylinder seals and other artifacts. A carved bone plaque from *Toumba tou Skourou* Tomb 1 portrays one individual in a short pleated kilt and another in a longer stippled robe (Vermeule and Wolsky 1977). Long spotted robes were also worn by the principal figures on Mycenaean pictorial craters. As Karageorghis (1959) has observed, long robes were the traditional garb of deities, kings, priests, and other elevated personnages in Near Eastern art, distinguishing persons of rank from commoners who appear nude or in shorter garments. These long robes were often decorated with elaborate flounces, borders, and other patterns which may be paralleled in the spotted fabrics worn by the figures on pictorial craters. Long robes seem to have been adopted in Crete as a form of priestly dress, but they were not otherwise common in the Aegean world, based on the evidence of artistic renderings. It is possible, therefore, that the robed figures in Mycenaean chariot scenes may actually represent Cypriot rather than Mycenaean nobles (Karageorghis 1959:193–5).

Whether Cypriot elites made routine use of chariots, a mode of conveyance which seems to have been the prerogative of Near Eastern and Mycenaean elites (Sasson 1966:175–6; Crouwel 1981:150), is unknown. However, it is noteworthy that in Amarna letter #34, the king of Alashiya asks the pharaoh for a chariot with gold fittings and two horses, and horse bones were found in Hala Sultan Tekke Tomb 2 (Karageorghis 1972b:72) and possibly in Morphou *Toumba tou Skourou* Tomb 1 (Vermeule and Wolsky 1990:164). The frequency of chariot scenes on the Mycenaean pictorial craters found in Cyprus, as well as their occasional occurrence on seal impressions from pithoi (Catling and Karageorghis 1960:122; Webb and Frankel 1994:13–4; Smith 1994; Hadjisavvas 1996:34–5) and carved ivories (such as the gaming box from Enkomi British Tomb 58) would certainly suggest that Cypriot elites wished to be associated with this form of transportation.

Over the course of the Late Cypriot period, styles and types of prestige goods were subject to occasional changes, yet there is also considerable continuity in theme. Bronze weapons re-emerged as important elements of elite paraphernalia in LCIIC–LCIII, and scenes involving warriors and warrior kings were depicted on new types of prestige goods. The elaborate carved ivory boxes and mirror standards from Enkomi and Kouklia, which may have been the products of local workshops (Poursat 1977; Barnett 1982; see also Maier and Wartburg 1985:148; Dikaios 1969:99–100) often portray themes in which a male protagonist is shown in domination over wild animals or supernatural beings. The faience rhyton recovered from the vicinity of the partially looted Kition Tombs 4 and 5 depicts an Asiatic king or hero in physical combat with a wild beast (Peltenburg 1974). These themes are also reiterated in the seal impressions on pithoi found at Alassa (Smith 1994; Hadjisavvas 1996; Herscher 1998:328). As in the past, various aspects of contemporary Near Eastern, Egyptian, and Aegean royal imagery seem to have been assimilated and reworked into the prestige complex of Cypriot elites. What is perhaps most distinctive about the later manifestations of this artistic blending is the heightened emphasis on the control exerted by heroes or kings over the cosmic or natural order. This in turn suggests a closer identification with, or perhaps a more sophisticated manipulation of, the Near Eastern ideology of kingship and political legitimacy in the wake of several centuries of prolonged sociopolitical interaction (Keswani 1989b:70).

## Conclusion: mortuary practices and social structure in the Late Bronze Age

At the beginning of this chapter several key problems in the transformation of LC mortuary practice were defined:

1. *What possible explanations may be offered for changes in the spatial disposition, ritual treatment, and grouping of the dead during the LC period?*

In comparing LC mortuary practices with those of the preceding EC–MC era, both changes and continuities are apparent. In urban areas, breaking with earlier traditions, the tombs of the dead were often located in close proximity to the habitations of the living. The use of unreserved burial space may to some extent reflect the heterogeneous social origins of the settlers who migrated to urban centers in regions that were previously sparsely inhabited; lacking a strong sense of communal identity at the outset, they 'privatized' their ancestors in domestic living areas rather than committing them to publicly defined, shared, and segregated spaces for the dead. However, as the EC–MC agricultural economy was transformed by the rise of specialized copper-producing enterprises, craft activities, and long-distance trade, the use of intramural burial space may also have become a more general phenomenon, serving to emphasize the ancestral rights of households and kin groups to specific productive, administrative, and residential complexes. The striking contrast between the relatively simple, mostly single-chambered tombs found within the LC settlement at Kourion *Bamboula* and the elaborate multi-chambered tomb complexes opening off the long trench-type dromoi at the nearby EC–MC cemetery at Episkopi *Phaneromeni* (Weinberg 1956; Carpenter 1981) may serve to illustrate the changing scale of the kin groups whose identities were commemorated in mortuary ritual.

Meanwhile, in the rural hinterland, older extramural cemeteries remained in use and new ones continued to be established. The persistence of this burial tradition, along with the intensification of EC–MC practices of secondary treatment and collective burial manifested by the so-called 'mass burials', suggests that the social changes associated with the expansion of copper production and trade were experienced differently in urban and non-urban areas. Outside the major towns, the expression of communal identity in funerary ritual may have remained as important as in the past, while participation in copper exchange networks led to a heightening of competitive social displays in the context of ritual celebrations. It may also be speculated that if urban immigrants maintained close ties with their ancestral villages in the countryside, the scale of collective burial events might have increased as the bodies of urban settlers were periodically returned to their native communities.

Within the emerging towns, mortuary rituals involving secondary treatment and collective burial also continued, but collective burials may have involved fewer individuals and it appears that their frequency diminished over the years. Over time, too, sequences of extended, multi-stage ritual treatment of the dead may have given way to a higher frequency of terminal primary interments that were 'incomplete' by traditional standards; this may account for the appearance of shaft graves in close proximity to some chamber tombs in LCIIC and LCIII. In earlier periods such low-energy burial features would have been emptied periodically for the reburial of the dead, rendering them more or less invisible archaeologically. Later on, as mortuary expenditure declined overall, the ritual sequence was abbreviated, thus preserving the interments in the archaeological record.

Although until recently the human skeletal remains from Late Cypriot chamber tombs were seldom completely analyzed and reported, the existing information from both older and current excavations reveals important changes and variations in tomb group composition, chronology, and size. The proportion of infants and children buried in LC chamber tombs relative to the proportion of adults, while less than might be expected in terms of actual deaths in a pre-industrial population, nevertheless represents a higher frequency of infant/child burials than was characteristic of the EC–MC periods. This might be viewed as an indication of an increasingly ascriptive set of criteria for the bestowal of mortuary honors, as opposed to a system in which those honors were largely achieved through social deeds as an adult. With regard to gender representation, it appears that women may have received chamber tomb burial less often than men in some communities or segments of communities, but among high status groups they were frequently present and richly equipped, suggesting that in both life and death they might occupy positions of prestige that were heritable by their descendants or other kinsmen. The social importance of high status females may be further attested by the depictions of women on Mycenaean pictorial craters from Kalavasos and Kourion.

Finally, while the protracted reuse of chamber tombs over generations and sometimes centuries bespeaks the ongoing importance and affirmation of lineal identity in the context of urbanization, concurrent differences in patterns of tomb use attest to conflicting tendencies or ongoing social contradictions. The variation in tomb group size is considerable, and the number of burials relative to the duration of tomb use is generally less than might be expected from a founding pair and their descendants over several generations. Tombs used for relatively short periods occur along with those used over very extended periods of time. These anomalies of tomb group size and chronology may be indicative of household and kin group fissioning associated with economic competition, compounded with other changes in social structure that led to the declining importance of mortuary displays with the passage of time.

2. *How are the long-term changes and synchronic variations in LC tomb architecture to be interpreted?*

As in the EC–MC periods, both regional peculiarities and cross-cutting similarities in tomb architecture are evident in the LC period. Multi-chambered tomb complexes opening from a central, shaft-like dromos were found at Morphou *Toumba tou Skourou* in the northwest and also at the distant site of Maroni *Kapsaloudhia* along the south coast. At Ayios Iakovos *Melia*, the cutting of elaborate, niched chambers with long tapering dromoi continued at least through LCI, extending the distinctively northeastern tradition established at Ayios Iakovos and Korovia *Paleoskoutella* during the Middle Cypriot period. Smaller variants of these tombs were also cut during LCI at Milia *Vikla Trachonas*, although much simpler ovoid tombs with relatively short dromoi were found in the enclosed cemetery by the fortress at Korovia *Nitovikla*. In the distant southwest, low vaulted chambers with deep, narrow central cists were common in localities around Kouklia and Yeroskipou *Asproyia*, and other examples of this tomb type have been found further east at Kalavasos *Ayios Dhimitrios*, although they are curiously lacking to date at Kourion *Bamboula*, located in between. The use of pit tombs, frequently observed along the northern foothills of the Troodos in the EC–MC period, persisted during LCI–LCII at Akhera, but also occurred farther afield at Hala Sultan Tekke. Other forms of chamber tombs—bilobate, buttressed types, circular or ovoid shapes, and variably rounded, asymmetrical plans—seem to have had a broad distribution throughout the island. Thus it would seem that what may initially have constituted rather localized practices of tomb construction tended to spread over wide geographical areas through social contacts, affiliations, alliances, and the migration of groups with heterogeneous origins to emergent centers of urban development.

Along with the diversity of tomb forms observed within and between geographical regions, there was also considerable variation in tomb architecture within particular settlements, most notably attested in the large sample from Enkomi. Differences in intended mortuary programs, regional affiliations, building constraints, and prestige considerations may explain the diversification of tomb types at this site. In addition to several of the tomb forms mentioned above, two types of built tomb, the rounded tholos tombs and the rectangular ashlar tombs, as well as rectangular chamber tombs that may have imitated the ashlar tombs, have been found at various locations throughout this town. The architecture of the built tombs suggests an emulation of or derivation from earlier and contemporary elite tombs in Syria and Palestine, particularly in the case of the ashlar tombs. Given the difficulties of constructing very large chamber tombs in already built-up urban areas, the adoption of prestigious forms of Near Eastern burial facilities may have afforded a new mode of status display.

Although the richest chamber tombs were often among the largest (and thus most effort-consuming to construct) tombs found at any particular site, the size of these tombs (e.g. Enkomi Swedish Tomb 18 at 8.5 sq m, Kition Tomb 9 at 12 sq m, Kalavasos *Ayios Dhimitrios* Tomb 11 at 10 sq m, and Kourion *Bamboula* Tombs 17/17A at 7.5 sq m and 8.2 sq m respectively) are not particularly large compared to some of the extraordinary Middle Cypriot chamber tombs with floor areas of over 20 sq m at Lapithos *Vrysi tou Barba* and Ayios Iakovos *Melia*. Once again, local building conditions may have precluded the digging of very large chamber tombs in urban areas, but it is noteworthy that, except for the large LCI tombs constructed at Ayios Iakovos, the practice of 'mega-tomb' construction also seems to have declined in rural areas as well during the Late Cypriot period. The mean size of LC chamber tombs overall was 6 sq m compared to 8.3 sq m for the MC period, marking a substantial decrease in energy expended in tomb construction. The use of pit or shaft graves at Enkomi, Kouklia *Kaminia*, Hala Sultan Tekke, Kourion, and Kition represents an even more pronounced decrease in building effort. Given the presumably 'heterarchical' as opposed to highly centralized or pyramidal distribution of political and economic power within towns such as Enkomi and Hala Sultan Tekke (Keswani 1996), it is difficult to equate the overall decline in chamber tomb size or the appearance of shaft grave burials with any official policy of sumptuary suppression. Rather, it seems likely that these developments are further indications of a decreasing scale of social participation and resource outlays in mortuary festivities over the course of the Late Cypriot period.

3. *What do LC burial assemblages reveal about contemporary prestige symbolism and prevailing forms of social hierarchy?*

The burial assemblages of LCI–LCII date testify to the development of a stratified social order characterized not only by quantitative disparities in the distribution of gold and other wealth objects between groups, but also by a hierarchy of symbolic goods, with the highest order luxury items and objects of high iconographic content occurring in the richest tombs overall. Members of the elite distinguished themselves through the use of lavish ornamentation and dress (e.g. the spotted robes of the chariot craters), elaborate ceremonial involving chariot processions and the display of ritual objects (asserting an instrumental or mediating influence *vis-à-vis* the cosmic order), paraphernalia associated with the control of the metals trade, and identification with the elites of neighboring 'peer' polities using similar prestige goods and cosmic symbols. The notable concentrations of higher order prestige goods such as richly worked and heavy gold jewelry, Mycenaean pictorial craters, and, from LCII onwards, bronze vessels and weaponry, accumulating in some tombs over a period of several generations, suggest that status differences were closely linked with descent group, or at least tomb group, affiliation and hereditary social rank. It is also important to note that many of the items of gold jewelry and Mycenaean pottery found in LC tombs show clear-cut evidence of wear, indicating that they were used not only in the context of mortuary ritual but on other ceremonial occasions as well. Status differences must therefore have been naturalized by overt recognition and constant reference in daily life as well as in death.

The distribution of elite burial goods may also be indicative of the types of social relationships prevailing between as well as within contemporaneous settlements. While the mortuary samples from most Late Cypriot south coast centers are generally smaller and less informative than the sample from Enkomi, it would nevertheless appear that these towns were dominated by elite groups making use of similar complements of prestige symbolism. Inasmuch as the palimpsest of tomb finds and/or the most outstanding individual tomb assemblages from towns such as Kition, Hala Sultan Tekke, Kalavasos, Kourion, and Kouklia tended to contain the highest order symbolic goods observed in the Enkomi sample, it is difficult to argue that the elite in any one of these settlements exercised paramount authority over the others. Indeed, the unique characteristics of the jewelry and Mycenaean pictorial pottery from some of these sites suggest that local elites maintained independent exchange contacts with foreign allies and traders rather than obtaining their valuables through any network of centralized or 'down-the-line' distribution (as is also argued by Manning and De Mita 1997). This is apparent in tombs of the LCIIA–B period, prior to the rise of the ashlar complexes in LCIIC, and supports the argument that these sites were not subordinate to Enkomi even at this relatively early date (cf. the positions of Muhly 1989; Knapp 1994a; 1994b; Peltenburg 1996; Webb 1999 discussed at the beginning of this chapter). Similarly, considering the finds associated with LCI burials at the north coast sites of Morphou *Toumba tou Skourou* and Ayia Irini *Palaeokastro*, it is likely that these communities were also peers rather than subordinates of Enkomi at the onset of the Late Bronze Age. Later developments on the north coast are unfortunately very scantily attested.

A different situation appears to have prevailed in the relationship between the coastal settlements and inland communities. After LCI, the range of valuable and/or imported goods recovered from tombs in the mining areas (e.g. Akhera, Politiko, Katydhata) and other rural regions that formed the hinterland or zones of transit to the major coastal centers (e.g. Ayios Iakovos, Angastina, Nicosia *Ayia Paraskevi*) seems to have been limited mostly to the lower tier of valuables found in elite tombs in the principal towns. This is a dramatic change relative to the distribution of status goods such as shafthole axes, Near Eastern cylinder seals, and ornaments of precious metal observed at the end of the Middle Cypriot period. It is probable that many of the inland sites that supplied the coast with copper or possibly agricultural produce were no longer 'peers' in exchange relationships, but had become instead subordinate economically and politically to the major centers (Keswani 1989a:603–13, 623–4). However, finds from the Politiko and *Ayia Paraskevi* areas suggest that some individuals occasionally obtained higher order valuables, whether through sociopolitical or economic transactions with coastal elites or more direct exchanges with foreigners.

4. *What are the implications of declining mortuary consumption over the course of the Late Cypriot period?*

It is probable that mortuary ritual continued to be *an* important arena for the affirmation and reproduction of social hierarchy throughout the Late Bronze Age. The sheer quantities of gold and other valuable goods disposed of in some tombs is suggestive of the need for constant reaffirmations of social status. Nevertheless, the 'privatization' of tombs in domestic contexts, the decreasing elaboration of ritual treatments especially evident in the shaft grave burials, the decreasing energy investment in chamber tomb architecture and shaft graves, the reduced consumption of pottery and copper relative to earlier periods, and the declining consumption of exotic goods in mortuary contexts during LCIII (when they are most plentiful in settlement deposits, at least at Enkomi; see Figs. 5.7, 5.8) cumulatively suggest that over time mortuary ritual may have ceased to be *the* central arena for the creation and negotiation of status within the community. As social competition stimulated the demand for foreign prestige goods, the status differentials expressed by those goods were increasingly underwritten by material differences in access to or control of copper production, trade, and other forms of economic activity. The development of political and religious institutions and other elite productive estates offered new contexts for the aggrandizement of social status and the accumulation of wealth, beyond the traditional matrix of competitive kin groups and their ceremonial displays. Thus the elaborate ritual practices that helped to promote crucial economic and political transformations at the onset of the Late Cypriot period were ultimately transformed by the sociopolitical order which their practitioners had created.

# 6 Mortuary Ritual, Social Structure, and Macro-Processual Change in the Cypriot Bronze Age

## Introduction

Over the course of the Bronze Age, Cypriot mortuary rituals underwent a series of transformations and reversals. With the appearance of the Philia Culture complex or 'facies' and the onset of the Early Cypriot period, Chalcolithic mortuary traditions, characterized for the most part by pit grave burials in unreserved or intra-settlement contexts, gave way to the widespread use of extramural cemeteries of rock-cut chamber tombs. The burials in these tombs were sometimes subjected to complex rituals involving secondary treatment, collective burial, and considerable outlays of material wealth. Mortuary expenditure, as quantified by calculations of chamber tomb floor areas and the mean number of ceramic and copper items per burial, seems to have increased over time, as funerals became ever more important occasions for prestige competition and display. At the transition between the Middle and Late Cypriot periods, elaborate mortuary celebrations involving collective secondary treatment appear to have reached their apogee with the so-called 'mass burials' at rural sites such as Pendayia *Mandres*, Myrtou *Stephania*, and Ayios Iakovos *Melia*. Yet at the same time, significant departures from EC–MC mortuary practices were underway. In the numerous coastal towns that emerged over the course of the Late Cypriot period, tombs were frequently located within or immediately adjacent to habitation areas rather than in spatially distinct cemeteries, and mean chamber floor area decreased relative to the preceding era. Large quantities of valuables including gold jewelry and luxury goods made of ivory, faience, glass, and other exotic materials were deposited in LC burial contexts, but the consumption of these items declined in LCIII, while the disposal rates for copper and ceramic goods were low in comparison to the EC–MC throughout the Late Cypriot period. Ritual practices involving secondary treatment and collective burial continued, but the growing use of shaft graves in the later part of the LC period suggests a decline in ritual elaboration and a shift away from traditional mortuary programs.

The patterns of mortuary transformation outlined above are in a number of ways consistent with the mortuary developments that other ethnographic and archaeological studies of societies experiencing economic intensification, urbanization, and increasing sociopolitical complexity would lead us to expect (see Chapter 2). The formalization of complex funerary practices observed in the EC–MC periods calls to mind Woodburn's (1982) discussion of differing mortuary complexes in 'immediate' and 'delayed' return societies, albeit within the context of a transition from hoe to plow agriculture in the Cypriot case, as opposed to the contrasting practices of mobile versus sedentary of intensive food collectors that Woodburn explores. The widespread appearance of extramural cemeteries in conjunction with the adoption of a subsistence strategy such as plow agriculture, requiring more land and higher labor inputs, is also unsurprising in light of archaeological analyses of linkages between economic intensification, territorial concerns, and the hereditary or lineal transmission of property rights (e.g. Saxe 1970; Goldstein 1976; Chapman 1995). And the elaboration of 'dual obsequies' involving secondary treatment and reburial of the dead, sometimes on a collective scale, is highly reminiscent of Kan's (1989) discussion of Tlingit mortuary practices and their relationship to increasing sedentism, population growth, and the development of lineal groups and social rank. Even the reversal of particular mortuary practices in the context of LC urbanism has interesting historical parallels; the return to intra-settlement burial in association with increasing intra-group competition in tenth century BC Greece (Morris 1987:181–2) is one example, and the diminution of mortuary expenditure following the establishment of a stratified social order in the Egyptian Old Kingdom is another (cf. Childe 1945; Trigger 1990; Cannon 1989). But although these cases contribute to our understanding of developments in Cyprus in a broad sense, the explication of changing mortuary practices—their local significance, causes, and consequences—requires a more specific contextual analysis.

In this final chapter, I turn to a consideration of the broader regional processes that impinged upon local systems of meaning and ritual, and of the ways in which those processes, not always fully perceived by individuals or communities, were experienced and influenced in the course of mortuary celebrations. I briefly review the principal developments in areas of demography, settlement patterns, economic production, and household and community organization that may be inferred from non-mortuary archaeological data, along with their implications for social structure and ideology. I then examine the changing role of mortuary ritual in the social life of Cypriot communities, and the issue of how local ideologies and ritual practices were affected by, and may reflexively have acted upon, supralocal or macro-processes of economic and sociopolitical change. I argue that as mortuary ritual became the central arena for prestige construction in the Early– Middle Bronze Age, the social envaluation of metal and, towards the end of the period, the display of exotic prestige goods, led to major transformations in productive organization and social structure in Cypriot communities. In turn, within the more stratified social order and associated elite institutions that emerged in the Late Bronze Age, the centrality of mortuary ritual gave way to other arenas of achievement, power, and display, accounting for the eventual decline in mortuary elaboration and expenditure. These hypotheses, along with a number of related observations, form the basis for a provisional model of the dynamic articulation of mortuary ritual with social change during the Cypriot Bronze Age.

## The social context of EC–MC mortuary ritual

The social landscape of Cyprus was radically altered during the second half of the third millennium BC. New colonists seem to have arrived from the mainland, bringing new technologies of subsistence and metallurgical production. Population increased, settlements expanded into previously unoccupied areas, household and community organization took on a dramatically different appearance relative to the Chalcolithic, and new dynamics of social interaction, exchange, and competition most likely developed both within and between communities. The transformations in mortuary ritual that took place concurrently would no doubt have been affected by the varying traditions of funerary practice in indigenous and immigrant communities, and, most importantly, by the reinterpretation of those traditions within the ritual forum, based on local experience of broad cultural changes.

Recent discussions of the Philia 'facies' and associated changes in material culture in Early Bronze Age Cyprus make a strong case for the occurrence of renewed migrations from mainland regions such as southwestern Anatolia during the mid-third millennium BC (Webb and Frankel 1999; Frankel *et al.* 1996; Peltenburg 1996). It has been suggested that the quest for copper sources may have been the impetus for these migrations (Frankel *et al.* 1996:49; Mellink 1991), but it is also possible that political developments in northwestern Syria and Anatolia were involved, notably the development of hierarchically organized polities prone to local conflicts and perhaps external aggression in attempts to control the rich metal resources of the Taurus mountains (Mellink 1993:506). Either of these 'push factors' (see Anthony 1990) may have been intensified by a period of destabilizing environmental degradation (Weiss *et al.* 1993) that prompted affected communities to seek more favorable living conditions elsewhere. The 'pull factors' that drew the immigrants to Cyprus are easy to imagine, for under these or any other circumstances the island must have seemed like an attractive haven, one that was located beyond the reach of local political authority and graced with ecologically benevolent conditions, a low density of indigenous settlement that presented few obstacles to colonization, and of course, abundant sources of copper.

The modes of interaction that developed between immigrant and indigenous communities during the Philia and Early Cypriot periods are open to conjecture, but the literature on other 'frontier' societies (Kopytoff 1987; Nyerges 1992) suggests a number of possibilities. In some regions such as northwestern Cyprus where previous Chalcolithic settlement was sparse, the founding of new 'Philia' settlements may initially have proceeded with minimal contact between colonists and natives. Over time, however, new social relationships would have developed as Philia communities expanded into adjacent areas where Chalcolithic communities had already established claims to land and other resources. These relationships may have oscillated between an emphasis on boundary maintenance and competition to strategies of alliance and exchange. In other regions such as southwestern Cyprus, where there was a long sequence of prior Chalcolithic settlement, colonists and locals must have interacted more intensively from the outset. Given the strong emphasis on fertility or pro-natalism apparent in Chalcolithic stone figurines and other artifacts such as the 'birthing shrine model' from Kissonerga *Mosphilia* (Peltenburg 1991c; Bolger 1992; Goring 1991), the low average life expectancy for females reconstructed from Chalcolithic mortuary samples (Bolger 1993:37), and the high rates of infant and child mortality (Lunt 1985: 246; 1995:58, Table 10.1), it is conceivable that newcomers were welcomed into existing communities and integrated as junior kinsmen within a prevailing ideology of 'wealth-in-people' (cf. Nyerges 1992). This would help to explain the admixture of Philia and traditional Chalcolithic or Erimi cultural elements in the late phases of the settlement at Kissonerga *Mosphilia* (Peltenburg 1991b:31–32) and at other southwestern sites such as Sotira *Kaminoudhia* (Manning and Swiny 1994:164). Alternatively, newcomers may have been welcomed and assimilated in relatively higher status positions as bearers of metallurgical expertise and valued goods that included cattle and other livestock. In any of these scenarios, 'ethnic' identities may have been maintained and symbolized through stylistic variation and other cultural practices for some period of time, but eventually, as exchange partnerships and patron-client relationships were formalized through intermarriage and the extension of kinship ties, the distinctions of ethnic origin are likely to have become increasingly ambiguous.

The arrival of new settlers in the Philia and subsequent EC–MC periods would have had many important ramifications for cultural change in Cyprus. In addition to creating new dynamics of ethnic diversity, inter-communal competition, and inter-communal exchange as discussed above, the infusion of immigrants may also have contributed to a significant increase in the island's population. Recent estimates put the number of known EC–MC settlements, dating roughly between 2400–1700 BC at 270, more than double the number of known Chalcolithic or Erimi

Culture sites assigned to the much longer timespan of 3800–2400 BC (Swiny 1989:16–7; Held 1992:122–3; Knapp 1994a:410). Furthermore, site size appears to have increased considerably relative to the Chalcolithic (Swiny 1989:16; Peltenburg 1991a:108; 1993:17). Population growth may have taken place not merely because migration swelled existing population levels, but also because the establishment of a larger pool of potential marriage partners would have reduced the demographic vulnerability of both native and immigrant communities over time (Wobst 1974; Cherry 1985). Population increases were accompanied by an expansion of settlement into previously unoccupied, lowland alluvial zones such as the region between the Aloupos river and the west end of the Kyrenia range and the eastern Mesaoria between the Kyrenia range and Enkomi, as well as the Karpass peninsula in the far northeast (Catling 1963:139–41), areas whose cultivation was presumably facilitated by the use of plow agriculture. And the use of this new subsistence technology, along with the expansion of copper metallurgy, had further ramifications for the social structure and lifeways of Cypriot communities.

Some of the most striking contrasts in the material culture of the EC–MC periods relative to the Chalcolithic and earlier periods of Cypriot prehistory are evident in household architecture and settlement structure. Chalcolithic settlements such as Erimi *Pamboules*, Lemba *Lakkous*, and Kissonerga *Mosphilia* were characterized by 'roundhouses'—freestanding, circular, single-roomed structures surrounded by outdoor work areas (e.g. Peltenburg 1993:14) and occasional indications of communal storage features (Peltenburg *et al.* 1985:326). The domestic buildings uncovered at EC–MC sites such as Ambelikou *Aletri* (Dikaios 1960), Marki *Alonia* (Frankel and Webb 1996; 1997; 1999; 2000b; 2001), Alambra *Mouttes* (Coleman 1996) and Sotira *Kaminoudhia* (Swiny 1985; 1989; Swiny *et al.* 2003) were rectangular, multi-roomed, accretive complexes. At the Philia Phase–Early Cypriot site of Sotira *Kaminoudhia* in the south, a series of distinct habitation units were observed, each consisting of a narrow corridor providing access to two rooms of seemingly multi-functional domestic purpose, equipped with hearths, benches, lime plaster bins, and assorted grain-grinding and pounding equipment (Swiny 1989:20). According to Swiny, 'The sizes and arrangements of the individual units, typically consisting of two rooms, one for habitation and the other for storage, which might also be subdivided, suggest that these discrete units correspond to the space requirements of a nuclear family' (1989:21). Early MC houses at Alambra *Mouttes* in the eastern Troodos foothills, for which Schaar (1985) has noted close parallels at EBII Tarsus, typically comprised a walled front courtyard and an inner room of similar size, sometimes subdivided into one or more square rooms which may have been used for storage or some other purpose. Private as opposed to communal storage facilities are characteristic of both of these sites, as is the tendency to enclose or seclude domestic activity areas. At *Kaminoudhia*, Swiny observed that so little space existed between houses that most activities must have been carried on inside, suggesting a strong concern with privacy that is further attested by the use of narrow corridors for access into individual compounds. Similarly, the walled courtyards of Alambra which overlook either the gorge or the walls of other houses may have been constructed with an interest in privacy.

Comparable shifts from circular to rectilinear architecture have occurred in other archaeologically known cultures, including those of the Neolithic Levant. Flannery (1972) has argued that the shift from circular to rectilinear architecture in ancient Near Eastern societies was related to the shift from extended polygynous family compounds to villages made up of monogamous nuclear or extended family households. Rectilinear architecture would have been favored in these redefined family groups because of the ease of adding or subtracting rooms as households expanded with the introduction of children's spouses and grandchildren, then contracted once again as married children moved away to establish households of their own. With the individual household replacing the compound group as the basic unit of production, the mandatory communal sharing characteristic of compound groups would no longer have been practiced, and each household would have had its own storage facilities inside or adjacent to the house (1972:38–9), affording a much greater potential for private accumulations of wealth (Flannery 1972:48; see also Bogucki 1993; Byrd 1994). In the Cypriot case, changes in household architecture were probably related to the transformations in kin and productive organization which indigenous communities underwent (and immigrants had already undergone) in conjunction with the adoption of plow agriculture.

As a number of researchers (e.g. Boserup 1970; Goody 1976; Sherratt 1981; Thomas 1987) have observed, systems of plow agriculture have a different logic *vis-à-vis* the deployment of labor and the sharing of its products relative to systems of hoe agriculture. In systems of hoe cultivation, the more wives, children, and other dependents a man has, the more gardens and fields he can have tilled and the more livestock he can maintain. However, he is obligated to share the produce all around. The use of the plow allows one man to cultivate up to ten times more area than is possible by hoe (Goody 1976:25), and with the assistance of other family members, even larger areas and greater returns may theoretically be achieved (assuming unlimited availability of land). But as the coterie of dependents increases in size, its members may consume more resources than their aggregate labor contributes, meaning that the primary food-producers must work relatively harder than they would in a system of hoe cultivation (see Sahlins 1972:89–95 on Chayanov's principle; also Chayanov 1986). Thus the basic household units of production may define their membership more narrowly and place greater restrictions on sharing than in the more communally focused compound or 'lineage' type societies that may have been characteristic of the Chalcolithic and earlier periods. Yet even with this major social and productive restructuring, EC–MC household units need not necessarily have been comprised of nuclear families. The labor requirements associated with extensive cultivation, delays in the inheritance of lands by children between marriage and death of their parents, and the cost or limited availability of mature, trained draft animals might have favored the continuing co-residence of extended families—that is, of parents and at least some of their married children. The agglomerative, multi-roomed construction of EC–MC houses may thus reflect patterns of intra-community relationships and cooperation.

Aside from the implications for household and productive organization, the use of plow agriculture would have further contributed to the transition from an immediate- to a delayed-return focus in Cypriot society, or more specifically to the heightened importance of hereditary transmission of property, both fixed and immovable. Gilman (1981) and others such as Gudeman (1977) have emphasized the significant increase in labor investment that attends the use of plow agriculture, particularly in the initial effort of land clearance and stump removal. In Cyprus, the heritability and enduring attachment to particular fields and territories in general may be evinced in the tendency for sites to cluster in particular areas over long periods of time (Swiny 1989:16–17; Knapp 1990a:158). Plow agriculturalists are also highly dependent on the transmission of draft animals as capital assets, and access to these may be a major criterion for the establishment of new households (Bogucki 1993). This emphasis on the inheritance of property may have extended to other personal possessions as well, perhaps explaining the remarkable differences in abandonment practices that differentiate Chalcolithic and EC–MC settlements—the former replete with pottery, tools, and other usable goods, the latter virtually swept clean, with few intact objects left behind (Frankel *et al.* 1996:47; Webb 1995; Coleman 1996:31).

The spread of copper metallurgy was another development with the potential for significantly transforming socioeconomic life. Given the scarcity of metal finds in EC–MC settlement contexts, copper-based tools and hardware are likely to have been highly valued and intensively curated at this time. However, the ceremonial uses of metal may have been even more important than its utilitarian consumption during the EC–MC period. Copper-based artifacts are found primarily in mortuary contexts, and both their composition and style often suggest that they were more valued as objects of display than as items for practical use (Swiny 1982; Philip 1991:68; Balthazar 1990: 430-2). With the increasing demand for copper generated by escalating prestige displays, social strategies—presumably based on kinship ties, alliances, and other types of exchange relation-ships—probably played an important role in the acquisition of copper, especially inasmuch as some of the richest accumulations of metal came from coastal sites such as Lapithos, *Vounous* and Vasilia, all located far away from the Troodos copper sources. It is unclear as to what coastal communities offered settlements in the mining zones in exchange for their metal output, but live-stock, pottery, other finished goods such as cloth, and marriage partners are among the possi-bilities. It is also conceivable that members of coastal kin groups established their own camps or settlements for mining, smelting, refining and the production of finished goods at sites such as Ambelikou *Aletri* (Dikaios 1960; Merrillees 1984) in the northern Troodos foothills and the recently excavated Pyrgos *Mavrorachi* in the south (Belgiorno 1999; 2000; Giardino 2000; Herscher 1998:321). But whether through alliance-based exchange networks or systems of direct procurement, a social infrastructure for copper production must have been established, and that infrastructure could later have been manipulated for other purposes such as production for overseas trade.

The dynamic social relationships between immigrants and indigenes, the changes in the domes-tic mode of production associated with plow agriculture, and variable access to sources of copper were all factors that may have promoted the emergence of wealth and status differences between household groups during the EC–MC periods. The successful accumulation and transmission of cattle and copper wealth might in principle have led to a spiraling process of social stratification with those households that were well-endowed in either finding themselves in a better position to accumulate more and more of each. Yet settlement excavations (Coleman 1996; Dikaios 1960; Frankel and Webb 1996; Overbeck and Swiny 1972; Swiny 1985; 1989; Todd 1993) thus far have not yielded any striking evidence for differences in household wealth within communities during this time (see also Frankel 2002:173). The only EC–MC building excavated to date that might reasonably be characterized as an 'elite residence' is an impressive complex of eleven rooms arranged around what may have been a central courtyard at the MC site of Kalopsidha *Tsaoudhi Chiflik* (Gjerstad 1926:27–37), but Frankel and Webb (1996:54) have recently questioned Gjer-stad's interpretation of this building as a unified complex. At other sites, intra-household com-parisons are difficult, in part because of the characteristic scarcity of artifacts and other refuse in primary contexts of discard (Frankel *et al.* 1996:47) and in part because of the difficulties of defin-ing individual household units (Frankel and Webb 1996:53). Largely on the basis of 'negative' evidence, however, it appears that any differences in household wealth that may have developed before the Late Bronze Age in Cyprus did not receive overt material expression in domestic architecture and household appurtenances. Nor are there as yet any indications of settlement hierarchies, public buildings, or specialized productive facilities under elite or official control. Present evidence thus suggests that the potential for social stratification in EC–MC society was as yet unrealized and perhaps even countered by other social factors.

## The EC–MC funeral and social change

As the social life of Cypriot communities was altered by migrations, changes in subsistence economy and household organization, and the increasing symbolic as well as technological importance of metal, mortuary rituals also changed. Some of the practices that arose—the use of extramural cemeteries and rock-cut tombs with multiple, sometimes secondary burials—may have been nascent both in the mortuary traditions of Cypriot Chalcolithic communities such as Souskiou and in the cultural heritage of the new settlers from the mainland. Yet the significance of the funerary programs that unfolded during the later third and early second millennia BC lay not so much in their novelty nor in their prior manifestations but in the elaboration, formalization, and quasi-ubiquity of practices that developed over this time period. The driving force behind these developments most likely emerged from the increasingly crucial importance of ancestors as guardians and legitimators of the contemporary social order. The use of spatially reserved burial precincts, the rise of 'dual obsequies', the use and reuse of pit and chamber tombs, and the material goods with which the deceased were provided effectively transformed the dead into ancestors, while affirming their connections to living relatives and descendants on the occasion of each successive funeral.

In the context of ongoing changes in productive systems and social structure, it is not difficult to imagine why EC–MC communities developed highly elaborated ancestral ideologies. The settlements of this period would most likely have required more territory than those of the Chalcolithic, not only because some, at least, were larger but also because of the land requirements associated with plow agriculture and pasturage of cattle. Given the labor invested in extensive land clearance and the fact that suitable agricultural land was not an inexhaustible supply on this mountainous, arid Mediterranean island, both newcomers and indigenes may have been concerned with asserting their claims to particular regions and resources *vis-à-vis* the claims of other social groups. The construction and reuse of durable, socially recognizable chamber tombs affirmed the long-term connections between particular kin groups and their ancestors from a household perspective, while the reuse of 'formal' cemeteries gave permanent testimonial to the links between the living, the dead, and their lasting connections to the land at the communal level. Other symbols, such as copper and stone axes, and the stone axe amulet from *Vounous* B Tomb 143 (Stewart and Stewart 1950), may also have expressed the 'ancestral' rights of firstcomers or latecomers to the land which community founders had to clear.

Another element of the cultural logic behind the formalization of ancestral ideologies and their ritual observances may relate to the powers with which the ancestors were invested as guarantors of fertility and prosperity. The intermediary role of the ancestors in insuring fertility may have been symbolized by the inclusion of ceramic models and pottery vessels with coroplastic 'genre' scenes depicting agricultural and food processing activities (plowing, harvesting, grinding) and images relating to human reproduction (women holding babies, kissing couples, man and woman in bed together, etc.) in Early–Middle Cypriot tombs (Morris 1985; Karageorghis 1991). Thus, the bestowal of appropriate honors upon the ancestors may not only have legitimized the rights of succession and social status claimed by the sponsors of their obsequies, but it may also have been deemed essential in securing fertility and prosperity.

The archaeological evidence discussed in Chapter 4 makes it possible to reconstruct the general schema of practices associated with EC–MC funerals for adults, beginning with the occasion of death. We cannot know exactly how the bodies were prepared, but it is evident that the corpses were arranged in a number of special positions, most commonly in a lateral, flexed, approximately fetal pose or in an upright seated posture with the knees drawn up to the chest. It is possible that the arms or legs were sometimes bound to hold them in particular positions, and in addition to being clothed (as the presence of dress pins would imply) the bodies may have been wrapped in shrouds or sacks, perhaps facilitating later episodes of mortuary treatment. The corpse might then be interred in a small pit grave, a new or existing chamber tomb, or a ceramic storage jar (possibly broken or halved), depending on local preference, status considerations, and the plans, if any, for future treatment. Some of these primary burial facilities were located within the same formal cemeteries where the local chamber or pit tombs were dug (as is apparent at Lapithos and *Paleoskoutella*), but it is not inconceivable that some burials were stored for a time within the settlement, as suggested by the recent excavations at Marki *Alonia* (Frankel and Webb 1997:88; 1999:90; 2000b:70). A variety of personal goods and gifts were placed with the deceased; these almost always included pottery, and many individuals were equipped with copper ornaments, weapons and/or other implements, along with occasional items of stone, bone, and, very rarely, exotic materials (for examples of individual, possibly primary burial complements see Alambra Tomb 102 in Coleman 1996 and the Karmi 'seafarer' in Stewart 1962b). We can only speculate about the mourning customs, lamentations, prayers, songs, dances, processions, animal sacrifices, feasts and other ceremonial undertakings that accompanied the primary funeral, but some combination of these activities was probably performed as well.

The decision to stage a secondary funeral may have been made on the occasion of death or at any time afterwards, based on the changing social calculations and aspirations of surviving relatives who were to sponsor such an event. Whether or not there was an interval of some specific and prescribed duration preceding exhumation is a matter for conjecture, but the variable preservation of individuals thought to represent secondary interments suggests that the timespan between obsequies may often have differed. This would certainly have been the case when a collective secondary celebration was held for several deceased ancestors at one time, as not all of those selected for treatment would necessarily have died in the same year. In any case, however, the interval must have been sufficient to allow the sponsors to make preparations for the next phase of mortuary rites: inviting guests from near and far to gather at a convenient time in the agricultural cycle, amassing the livestock, grain, beer, and other foodstuffs necessary for large scale feasting, gifts to the dead, and general hospitality, accumulating additional quantities of pottery, copper and other goods to be given to the dead and perhaps to honored invitees as well, and, in some instances, digging a new and suitably elaborate chamber tomb in which the remains of the ancestor(s) were to be placed. The choice of whether to reuse an old tomb or make a new one must have been governed by both practical and prestige considerations. On the one hand, cutting a new tomb required more labor and help from friends and kin, but it made an important statement about the immediate wealth and social resources of the kin group. On the other hand, reusing an older tomb would reinforce 'social memory' of the group's local forebears, their wealth and longstanding prestige. Either choice could therefore result in an enhancement of prestige.

The occasion of a secondary funeral must have been a dramatic and memorable event. One or more individuals had to be exhumed from temporary graves or the older chamber tombs in which they were originally placed. The odors and textures associated with handling partially decomposed corpses and/or disarticulated skeletal remains must have been remarkable, and we can only wonder whether this was a task that the bone diggers approached with dread and revulsion, sadness and respect, or joking and ribald enthusiasm. The uncovering of buried personal goods was another situation replete with possibilities. If the wealth was scanty, the heirs might lose face unless they provided ample new gifts for the reburial. If the finds were rich, they must have gained prestige from having the wealth of their ancestors redisplayed before an audience, but they may also have found it problematic to selectively remove and cache particular goods for future reuse—if they were so inclined. It would be interesting to know, therefore, whether the exhumation was undertaken in public or private. The reburial, however, was almost certainly a public event. The dead made their journey to what was ostensibly their final resting place, a collective tomb, sometimes decorated as a shrine (Frankel and Tamvaki 1973; Åström 1988; Stewart 1962a:216; Stewart 1962b:197; Stewart and Stewart 1950:152, 158, 162–3; Herscher 1978:705). The original grave goods would have been on view again in the course of the transfer, and fresh gifts may have been added to the collection by the funeral sponsors and their guests. The corpses or skeletons undergoing reburial often seem to have been arranged with considerable care, but in some instances they may have been inserted more casually, perhaps in the aftermath of other, more important rituals taking place outside the tombs. Skulls and other body parts from earlier burials may also have been removed or deliberately rearranged sometimes when new burials were introduced. Meanwhile, livestock such as cattle, sheep and goat were slaughtered as sustenance for the dead and, equally important, as feast food for the living (Keswani 1994), to be served along with whatever other types of food and drink were appropriate to the occasion.

The size and elaboration of EC–MC chamber tombs, along with the lavish consumption of livestock, metal, pottery, and other grave goods suggest that the associated mortuary celebrations were meant to be attended—and thus appreciated—by a large number of guests from the local community and farther afield. The island-wide similarities in metal weaponry and ornamentation, along with the broad similarities in 'genre scenes' and 'ritual' vessels found in tombs from communities as distant as *Vounous*, Marki, and Kalavasos may be indicative of the breadth of extra-local participation in mortuary observances; these were the foremost occasions on which a kin group's treasures might be displayed, admired, and subsequently emulated or outdone. But 'showing off' need not have been the sole purpose of the undertaking, for by staging lavish mortuary festivities to which members of other communities were invited, the sponsors not only asserted their own wealth and importance; they also demonstrated the desirability of continuing alliances and exchange relationships established in the past and perhaps attracted new partners for the future. Hospitality, largesse, convivial socializing, and quite possibly gaming, as suggested by the recently discovered gaming board cutting at at Dhenia (Herscher 1998:320), would all have served to create and reinforce these social bonds.

Given the dearth of evidence from settlement excavations (none of which, regrettably, have been contiguous with the major cemetery excavations) for obvious social differentiation in housing or other forms of quotidian display, it seems likely that mortuary ritual was the central arena for the affirmation and negotiation of social status differentials that were neither constantly nor overtly manifested in everyday life. At the north coast cemeteries of *Vounous* and Lapithos, all of the major indices of mortuary expenditure—estimated chamber floor area, 'pots per burial' and 'copper items per burial' rose over the course of the Early and Middle Cypriot periods. The increase in the median as well as the mean values of these variables is especially important because it implies that the propensity towards conspicuous consumption was broadly based within these communities and not merely limited to an hereditary elite. The fact that variations in grave assemblages became progressively more continuous and quantitative rather than discrete and qualitative further suggests that social status was more the product of competitive display than of a highly formalized system of ranking. Most kin groups, both richer and poorer, would have participated in competitive ritual displays, some more successfully than others, but relatively few may have been able to sustain such expenditures over time, as costly displays would have tended to serve as wealth depleting and levelling mechanisms.

Yet this does not mean that EC–MC mortuary ritual was the product—or producer—of an essentially static socioeconomic system. On the contrary, the multiplication of these prestige displays, repeated and exceeded with each successive funeral in communities throughout the island, had important cumulative effects that went well beyond the fluctuating ascendancy of competing kin groups or communities. Because of the demand for copper to be crafted into a variety of prestige goods, and because those goods were periodically removed from circulation, there had to have been a significant increase in the scale of metallurgical production, along with the spread of metallurgical technology and its practitioners. Concurrently, the production of surplus agricultural goods and a variety of local craft items—ceramics, worked stone, cloth—may have been increased for barter in the copper trade. Mortuary consumption was, in effect, a significant driver of economic intensification. In turn, as the scale of metallurgical production increased, the visibility and international reputation of Cypriot copper resources would also have been enhanced. Any visitor who happened to be present on the occasion of one of the more elaborate Early and Middle Cypriot funerals would certainly have been impressed by the quantities of copper discarded with the dead, and might have carried home anecdotes that inspired deliberate trading visits in the future. Cypriots travelling abroad may also have told tales of their island's wealth. Presumably this is one reason why the quantities of foreign valuables recovered from Cypriot sites began to increase at the time of the ECIIIB/MCI transition, when the widespread consumption of metal goods becomes evident in mortuary deposits. These developments approximately coincided with the breakdown of the older exchange networks through which Near Eastern and Egyptian polities had previously acquired their copper supplies (Klengel 1984; Knapp 1986a), prompting traders from these regions to seek alternative sources of copper in Cyprus.

The ways in which these new exotic goods were envalued and manipulated within local systems of mortuary ritual and exchange also had far reaching ramifications for productive and ideological transformations in Middle Cypriot society. In the prevailing environment of competitive display associated with mortuary ritual, the deployment of imported prestige goods (whose associations with foreign elites would have been conveyed through a combination of personal contacts and observations of iconography) offered a means for redefining status distinctions in terms that were now qualitative as well as quantitative. In turn, through a combination of strategies that included manipulating the supply of those exotic valuables, ambitious social entrepreneurs could parlay their long-distance trade contacts into enhanced, and to some extent exclusive, local exchange partnerships, maximizing their access to copper supplies. By the end of the Middle Cypriot period, some of these social entrepreneurs seem to have migrated to auspicious localities for the pursuit of long-distance trade, establishing metallurgical workshops that attracted repeated visits from representatives of foreign polities. These trading entrepôts in time grew into large and important towns. The social dynamics of mortuary ritual, involving the display and disposal of copper wealth, along with the rising envaluation of imported prestige goods, may therefore have prompted both the infrastructural and the ideological transformations leading to the emergence of politically complex and socially stratified urban communities in the Late Bronze Age.

## Urbanization and changes in social structure during the Late Bronze Age

The structure of settlement, economic production, and social relations underwent major transformations during the Late Bronze Age in Cyprus. Towns were established in coastal locations favorably situated for long-distance trade, the copper industry expanded, and numerous economic and craft specializations appear to have flourished. In many of the major towns, social heterogeneity must have increased as settlers from different communities took up residence and began competing with each other for economic and political hegemony. Institutionalized relations of hierarchy developed within towns and between the towns and hinterland regions that supplied copper and agricultural produce, as ambitious elite groups formulated new strategies for controlling critical resources and productive enterprises, along with new ideologies for legitimizing that control. Many of these social transformations are evident in the realm of mortuary ritual as well as in the archaeological evidence from settlement contexts. At the same time, it is apparent that the role of mortuary ritual in structuring social relationships was also undergoing important changes.

It is difficult to estimate the scale of population growth in the Late Cypriot period (over 300 sites have been ascribed to this period, compared to the 270 noted for the preceding EC–MC period), but it appears that a significant population increase and a major redistribution of population was underway, as high concentrations of settlers gathered in towns that ranged from 12–70 ha or more in area (Catling 1963; Swiny 1981:78; Knapp 1994b: Fig. 9.5; 1996a:61, 80, 1997:47–8). Current archaeological evidence suggests that the associated patterns of urbanization and internal organization were diverse. The earliest coastal centers at Enkomi and *Toumba tou Skourou* seem to have been established in areas where previous occupation was sparse. It is probable that these sites were settled by groups from a number of outlying communities, and consequently, there would have been a greater degree of social distance between residential groups than was characteristic of rural village communities. This 'social distance' may be reflected in the multiple mound configuration of *Toumba tou Skourou* (Vermeule and Wolsky 1990:14–5) and in the open spaces surrounding the earliest household and industrial complexes at Enkomi (Dikaios 1969: Pls 247–8, 267–8; Courtois 1986b:5)—contrasting with the agglomerative tendencies of EC–MC settlement architecture (Swiny 1989)—as well as in the co-occurrence at these sites of diverse regional ceramic types (e.g. Vermeule and Wolsky 1990:289).

Presumably the founding groups were drawn to these coastal locations by the prospects for overseas trade, led by elite entrepreneurs or 'aggrandizers' (Manning and De Mita 1997:108; see also Clark and Blake 1994) seeking to secure their own exclusive supplies of exotic prestige goods that were essential to the maintenance of their status and authority. Manning and De Mita have suggested that foreign, professional merchants who negotiated special relationships with local elites were the 'administrative master-minds' behind the organization of production in each region, yet they acknowledge that the political basis for this mobilization must have been 'embedded in the pre-existing regional alliances and kinship networks' (1997:114). In other words, longstanding social relationships may have facilitated the extraction and transport of copper from early production locations such as the LCI smelting site at Politiko *Phorades* (Knapp *et al.* 1998; 1999; 2002). Meanwhile, groups that established successful enterprises in the new towns would likely have attracted many 'clients'—kin or non-kin—to their urban estates, thus creating a local power base of labor, and perhaps in some cases (e.g. the LCI fortress at Enkomi) defensive, capabilities. Hierarchical social relationships based on real differences in access to productive resources, and perhaps legitimated in part by an ideology of first-comer versus late-comer privileges (Kopytoff 1987), would have arisen and intensified as more and more settlers flocked to the urban sites. Yet the overall structure of political organization within these settlements may have remained predominantly 'heterarchical' or oligarchic, as long as multiple kin groups were able to sustain their own independent supply networks and workshops that yielded copper and other commodities of exchange. In the case of Enkomi, this would account for both the presence of seemingly numerous elite residential areas and productive zones and the absence of any clearly paramount administrative complex within the site (Keswani 1996).

Other urban communities that developed in regions with prior sequences of occupation may have experienced variable forms and degrees of centralization and social hierarchy. Very large towns such as Kition (Karageorghis and Demas 1985) and Hala Sultan Tekke (Åström 1982; 1986; 1996), located in favorable harbor sites along Larnaca Bay, most likely expanded in part through immigration from outside areas, and the trajectory of sociopolitical development within these settlements may have resembled the one just outlined for Enkomi and *Toumba tou Skourou*. Other urban sites such as Maroni (Cadogan 1984; 1989; 1996; Manning *et al.* 1994; Manning 1998a; 1998b), Kalavasos (South 1984; 1989; 1996; Todd and South 1992), and Alassa (Hadjisavvas 1986; 1989; 1991; 1994; 1996) located further west may have developed from relatively localized processes of population nucleation. These sites appear to have had more pyramidal or centralized administrative structures, possibly because of their relative proximity to the copper mines, which facilitated a more monopolistic control over copper supplies and production. It is also possible that there were fewer elite groups competing for hegemony in these localities. Still other small centers such as Maa *Palaeokastro* (Karageorghis and Demas 1988) and Pyla *Kokkinokremos* (Karageorghis and Demas 1984) may conceivably have been founded as outposts of other larger urban settlements like Kouklia (Maier and Wartburg 1985) or Kition, thus constituting externally imposed foci of elite power within the surrounding countryside (cf. Karageorghis 1984; 1990b for an alternative interpretation involving Aegean colonists). This may also apply to various 'secondary' centers and sanctuaries established in hinterland regions to administer copper production, expedite the shipment of copper supplies, provision mining communities, and perhaps even supply some urban centers with essential subsistence commodities (Keswani 1993; Keswani and Knapp 2003; Webb and Frankel 1994).

In conjunction with changing settlement patterns, new forms of economic specialization were developing in urban areas, and these would over the long run have altered the traditional basis of household reproduction and kin-based modes of productive activity. The growth of the copper trade meant that copper production had to be expanded to accommodate the needs of exporting as well as local consumption. The scale of specialist activity in the fields of mining, smelting, refining, and production of finished goods would have to have increased. In addition, the ranks of specialists must have been further augmented by the need for merchants, sailors, builders, masons, and various craftsmen associated with ceramic production, seal cutting, ivory carving, gold working, and other forms of luxury and utilitarian production. Some of these specialists may have been attached to elite households, political institutions, or temples, while others may have worked independently, sometimes as itinerants. Yet regardless of their affiliations, their activities overall would have served to diversify a productive regime formerly based on agriculture, creating not only new commodities for exchange but also new requirements for surplus subsistence goods and objects of 'wealth finance' (D'Altroy and Earle 1985; Keswani 1993), resulting in further specialization and economic intensification. At the same time, institutional control and/or strategic manipulation of their products would have afforded members of the emerging political elite new means of establishing and extending their wealth and power.

The development of social stratification and power differentials within the major LC urban centers is abundantly attested in the architectural evidence from sites such as Enkomi, Kition, Hala Sultan Tekke, Maroni, Kalavasos, and Alassa. The construction of monumental ashlar buildings, some of which may be classified as elite residential and/or administrative complexes, others as temple complexes, indicates a deliberate differentiation of elite versus non-elite and institutional versus domestic domains, along with the requisite control over the labor needed to build such structures. These and other impressive buildings associated with elite groups were also frequently the loci of important productive and/or storage facilities. Some of the earliest copper workshops at Enkomi were associated with the large and well-built LCI fortress located at the north end of the site, a complex that was succeeded by other large buildings where copper working continued on a varying scale throughout the remainder of the site's occupation (see Pickles and Peltenburg 1998 on the significance of architectural changes in this part of the site). Other copper workshops were located in close proximity to imposing elite residences and temples, some built of ashlar masonry, found in the central part of the town during LCII and LCIII (Dikaios 1969; Courtois 1971; 1982; 1986b). Copper workshops were also directly affiliated with temple complexes at the town of Kition in LCII and LCIII (Karageorghis and Demas 1985) and possibly at Hala Sultan Tekke (Åström 2000:34). Elite administrative complexes with extensive storage facilities for agricultural produce (some of which may have been used for the provisioning of miners and other specialists) have been found at Maroni (Cadogan 1996), Kalavasos (South 1996), and Alassa (Hadjisavvas 1996). The functional characteristics and work activities associated with these buildings attest to elite control over forms of production that served as the material basis for social stratification.

The means by which this control was achieved and sustained must have depended upon a variety of new power strategies and modes of social legitimation. To some extent, traditional relationships of alliance and exchange that had facilitated the flow of copper during the Early–Middle Bronze periods may have persisted during the Late Bronze Age, but the appearance in the MCIII–LCI period of fortresses at Enkomi (Dikaios 1969; Fortin 1989), Korovia *Nitovikla* (Gjerstad *et al.* 1934; Hult 1992; Merrillees 1994), and various locations in the eastern Troodos and southern Kyrenia mountains (Catling 1963:140–1; Gjerstad 1926:37–47; Overbeck and Swiny 1972) may indicate the coercive subordination of interior regions by coastal towns such as Enkomi and *Toumba tou Skourou* at this time (Peltenburg 1996:30–5). Later, other forms of administration involving ritual regulation and formal systems of tribute collection may have been adopted. The sanctuary site at Athienou (Dothan and Ben-Tor 1983; Maddin *et al.* 1983), for example, may represent a secondary center that functioned as a transshipment point for copper between the mines and a coastal center such as Enkomi, Kition, or Hala Sultan Tekke. Other sanctuary sites such as Myrtou *Pighades* (Taylor 1957) and Ayios Iakovos *Dhima* (Gjerstad *et al.* 1934) may have facilitated different types of exchange between coastal and inland areas, along with providing ritual services for the surrounding countryside. A combination of official strategies involving wealth and staple finance may have served to underwrite the operations of sanctuary sites, primary copper production sites such as Apliki *Karamallos* (Taylor 1952), and other administrative centers such as Analiondas *Palioklichia* (Webb and Frankel 1994) and Aredhiou Vouppes (Knapp *et al.* 1994; Given and Knapp 2003:179–82) that may have provisioned mining communities (Keswani 1993; Knapp 1997).

The deployment of a new complex of ideology and prestige symbolism would have been crucial in naturalizing or legitimizing the now overt wealth and power differentials within and between communities as well as the extractive demands imposed by the elite. This complex was strongly influenced by external contacts that introduced not only new types of status goods but a familiarity with other political systems and politico-religious cosmologies. That elements of this knowledge were in turn incorporated and adapted into the ideological and symbolic systems of Cypriot communities is evident from the diverse Near Eastern, Egyptian, and Aegean affinities of Late Cypriot art and architecture. The use of personal adornments and ceremonial paraphernalia that posited a relationship or identity with powerful foreign elites and divinities was one means by which the local elites set themselves apart from and above the populace that labored on their behalf. Who could see them with their fine robes, gold jewelry, bronze weapons, and horse-drawn chariots and not be convinced that they were a special, exalted kind of human? Yet the distinctions of 'eliteness' must not only have been symbolized in physical appurtenances but tangibly enacted in public ceremonies. The performance of ritual duties allowed the elite to actively portray themselves as the new intermediaries between the people and the cosmic powers—intermediaries whose ministrations to various deities ensured the fertility, prosperity, and overall well-being of the community at large (Knapp 1986b; 1988a). The supernatural capabilities that had formerly been ascribed to the ancestors were now appropriated by the elite.

## The social dynamics of Late Bronze Age mortuary ritual

Urbanization and the attendant economic and sociopolitical developments of the LC period were accompanied by both continuity and changes in mortuary ritual. Continuity is most evident at rural sites where the use of extramural cemeteries persisted throughout the Late Bronze Age, perhaps because older forms of household organization and an agriculturally based mode of production persisted there as well. Moreover, at many of these localities elaborate celebrations involving secondary treatment and collective burial not only continued but also seem to have increased in scale, especially during LCI, as attested by the so-called 'mass burials' found at sites such as Ayios Iakovos, Pendayia, and Myrtou *Stephania*. These extraordinary ritual events may attest to an intensification of prestige competition and ritualized assertions of group identity in the face of the changing economic and political conditions that were gradually undermining traditional lifeways. It is possible that such celebrations were staged periodically when the bodies of several kinsmen who had migrated to the towns were returned to their ancestral villages for permanent burial. If so, the 'return' of the deceased would almost certainly have been an important occasion for the reaffirmation of alliances between coastal and inland, or urban and rural, communities, much as it was during the preceding era.

In the urban centers, the most immediate and obvious departure from EC–MC practice was the placement of tombs within settlement contexts. In part because of the heterogeneous origins of the urban settlers, and in part because of increasing competition between household groups associated with the growth of economic specializations and trade, the dead were no longer buried in spatially reserved cemeteries used by the community at large. Instead the ancestors were appropriated by smaller residence groups, who placed their tombs beneath houses and courtyards, or adjacent to houses and workshops in streets or other open areas, where they served as important symbols of group identity and of rights transmitted through descent over generations and sometimes centuries of reuse. Some of these may have been 'public' rather than private locations, but the association of the dead with specific social groups must nevertheless have been obvious. As social distance increased among the living, it also increased among the dead.

Despite the shift in burial location, many of the funerals which took place within the new towns may have been similar in format to those of the EC–MC period. Enkomi Swedish Tombs 2, 6, and 17 (Gjerstad *et al.* 1934), representing both middle and high status groups, would seem to attest to the continuation of ritual programs involving secondary treatment and collective reburial in LCI and LCII. Several of the burials in each of these tombs appear to have been placed in squatting positions around the perimeter of the chamber, in an arrangement also observed in Ayios Iakovos *Melia* Tomb 6 (Gjerstad *et al.* 1934) and Katydhata Tomb 42 (Gjerstad 1926:82). Interestingly, the latest burial in Tomb 17 was an extended, possibly primary burial of an older male who had been equipped with a very elaborate gold pin, a gold bowl, and the so-called 'Zeus' crater depicting a chariot scene, an individual in elite dress holding a pair of balance pans, and another person carrying what appears to be a copper oxhide ingot. It is possible that the person accorded this high status burial treatment had himself been a sponsor of one or more of the earlier secondary funerals that occurred in association with this tomb. The case of Tomb 6 is also noteworthy because it is a rather late (LCIIC) example of collective secondary treatment, taking place at a time when such ceremonies seem to have been on the wane. Perhaps this is an example of a deliberately 'archaizing' celebration, intended to mark the social ascendancy of a new or 'upstart' kin group, or to reassert the status of an older group seeking new prominence and legitimacy.

Other chamber tomb burials may have been interred in the context of grand primary funerals making use of new or existing tombs. In some cases, the use of a new tomb for such an event may have been inaugurated with the concurrent interment of one or more burials exhumed from an older tomb, symbolizing the continuity of the lineal group. This may have been the situation in Kalavasos *Ayios Dhimitrios* Tomb 11 (South 1997; 2000; Goring 1989), where the incomplete and almost certainly secondary remains of one adult female (Skeleton II) were found along with two more or less complete adult female skeletons. Skeleton III, which was found on the narrower east bench with some bones in anatomically impossible positions, might have been the original primary burial in the chamber, shifted from the larger west bench when the remains of the woman represented by Skeleton I were interred there. All of these burials were very richly equipped with gold jewelry and other luxury goods. The increasing frequency of extended, articulated and probably primary burials and of tombs with rock-cut benches intended to accommodate extended burials, may suggest a shift in LC mortuary ritual from an emphasis upon elaborate secondary funerals to the elaborate arraying and presentation of the dead for a single impressive viewing in the context of a primary funeral (see also Thomas 1991:129).

There can be no doubt that elite funerals of the Late Cypriot period were elaborate spectacles. One may imagine the deceased laid out and attired in their finest clothing and jewelry—diadems, mouthpieces, earrings, finger rings, bracelets, anklets, necklaces, dress pins and other ornaments— with combinations of these and other goods varying somewhat based on the gender of the burial and the fashions of the period. Perhaps the body was displayed before burial on some sort of bier, surrounded by personal possessions and gifts that included toilet articles (cosmetic jars and dishes made of ivory, faience and alabaster, glass perfume bottles, stone mortars and pestles, ivory combs and bronze mirrors, ceramic oil and unguent containers of Mycenaean and Red Lustrous wares), drinking sets and tableware (with Mycenaean pictorial craters and, in later years, bronze vessels among the most prized treasures), bronze weaponry (especially from LCIIC onwards), gaming equipment (ivory boxes and faience gaming pieces), and ceremonial paraphernalia (such as faience, Mycenaean and Red Lustrous libation vessels, scepters or batons, cylinder and stamp seals most likely worn as jewelry, and balance weights and scale pans symbolic of the copper trade). Such arrays of wealth affirmed both the status that the deceased had enjoyed during life and the status of the relatives who could afford to bury them away in a tomb, where only the dead would use them. Presumably family and household members would have been witnesses to these proceedings, along with representatives of other important families from this and other towns in which the family had social connections. Funerals taking place in the street outside the home of the deceased would doubtless have attracted other onlookers as well. But the frequent practice of interring the dead in tombs within the family compound suggests that the viewing of the dead may generally have required a direct invitation; it was no longer an open community affair.

The disparities in the distribution of gold objects and other valuables of exceptional workmanship and high informational content (inscriptions, pictorial decoration), and the accumulations of such goods in certain tombs over long periods of time, unquestionably attest to the emergence of a hierarchical social order dominated by a hereditary elite during the Late Cypriot period. However, the material indices of mortuary expenditure also display some interesting contradictory tendencies at the same time. On the one hand, the disposal of large quantities of gold and other exotic goods suggests that mortuary ritual continued to be an important occasion for demonstrations of wealth and social prestige. On the other hand, the size and elaboration of chamber tomb architecture does not begin to compare with some of the extraordinary Middle Cypriot constructions at various sites in the north, from which some of the settlers of Enkomi and *Toumba tou Skourou* may have originated. Undoubtedly the constraints of urban space and the quality of the local bedrock may have limited the sizes of tombs that could be cut in urban centers, but if town dwellers were inclined to construct elaborate chamber tombs, they might have sought more appropriate locations beyond the bounds of their settlements. At present it does not appear that they did so. Taken in conjunction with the decreasing quantities of tomb pottery (possibly representing contributed grave goods as well as personal effects), the apparent scarcity of large animal sacrifices that may have been associated with funeral feasting in earlier periods, and the tendency towards 'privatization' of tomb locations, this implies a relative diminution in the scale and centrality of mortuary ritual. Social status differentials were no longer produced primarily in the context of grandiose mortuary events and the associated disbursement of hospitality. They were founded in differential access to productive resources and exchange networks, and manifested on a continuous basis in the differentiation of settlement architecture and in a broader range of ceremonial events. They were a part of the *habitus* (Bourdieu 1977) created by the built environment and every day experience.

The prolonged reuse of LC chamber tombs, whether continuous or intermittent, attests to the enduring importance of lineal identity as the basis for status and social legitimacy throughout much of the LC period. Yet at the transition between LCIIC and LCIIIA, when trade, copper production, and monumental and civic construction expanded to an unprecedented degree, important changes in mortuary practice were underway. Single burials or small burial groups of two to three persons interred in earthen or stone-lined shaft graves became extremely common. Considerable variation in wealth is evident amongst these burials, and while many were quite poor, at least a few would seem to represent individuals of relatively high rank. This break with past mortuary traditions may have been related to the growth of court and temple institutions and an associated rise in the numbers of officials, dependents, and functionaries whose status was more and more independent of membership, or claims to membership, in the ancient kin groups of the city. Some of these persons may have come from communities elsewhere on the island or even beyond, and were thus detached from their ancestral descent and tomb groups. Others may have been local residents whose mortuary rites were truncated, terminating in primary burial without subsequent exhumation and reburial in collective family chamber tombs. Such an abbreviation of traditional mortuary practices may have become acceptable as the increasingly complex political and economic environment created new contexts for the establishment of social status and the accumulation of wealth.

It is important to keep in mind, however, that older customs may have persisted at rural sites about which we have less information, and these traditions were reworked once again after the collapse of the LC urban system. In LCIIIB and the subsequent Cypro-Geometric era, a reversion to the use of extramural cemeteries is evident throughout the island, new and elaborate forms of chamber tombs were constructed, complex mortuary treatments involving cremation as well as inhumation were sometimes undertaken, and large collections of ceramic and metal grave goods were again disposed of with the dead (Steel 1995). Once more, in the context of migrations from the Aegean and other outlying regions, communal burial grounds may have taken on renewed importance as symbols of communal identity and ancestral rights, and mortuary display may have re-emerged as a critical arena for the negotiation of a new sociopolitical order.

## Epilogue: realizing the potential of mortuary analysis— methodological imperatives and directions for future research

The death of family members, spouses, friends, and acquaintances is an inescapable phenomenon with which every human being and every community must come to terms, physically (in the exigency of disposing of the body), emotionally, and socially. Ideologies of death and customary practices may dictate the broad outlines of the treatment owed to the dead, as well as the conduct that is expected of the mourners, but those ideologies and practices are interpreted by individuals acting within specific contexts of local meaning. Each instantiation of mortuary ritual is therefore replete with dynamic potential, for each successive funeral may be executed as the perpetuation, enhancement, or repudiation of previous traditions. And to whatever extent the funeral is witnessed by others, its enactment will serve as a model of how, or how not, to stage such events in the future. Thus each episode of mortuary practice can influence the trajectory of mortuary ritual overall, as it proceeds not simply according to a set of static social prescriptions, but rather unfolds as the cumulative outcome of many individual decisions and actions, producing and produced by cultural meanings, both old and new.

In most societies, the details of the funeral are influenced to some extent by the social status and persona or 'individuality' of the deceased. Yet the dead do not bury themselves, and even when their specific wishes are carefully honored, their funerals are ultimately social presentations devised by the living. Even as the survivors grieve for, honor, and celebrate the deceased, they may also embroider, deny, and reinterpret certain facets of the dead person's identity as a living human, for the presentation of the dead inevitably reflects upon the social status of the heirs and funeral sponsors. In turn, the choices made in designing funerals as social events may have ramifications that extend well outside the mortuary arena. Both the personal status of the sponsors and the network of social relationships that they participate in may be affected, for better or for worse. Many funerals, moreover, entail material outlays and monumental constructions which may alter the sociopolitical landscape, and in some preindustrial cases, the economic infrastructure of an entire society. It has been argued in this book that the expansion of the copper industry and the onset of the urbanization process in Bronze Age Cyprus may be attributed in significant measure to the central importance of mortuary ritual and associated prestige displays—presentations of the dead that served, cumulatively, as incentives to intensified metallurgical production and the growth of long-distance trade.

The archaeological study of mortuary remains has a vast potential for illuminating the ideologies, ritual practices, and social structures of past societies. Where long-term sequences of mortuary data and contemporary evidence from settlement contexts are available, it becomes possible not only to describe the details of burial practices but also to theorize about how those practices reflexively expressed and influenced the dynamics of social life in a broader historical or diachronic perspective. Here I have made a lengthy attempt to interpret mortuary ritual and society in ancient Cyprus, but I would like to stress that despite the size and scope of this work, it remains a preliminary endeavor—a provisional model that new data, analytical methodologies, theories and philosophical perspectives will undoubtedly revise. However, future progress in Cypriot mortuary studies will be limited until all excavators come to regard burial features not merely as repositories of interesting objects, but as invaluable and perilously fragile records of the past, in every aspect of their content and their context. In closing, therefore, I would like to outline briefly a few of the archaeological procedures that are critical to eliciting social and ritual information, and ultimately cultural meanings, from the mortuary record. Some of these procedures are standard in other parts of the world and have begun to be applied in Cyprus; others emerge from the methodology for studying collective burials that was developed in Chapter 3 of this book.

First and foremost, all skeletal remains should be meticulously excavated and mapped, preferably by a qualified physical anthropologist, with careful observation of the articulation, disarticulation, and rearrangements of body parts. Subsequently, all of the bone material (not merely the cranial remains) should be described, quantified, and published in full. Osteological analyses should endeavor to establish not only the age-sex distribution of the individuals buried within the tomb, but also to distinguish relatively complete from anomalously incomplete skeletons, and to identify isolated body parts. In this way it may be possible to establish the frequency of secondary burials, to determine the body parts considered essential for completed secondary burials, and to establish patterns of dismemberment linked to the symbolism of particular body parts. Further analyses of individual health conditions along with patterns of dietary and physical stress are essential to understanding the relationships between social status, nutrition, work and other forms of activity, as is the study of hereditary non-metrical traits to understanding the biological and social relationships among individuals within particular tombs and cemeteries (e.g. Gamble *et al.* 2001).

Secondly, all tomb excavations must be undertaken with greater concern for the formation processes associated with burial deposits—even (and sometimes especially) in cases where tombs appear to have been 'disturbed'. Through geoarchaeological studies of tomb sediments and their stratigraphy, it may be possible to distinguish water-laid deposits from the mixed soil matrices of redeposited burial remains, and in turn to differentiate ritual practices from ancient or modern acts of 'looting', as well as from natural post-depositional disturbances. More detailed analyses of the depositional sequences of burials are also needed to distinguish incremental from 'mass' or collective burials and to differentiate the effects of post-depositional disturbances from the condition of burial remains at the time of deposition.

Thirdly, all faunal and botanical remains from non-intrusive deposits should be recovered, analyzed, and published, so that ideologies of provisioning the dead (and feasting among the living) can begin to be understood. The analysis of organic residues from pottery and other containers, where feasible, is also highly desirable.

Fourthly, the condition of the grave goods—fragmentary versus complete, new versus used etc.—recovered from each tomb is another matter which deserves more attention. Where grave goods are fragmentary or damaged, it is important to ascertain whether this was the result of sustained reuse of the tomb, other post-depositional processes relating to flooding, roof collapse, or looting, or redeposition from another context. The incidence of joins between ceramic vessels from different mortuary features requires particular attention, for this may offer direct proof for the movement of burial remains between various mortuary facilities. Where the grave goods recovered are largely complete, it will be useful to determine whether they had been used for an extended period of time prior to deposition, or whether they were newly made for mortuary use, thus shedding light on whether the objects were more likely personal possessions or gifts to the dead.

Fifthly, it would be extremely valuable to calculate three-dimensional estimates of tomb size at the time of excavation, as the determination of the volume of soil or *havara* that had to be removed when each tomb was cut may make possible the computation of the time and human labor needed for construction.

Sixthly, more intensive investigations of all features associated with mortuary localities are essential to the reconstruction of multi-phase ritual systems involving secondary treatment. Excavators need to pay more attention to features such as 'empty' or 'unused' tombs that may indeed have been used as temporary burial facilities, and to other features that may have been associated with the processing of corpses and/or with the celebrations (feasting, sacrifices, gaming, etc.) accompanying the burial or reburial of the dead. Concomitantly, the occurrence of human bone in non-mortuary contexts such as middens, wells, and other features where some body parts (or even complete individuals, especially infants) may have been discarded should be carefully monitored.

Finally, at the more general level of research design, there are a number of sampling issues that might profitably be addressed through survey and excavation. Extensive, systematic investigation of tombs and other mortuary features at multiple localities within a given region are essential to reconstructing the social and demographic variability in mortuary practice within and between different regions. Even in areas where extensive looting has taken place, the intensive survey of cemetery sites, with detailed mapping and quantification of mortuary features (as in Swiny 1981; Frankel and Webb 1996; Sneddon 2002) may contribute to a more complete understanding of the spatial distribution of cemeteries and tombs relative to settlement sites, and to the scale of the burying groups associated with each settlement. Furthermore, the interpretation of data from tomb excavations must be balanced by a consideration of the evidence from adjacent settlements. Variability in the treatment of the dead, after all, must be understood in the context of variability in the households of the living community.

These are, of course, only a few of the directions that might be taken in future research. There is no shortage of work for those who would challenge or refine the construct of mortuary ritual and social change in Bronze Age Cyprus that has been presented in this book.

# Bibliography

Acsádi, G. and J. Nemeskéri (1970) *History of Human Life Span and Mortality*. Budapest: Akadémiai Kiadó.

Admiraal, S. (1982) Late Bronze Age tombs from Dromolaxia. *Report of the Department of Antiquities of Cyprus*, pp. 39–59.

Amiran, R. (1971) The much-discussed Vounous foreign vessel again. *Report of the Department of Antiquities of Cyprus*, pp. 1–6.

Amiran, R. (1973) More about the Vounous jar—some EBIV antecedents. *Bulletin of the American Schools of Oriental Research* 210:63–6.

Angel, J. L. (1972) Late Bronze Age Cypriotes from Bamboula: the skeletal remains. Appendix B in J. L. Benson, *Bamboula at Kourion*. Philadelphia: University of Pennsylvania Press, pp. 148–58.

Anthony, D.W. (1990) Migrations in archeology: the baby and the bathwater. *American Anthropologist* 92(4):895–914.

Ariès, P. (1974) *Western Attitudes toward Death from the Middle Ages to the Present*. Baltimore: Johns Hopkins Press.

Ariès, P. (1977) *L'homme devant la mort*. Paris: Editions du Seuil.

Åström, L. and P. Åström (1972) *The Swedish Cyprus Expedition*. Vol. IV, Part ID. *The Late Cypriote Bronze Age, Other Arts and Crafts*. Lund: Swedish Cyprus Expedition.

Åström, P. (1957) *The Middle Cypriote Bronze Age*. Lund: Håkan Ohlssons Boktryckeri.

Åström, P. (1960) A Middle Cypriote tomb from Galinoporni. *Opuscula Atheniensa* 3:123–33.

Åström, P. (1962) Supplementary material from Ayios Iakovos Tomb 8. *Opuscula Atheniensa* 4:207–24.

Åström, P. (1966) *Excavations at Kalopsidha and Ayios Iakovos*. Studies in Mediterranean Archaeology 2. Lund. Paul Åström's Förlag.

Åström, P. (1972a) *The Swedish Cyprus Expedition*. Vol. IV, Part IB. *The Middle Cypriote Bronze Age*. Lund: Berlingska Boktryckeriet.

Åström, P. (1972b) *The Swedish Cyprus Expedition*. Vol. IV, Part IC. *The Late Cypriote Bronze Age, Architecture and Pottery*. Lund: Berlingska Boktryckeri.

Åström, P. (1974) Livadhia 'Kokotes' Tomb 1. *Report of the Department of Antiquities of Cyprus*, pp. 51–60.

Åström, P. (1977) The Pera bronzes. In *Scripta Minora 1977–1978 in honorem Einari Gjerstad*. Lund: CWK Gleerup, pp. 5–43.

Åström, P. (1980) Cyprus and Troy. *Opuscula Atheniensa* 13:23–28.

Åström, P. (1982) The bronzes of Hala Sultan Tekke. In *Acta of the International Archaeological Symposium: Early Metallurgy in Cyprus, 4000–500 B.C.*, edited by J. D. Muhly, R. Maddin and V. Karageorghis. Nicosia: Pierides Foundation, pp. 177–84.

Åström, P. (1983) Chamber tombs. In *Hala Sultan Tekke* VIII, edited by P. Åström, E. Åström, A. Hatziantoniou, K. Niklassen and U. Obrink. Studies in Mediterranean Archaeology 45. Göteborg: Paul Åström's Förlag, pp. 145–68.

Åström, P. (1986) Hala Sultan Tekke—an international harbour town of the Late Cypriote Bronze Age. *Opuscula Atheniensa* 16:7–17.

Åström, P. (1987) Intentional destruction of grave goods. In *Thanatos: les coutumes funéraires en Egée à l'âge du Bronze: actes du colloque de Liège, 21-23 avril 1986*, edited by R. Laffineur. Aegaeum I. Liège: Université de l'Etat, Histoire de l'art archéologie de la Grèce antique, pp. 213–17.

Åström, P. (1988) A Cypriote cult scene. *Journal of Prehistoric Religion* 2:5–11.

Åström, P. (1989) *Katydhata. A Bronze Age Site in Cyprus*. Studies in Mediterranean Archaeology 86. Göteborg: Paul Åström's Förlag.

Åström, P. (1996) Hala Sultan Tekke—a Late Cypriote harbour town. In *Late Bronze Age Settlement in Cyprus: Function and Relationship*, edited by P. Åström and E. Herscher. Studies in Mediterranean Archaeology and

Literature, Pocketbook 126. Jonsered: Paul Åström's Förlag, pp. 9–14.

Åström, P. (2000) A coppersmith's workshop at Hala Sultan Tekke. In *Periplus. Festschrift für Hans-Günter Buchholz zu seinem achtzigsten Geburtstag am 24. Dezember 1999*, edited by P. Åström and D. Sürenhagen. Studies in Mediterranean Archaeology and Literature, Pocketbook 127. Jonsered: Paul Åström's Förlag, pp. 33–5.

Åström, P. and G.R. Wright (1962) Two Bronze Age tombs at Dhenia in Cyprus. *Opuscula Atheniensa* 4:225–76.

Bailey, D. M. (1972) The British Museum excavations at Hala Sultan Tekke in 1897 and 1898. The material in the British Museum. In *Hala Sultan Tekke I. Excavations 1897–1971*, edited by P. Åström, D. M. Bailey and V. Karageorghis. Studies in Mediterranean Archaeology 45(1). Göteborg: Paul Åström's Förlag, pp. 1-32.

Balthazar, J. W. (1990) *Copper and Bronze Working in Early through Middle Bronze Age Cyprus*. Studies in Mediterranean Archaeology and Literature, Pocketbook 84. Göteborg: Paul Åström's Förlag.

Barnett, R. D. (1982) *Ancient Ivories in the Middle East and Adjacent Countries*. Jerusalem: Institute of Archaeology, Hebrew University of Jerusalem.

Bartel, B. (1982) A historical review of ethnological and archaeological analyses of mortuary practice. *Journal of Anthropological Archaeology* 1:32–58.

Bass, G. F. (1986) A Bronze Age shipwreck at Ulu Burun (Kas):1984 campaign. *American Journal of Archaeology* 90:269–96.

Bass, W. M. (1981) *Human Osteology: A Laboratory and Field Manual of the Human Skeleton*. Second edition. Columbia, MO: The Missouri Archaeological Society.

Baxevani, E. (1997) From settlement to cemetery burial: the ideology of death in the Early Bronze Age societies of Cyprus and Crete. In *Proceedings of the International Archaeological Conference 'Cyprus and the Aegean in Antiquity'. From the Prehistoric Period to the 7th Century AD Nicosia 8–10 December 1995*, edited by D. Christou. Nicosia: Department of Antiquities, pp. 57–68.

Belgiorno, M. R. (1997) A coppersmith tomb of Early–Middle Bronze Age in Pyrgos (Limassol). *Report of the Department of Antiquities of Cyprus*, pp. 119–46.

Belgiorno, M. R. (1999) Preliminary report on Pyrgos excavations 1996, 1997. *Report of the Department of Antiquities of Cyprus*, pp. 71-86.

Belgiorno, M. R. (2000) Project 'Pyrame' 1998-1999: archaeological, metallurgical and historical evidence at Pyrgos (Limassol). *Report of the Department of Antiquities of Cyprus*: 1–17.

Belgiorno, M. R. (2002) Rescue-excavated tombs of the Early and Middle Bronze Age from Pyrgos (Limassol). Part I. *Report of the Department of Antiquities of Cyprus*: 1–32.

Bendann, E. (1930) *Death Customs: An Analytical Study of Burial Rites*. New York: Knopf.

Benson, J. L. (1972) *Bamboula at Kourion. The Necropolis and the Finds*. Philadelphia: University of Pennsylvania Press.

Benson, J. L. (1973) *The Necropolis at Kaloriziki*. Studies in Mediterranean Archaeology 36. Göteborg: Paul Åström's Förlag.

Biggar, H. P. (ed.) (1929) *The Works of Samuel de Champlain*. Vol. III. Toronto: The Champlain Society, 1922–1936.

Binford, L. R. (1972) Mortuary practices: their study and their potential. In *An Archaeological Perspective*, by L. R. Binford. New York: Seminar Press, pp. 208–43. Originally published in 1971 in *Approaches to the Social Dimensions of Mortuary Practices*, edited by J. A. Brown. Society for American Archaeology Memoir 25, pp. 6–29.

Bloch, M. (1971) *Placing the Dead*. New York: Seminar Press.

Bloch, M. (1982) Death, women and power. In *Death and the Regeneration of Life*, edited by M. Bloch and J. Parry. Cambridge: Cambridge University Press, pp. 211–30.

Bogucki, P. (1993) Animal traction and household economies in Neolithic Europe. *Antiquity* 67:492–503.

Bolger, D. (1983) Khrysiliou-*Ammos*, Nicosia-*Ayia Paraskevi* and the Philia Culture of Cyprus. *Report of the Department of Antiquities of Cyprus*, pp. 60–73.

Bolger, D. (1991) Early Red Polished ware and the Origin of the 'Philia Culture'. In *Cypriot Ceramics: Reading the Prehistoric Record*, edited by J. A. Barlow, D. L. Bolger and B. Kling. University Museum Monograph 74. Philadelphia: University Museum, pp. 29–35.

Bolger, D. (1992) The archaeology of fertility and birth: a ritual deposit from Chalcolithic Cyprus. *Journal of Anthropological Research* 48:145–64.

Bolger, D. (1993) The feminine mystique: gender and society in prehistoric Cypriot studies. *Report of the Department of Antiquities of Cyprus*, pp. 29–41.

Bolger, D. (2002) Gender and mortuary ritual in Chalcolithic Cyprus. In *Engendering Aphrodite: Women and Society in Ancient Cyprus*, edited by D. Bolger and N. Serwint. American Schools of Oriental Research Archaeological Reports, Volume 7. Cyprus American Archaeological Research Institute Monographs, Volume 3. Boston: American Schools of Oriental Research, pp. 67–86.

Boserup, E. (1970) *Women's Role in Economic Development*. London: Allen & Unwin.

Bourdieu, P. (1977) *Outline of a Theory of Practice*. Translated by Richard Nice. Cambridge: Cambridge University Press.

Branigan, K. (1966) Byblite daggers in Cyprus and Crete. *American Journal of Archaeology* 70: 123–26.

Branigan, K. (1967) Further light on prehistoric relations between Crete and Byblos. *American Journal of Archaeology* 71:117–22.

Braun, D. (1981) A critique of some recent North American mortuary studies. *American Antiquity* 46:398–415.

Bright, L. (1995) Approaches to the archaeological study of death with particular reference to ancient Cyprus. In *The Archaeology of Death in the Ancient Near East*, edited by S. Campbell and A. Green. Oxbow Monograph 51. Oxford: Oxbow, pp. 62–74.

Brown, J.A. (ed.) (1971) *Approaches to the Social Dimensions of Mortuary Practices*. Society for American Archaeology Memoir 25.

Brown, J.A. (1981) The search for rank. In *The Archaeology of Death*, edited by R. Chapman, I. Kinnes, and K. Randsborg. New York: Cambridge University Press, pp. 25–37.

Brumfiel, E. M. and T. K. Earle (1987) Specialization, exchange, and complex societies: an introduction. In *Specialization, Exchange, and Complex Societies*, edited by E. M. Brumfiel and T. K. Earle. Cambridge: Cambridge University Press, pp. 1–9.

Buchholz, H. G. (1973) Tamassos, Zypern, 1970–1972. *Archäologischer Anzeiger* 3:295–387.

Buchholz, H. G. (1979) Bronzen Schaftrohräxte aus Tamassos und Ungebung. In *Studies Presented in Memory of Porphyrios Dikaios*, edited by V. Karageorghis *et al*. Nicosia: Lions Club of Nicosia (Cosmopolitan), pp. 76–88.

Buchholz, H. G. and K. Untiedt (1996) *Tamassos: Ein Antikes Königreich auf Zypern*. Studies in Mediterranean Archaeology Pocketbook 136. Jonsered: Paul Åström's Förlag.

Bushnell, D. I. Jr (1920) *Native Cemeteries and Forms of Burial East of the Mississippi*. Bureau of American Ethnology Bulletin 71. Washington, DC: Bureau of American Ethnology.

Buxton, L. H. D. (1920) The anthropology of Cyprus. *Journal of the Royal Anthropological Institute* 50:183–235.

Buxton, L. H. D. (1931) Künstlich deformierte Schädel von Cypern. *Anthropologischer Anzeiger* 7:236–40.

Byrd, B. F. (1994) Public and private, domestic and corporate: the emergence of the southwest Asian village. *American Antiquity* 59(4):639–66.

Byrd, B. F. and C. M. Monahan (1995) Death, mortuary ritual, and Natufian social structure. *Journal of Anthropological Archaeology* 14(3):251–87.

Cadogan, G. (1984) Maroni and the Late Bronze Age of Cyprus. In *Cyprus at the Close of the Late Bronze Age*, edited by V. Karageorghis and J. D. Muhly. Nicosia: Leventis Foundation, pp. 1–10.

Cadogan, G. (1989) Maroni and the monuments. In *Early Society in Cyprus*, edited by E. J. Peltenburg. Edinburgh: Edinburgh University Press, pp. 43–51.

Cadogan, G. (1996) Maroni: change in Late Bronze Age Cyprus. In *Late Bronze Age Settlement in Cyprus: Function and Relationship*, edited by P. Åström and E. Herscher. Studies in Mediterranean Archaeology and Literature, Pocketbook 126. Jonsered: Paul Åström's Förlag, pp. 15–22.

Cannon, A. (1989) The historical dimension in mortuary expressions of status and sentiment. *Current Anthropology* 30(4):437–58.

Carpenter, J. R. (1981) Excavations at Phaneromeni, 1975–1978. In *Studies in Cypriote Archaeology*, edited by J.C. Biers and D. Soren. UCLA Institute of Archaeology, Monograph 18. Los Angeles: UCLA Institute of Archaeology, pp. 59–78.

Carter, E. and A. Parker (1995) Pots, people and the archaeology of death in northern Syria and southern Anatolia in the latter half of the third millennium BC. In *The Archaeology of Death in the Ancient Near East*, edited by S. Campbell and A. Green, Oxbow Monograph 51. Oxford: Oxbow, pp. 96–119.

Cassimatis, H. (1973) Les rites funéraires à Chypre. *Report of the Department of Antiquities of Cyprus*, pp. 116–66.

Catling, H. W. (1956) Bronze cut-and-thrust swords in the eastern Mediterranean. *Proceedings of the Prehistoric Society* 22, pp. 102–25.

Catling, H. W. (1963) Patterns of settlement in Bronze Age Cyprus. *Opuscula Atheniensa* 4:129–69.

Catling, H. W. (1964) *Cypriot Bronzework in the Mycenaean World*. London: Oxford University Press.

Catling, H. W. (1968) Kouklia Evreti Tomb 8. *Bulletin de Correspondance Hellénique* 92:162–69.

Catling, H. W. (1971) Cyprus in the Early Bronze Age. In *The Cambridge Ancient History,* Vol. I, Part II, third edition, edited by I. E. S. Edwards, C. J. Gadd and N. G. L. Hammond. Cambridge: Cambridge University Press, pp. 802–23.

Catling, H. W. (1979) The St. Andrews-Liverpool Museum Kouklia tomb excavations 1950–1954. *Report of the Department of Antiquities of Cyprus*, pp. 270–75.

Catling, H. W. (1984) Workshop and heirloom: prehistoric bronze stands in the eastern Mediterranean. *Report of the Department of Antiquities of Cyprus*, pp. 69–91.

Catling, H. W. and V. Karageorghis (1960) Minoika in Cyprus. *Annual of the British School at Athens* 55:109-27.

Cesnola, L. Palma di (1877) *Cyprus: Its Ancient Cities, Tombs and Temples*. New York: Harper & Brothers.

Chapman, R. (1977) Burial practices: an area of mutual interest. In *Archaeology and Anthropology: Areas of Mutual Interest*, edited by M. Spriggs. British Archaeological Reports, Supplementary Series 19. Oxford: British Archaeological Reports, pp. 19–33.

Chapman, R. (1981a) Archaeological theory and communal burial in prehistoric Europe. In *Pattern of the Past: Studies in Honour of David Clarke*, edited by I. Hodder, G. Isaac and N. Hammond. Cambridge: Cambridge University Press, pp. 387–411.

Chapman, R. (1981b) The emergence of formal disposal areas and the 'problem' of megalithic tombs in prehistoric Europe. In *The Archaeology of Death*, edited by R. Chapman, I. Kinnes and K. Randsborg. New York: Cambridge University Press, pp. 71–82.

Chapman, R. (1995) Ten years after—megaliths, mortuary practices, and the territorial model. In *Regional Approaches to Mortuary Analysis*, edited by L. A. Beck. New York and London: Plenum Press, pp. 29–51.

Chapman, R., I. Kinnes and K. Randsborg (eds) (1981) *The Archaeology of Death*. New York: Cambridge University Press.

Charles, D. K. (1995) Diachronic regional social dynamics: mortuary sites in the Illinois Valley/American Bottom region. In *Regional Approaches to Mortuary Analysis*, edited by L. A. Beck. New York and London: Plenum Press, pp. 77–99.

Charles, R.-P. (1960) Observations sur les crânes de Chrysopolitissa. In V. Karageorghis, Fouilles de Kition 1959, pp. 584–8. *Bulletin de Correspondance Hellénique* 84:504–88.

Charles, R.-P. (1965a) Étude des crânes d'Akhera. In *Nouveaux Documents pour l'Étude du Bronze Récent à Chypre*, by V. Karageorghis, Études Chypriotes 3. Paris: Boccard, pp. 139–49.

Charles, R.-P. (1965b) Note sur les restes humaines de Pendayia. In *Nouveaux Documents pour l'Étude du Bronze Récent à Chypre*, by V. Karageorghis, Études Chypriotes 3. Paris: Boccard, pp. 65-70.

Chayanov, A. V. (1986) *The Theory of Peasant Economy*. Madison, WI: University of Wisconsin Press.

Cherry, J. F. (1985) Islands out of the stream: isolation and interaction in early east Mediterranean insular prehistory. In *Prehistoric Production and Exchange: the Aegean and the Eastern Mediterranean*, edited by A. B. Knapp and T. Stech. UCLA Institute of Archaeology, Monograph 25. Los Angeles: UCLA Institute of Archaeology, pp. 12–29.

Chesson, M. S. (1999) Libraries of the dead: Early Bronze Age charnel houses and social identity at urban Bab edh-Dhra', Jordan. *Journal of Anthropological Archaeology* 18:137–64.

Chesson, M. S. (ed.) (2001) *Social Memory, Identity, and Death: Anthropological Perspectives on Mortuary Rituals*. Archaeological Papers of the American Anthropological Association 10.

Childe, V. G. (1945) Directional changes in funerary practices during 50,000 years. *Man* 45:13–19.

Christou, D. (1989) The Chalcolithic cemetery of Souskiou-*Vathyrkakas*. In *Early Society in Cyprus*, edited by E. J. Peltenburg. Edinburgh: Edinburgh University Press, pp. 82–94.

Clark, J. E. and M. Blake (1994) The power of prestige: competitive generosity and the emergence of rank societies in lowland Mesoamerica. In *Factional Competition and Political Development in the New World*, edited by E. M. Brumfiel and J. W. Fox. Cambridge: Cambridge University Press, pp. 17–30.

Clarke, D. L. (1973) Archaeology: the loss of innocence. *Antiquity* 47:6–18.

Coleman, J. E. (1996) *Alambra: A Middle Bronze Age settlement in Cyprus: archaeological investigations by Cornell University 1975–1985.* Studies in Mediterranean Archaeology 118. Jonsered: Paul Åström's Förlag.

Courtois, J.-C. (1969) Enkomi-Alasia: Glorious capital of Cyprus. *Archaeologia Viva* 2:93–100.

Courtois, J.-C. (1971) Le Sanctuaire du Dieu au Lingot d'Enkomi-Alasia (Chypre). In *Alasia*, Vol. I, edited by C. F. A. Schaeffer. Mission archéologique d'Alasia IV. Paris: Klincksieck, pp. 151–362.

Courtois, J.-C. (1981) *Alasia II. Les Tombes d'Enkomi: Le Mobilier Funéraire.* Paris: Boccard.

Courtois, J.-C. (1982) L'activité métallurgique et les bronzes d'Enkomi au bronze récent (1650–1100 avant J.-C.). In *Acta of the International Archaeological Symposium: Early Metallurgy in Cyprus, 4000–500 BC*, edited by J. D. Muhly, R. Maddin and V. Karageorghis. Nicosia: Pierides Foundation, pp. 155–76.

Courtois, J.-C. (1983) Le trésor de poids de Kalavasos-*Ayios Dhimitrios*. *Report of the Department of Antiquities of Cyprus*, pp. 117–30.

Courtois, J.-C. (1984a) *Alasia III. Les Objets des Niveaux Stratifiés d'Enkomi. Fouilles C.F.A. Schaeffer (1947–1970)*. Éditions Recherche sur les Civilisations, Mémoire 32. Paris: Éditions Recherche sur les Civilisations.

Courtois, J.-C. (1984b) Les poids de Pyla-*Kokkinokremos*. Étude métrologique. Appendix II in *Pyla-Kokkinokremos*, by V. Karageorghis and M. Demas. Nicosia: Department of Antiquities, pp. 80–85.

Courtois, J.-C. (1986a) À propos des apports orientaux dans la civilisation du bronze récent à Chypre. In *Acts of the International Archaeological Symposium, Cyprus between the Orient and the Occident*, edited by V. Karageorghis. Nicosia: Department of Antiquities, pp. 69–87.

Courtois, J.-C. (1986b) Bref historique des recherches archéologiques à Enkomi. In *Enkomi et le bronze récent à Chypre*, by J.-C. Courtois, J. Lagarce and E. Lagarce. Nicosia: Leventis, pp. 1–50.

Courtois, J.-C. and J. M. Webb (1987) *Les cylindres-sceaux d'Enkomi (Fouilles françaises 1957–1970)*. Nicosia: Mission archéologique française d'Alasia.

Croft, P. (1989) Animal bones. In *Vasilikos Valley Project III: Kalavasos-Ayios Dhimitrios II: Ceramics, Objects, Tombs, Specialist Studies*, by A. K. South *et al.*. Studies in Mediterranean Archaeology 71(3). Göteborg: Paul Åström's Förlag, pp. 70–72.

Croft, P. (n.d.) Faunal remains. In *Vasilikos Valley Project IV: Kalavasos-Ayios Dhimitrios III. Tombs 8,9, 11-20*, by A. K. South *et al*. Studies in Mediterranean Archaeology 71(4). Jonsered: Paul Åström's Förlag.

Crone, P. (1989) *Pre-Industrial Societies*. Oxford: Basil Blackwell.

Crouwel, J. H. (1981) *Chariots and Other Means of Land Transport in Bronze Age Greece*. Amsterdam: Pierson.

D'Altroy, T. and T. K. Earle (1985) State finance, wealth finance, and storage in the Inka political economy. *Current Anthropology* 26:187–206.

Danforth, L. M. (1982) *The Death Rituals of Rural Greece*. Princeton: Princeton University Press.

Daniel, J. F. (1937) Two Late Cypriote III tombs from Kourion. *American Journal of Archaeology* 41:51–85.

Davies, P. (1995) Mortuary Patterning and Social Complexity in Prehistoric Bronze Age Cyprus. Unpublished Masters thesis, University of Melbourne.

Davies, P. (1997) Mortuary practices in prehistoric Bronze Age Cyprus. Problems and potential. *Report of the Department of Antiquities of Cyprus*, pp. 11–26.

Decker, L. F. and J. A. Barlow (1983) Study of tombs. In Cornell excavations at Alambra, 1982, by J. E. Coleman *et al.*, pp. 82–89. *Report of the Department of Antiquities of Cyprus*, pp. 76–91.

Dikaios, P. (1940) *The Excavations at Vounous-Bellapais in Cyprus, 1931–1932*. Archaeologia 88. Oxford: Society of Antiquaries of London.

Dikaios, P. (1946) Early Copper Age discoveries in Cyprus:3rd millennium BC coppermining. *Illustrated London News* 2 March.

Dikaios, P. (1960) A conspectus of architecture in ancient Cyprus. *Kypriaki Spoudai*. Volume D, pp. 3–30.

Dikaios, P. (1962) The Stone Age. In *The Swedish Cyprus Expedition,* Vol. IV, Part IA. Lund: The Swedish Cyprus Expedition.

Dikaios, P. (1969–71) *Enkomi 1948–1958*. Vols. I–IIIb. Mainz am Rhein: Verlag Philip von Zabern.

Dillehay, T. D. (1990) Mapuche ceremonial landscape, social recruitment and resource rights. *World Archaeology* 22(2):223–41.

Domurad, M. R. (1986) The Populations of Ancient Cyprus. Unpublished PhD dissertation, University of Cincinnati.

Domurad, M. R. (1989) Whence the first Cypriots? In *Early Society in Cyprus*, edited by E. J. Peltenburg. Edinburgh: Edinburgh University Press, pp. 66–70.

Domurad, M. R. (1996) The human remains from Alambra. In *Alambra: a Middle Bronze Age Settlement in Cyprus: Archaeological Investigations by Cornell University 1975–1985*, by J. E. Coleman. Studies in Mediterranean Archaeology 118. Jonsered: Paul Åström's Förlag, pp. 515–8.

Doran, J. E. and F. R. Hodson (1975) *Mathematics and Computers in Archaeology*. Cambridge, MA: Harvard University Press.

Dossin, G. (1939) Les archives économiques du palais de Mari. *Syria* 20:97–113.

Dossin, G. (1965) Les découvertes épigraphiques de la Xve campagne de fouilles de Mari au printemps de 1965. *Académie des Inscriptions et Belles-Lettres. Comptes Rendus des Séances de l'Année 1965*. Paris: Klincksiek, pp. 400–406.

Dothan, T. K. and A. Ben-Tor (1983) *Excavations at Athienou, Cyprus, 1971–1972*. Jerusalem: Institute of Archaeology, Hebrew University of Jerusalem.

Douglas, W. A. (1969) *Death in Murelaga: Funerary Ritual in a Basque Village*. Seattle: University of Washington Press.

Dugay, Laurinda (1996) Specialized pottery production on Bronze Age Cyprus and pottery use-wear analysis. *Journal of Mediterranean Archaeology* 9(2):167–92.

Dunn-Vaturi, A.-E. (2003) *Vounous: C.F.A. Schaeffer's Excavations in 1933. Tombs 49-79*. Studies in Mediterranean Archaeology 130. Jonsered: Paul Åström's Förlag.

Dussaud, R. (1952) Note préliminaire. Identification d'Enkomi avec Alasia. In *Enkomi-Alasia*, by C. F. A. Schaeffer. Paris: Klincksieck, pp. 1-10.

Earle, T. K. (1977) A reappraisal of redistribution: complex Hawaiian chiefdoms. In *Exchange Systems in Prehistory*, edited by T. K. Earle and J. Ericson. New York: Academic Press, pp. 213–29.

Earle, T. K. (1987) Chiefdoms in archaeological and ethnohistorical perspective. *Annual Review of Anthropology* 16:279–308.

Eriksson, K. (1991) Red Lustrous Wheelmade Ware: a product of Late Bronze Age Cyprus. In *Cypriot Ceramics: Reading the Prehistoric Record*, edited by J. A. Barlow, D. L. Bolger and B. Kling. University Museum Monograph 74. Philadelphia: University Museum, pp. 81–96.

Fischer, P. M. (1986) *Prehistoric Cypriot Skulls*. Studies in Mediterranean Archaeology 75. Göteborg: Paul Åström's Förlag.

Flannery, K. V. (1968) The Olmec and the valley of Oaxaca: a model for interregional interaction in Formative times. In *Dumbarton Oaks Conference on the Olmec*, edited by E. P. Benson. Washington, DC: Dumbarton Oaks, pp. 79–110.

Flannery, K. V. (1972) The origins of the village as a settlement type in Mesoamerica and the Near East: a comparative study. In *Man, Settlement, and Urbanism*, edited by P. J. Ucko, R. Tringham and G.W. Dimbleby. London: Duckworth, pp. 23–53.

Flourentzos, P. (1988) Tomb discoveries from the necropolis of Ayia Paraskevi, Nicosia. *Report of the Department of Antiquities of Cyprus*, pp. 121–5.

Flourentzos, P. (1989) A group of tombs of Middle Bronze Age date from Linou. In *Katydhata. A Bronze Age Site in Cyprus*, by P. Åström. Göteborg: Paul Åström's Förlag, pp. 61–8.

Forman, S. (1980) Descent, alliance, and exchange ideology among the Makassae of East Timor. In *The Flow of Life: Essays on Eastern Indonesia*, edited by J. J. Fox. Cambridge, MA: Harvard University Press, pp. 152–77.

Fortin, M. (1989) La soi-disante forteresse d'Enkomi (Chypre) à la fin du bronze moyen et au début du bronze récent. In *Transition. Le monde égéen du bronze moyen au bronze récent. Actes de la deuxiéme rencontre égéenne internationale de l'Université de Liège (18-20 avril 1988)*, edited by R. Laffineur. Aegaeum 3. Liège: Université de l'État à Liège, pp. 239–48.

Frankel, D. (1974) *Middle Cypriot White Painted Pottery. An Analytical Study of the Decoration*. Studies in Mediterranean Archaeology 42. Göteborg: Paul Åströms Förlag.

Frankel, D. (2002) Social stratification, gender and ethnicity in third millennium Cyprus. In *Engendering Aphrodite: Women and Society in Ancient Cyprus*, edited by D. Bolger and N. Serwint. American Schools of Oriental Research Archaeological Reports, Volume 7. Cyprus American Archaeological Research Institute Monographs, Volume 3. Boston: American Schools of Oriental Research, pp. 171–9.

Frankel, D. and A. Tamvaki (1973) Cypriote shrine models and decorated tombs. *Australian Journal of Biblical Archaeology* 2:39–44.

Frankel, D., J. M. Webb and C. Eslick (1996) Anatolia and Cyprus in the third millennium

BCE. A speculative model of interaction. In *Cultural Interaction in the Ancient Near East. Papers Read at a Symposium Held at the University of Melbourne, Department of Classics and Archaeology (29–30 September 1994)*, edited by G. Bunnens. Abr-Nahrain Supplement 5. Louvain: Peeters Press, pp. 37–50.

Frankel, D. and J. M. Webb (1996) *Marki Alonia. An Early and Middle Bronze Age Town in Cyprus. Excavations 1990–1994*. Studies in Mediterranean Archaeology 123(1). Jonsered: Paul Åström's Förlag.

Frankel, D. and J. M. Webb (1997) Excavations at Marki-*Alonia*, 1996–7. *Report of the Department of Antiquities of Cyprus*, pp. 85–109.

Frankel, D. and J. M. Webb (1999) Excavations at Marki-*Alonia*, 1998–9. *Report of the Department of Antiquities of Cyprus*, pp. 87–110.

Frankel, D. and J.M. Webb (2000a) Marki Alonia: a prehistoric Bronze Age settlement in Cyprus. *Antiquity* 74: 763–5.

Frankel, D. and J.M. Webb (2000b) Excavations at Marki-*Alonia*, 1999–2000. *Report of the Department of Antiquities of Cyprus*: 65–94.

Frankel, D. and J.M. Webb (2001) Excavations at Marki-*Alonia*, 2000. *Report of the Department of Antiquities of Cyprus*: 15–44.

Frankenstein, S. and M. Rowlands (1978) The internal and regional context of early Iron Age society in southwestern Germany. *University of London Institute of Archaeology Bulletin* 15: 73–112.

Frazer, J. G. (1886) On certain burial customs as they illustrate the primitive theory of the soul. *Journal of the Royal Anthropological Institute* 15:64–104.

Freedman, M. (1966) *Chinese Lineage and Society*. London: Athlone.

Fürst, C. M. (1933) *Zur Kenntnis der Anthropologie der Prähistorischen Bevolkerung der Inseln Cypern*. Lunds Universitets Arsskrift, NF, Avd. 2, Bd. 29, no. 6.

Fustel de Coulanges, N. D. (1980) *The Ancient City: a Study on the Religion, Laws and Institutions of Greece and Rome*. Baltimore and London: Johns Hopkins University Press. Originally published in French in 1864 as *La cité antique; étude sur le culte, le droit, les institutions de la Grèce et de Rome*. Paris: Hachette et Cie.

Gale, N., Z. Stos-Gale, and W. Fasnacht (1996) Copper and copper working at Alambra. Appendix 2 in *Alambra: A Middle Bronze Age Settlement in Cyprus: Archaeological Investiga-*

tions by Cornell University 1975–1985, edited by J.E. Coleman *et al.* Studies in Mediterranean Archaeology 118. Jonsered: Paul Astrom's Forlag, pp. 359–426.

Gamble, L. H., P. L. Walker and G. S. Russell (2001) An integrative approach to mortuary analysis: social and symbolic dimensions of Chumash burial practices. *American Antiquity* 66(2):185–212.

Georgiou, G. (2000) An Early Bronze Age tomb at Psematismenos-*Trelloukkas*. *Report of the Department of Antiquities of Cyprus*: 47–63.

Georgiou, G. (2001) Three Early Bronze Age tombs from the hills of the Larnaka District. *Report of the Department of Antiquities of Cyprus*: 49–71.

Georgiou, G. (2002) The necropolis of *Agia Paraskevi* revisited. *Report of the Department of Antiquities of Cyprus*: 49–63.

Georgiou, H. (1979) Relations between Cyprus and the Near East in the Middle and Late Bronze Age. *Levant* 11:84–100.

Giardino, C. (2000) Prehistoric copper activity at Pyrgos. *Report of the Department of Antiquities of Cyprus*: 19–31.

Giardino, C., G. E. Gigante and S. Ridolfi (2002) Archaeometallurgical investigations on the Early–Middle Bronze Age finds from the area of Pyrgos (Limassol). *Report of the Department of Antiquities of Cyprus*: 33–48.

Giddens, A. (1979) *Central Problems in Social Theory*. Berkeley: University of California Press.

Giddens, A. (1984) *The Constitution of Society: Outline of a Theory of Structuration*. Cambridge: Polity Press.

Gilman, A. (1981) The development of social stratification in Bronze Age Europe. *Current Anthropology* 22:1–23.

Gilmour, G. (1995) Aegean influence in Late Bronze Age funerary practices in the southern Levant. In *The Archaeology of Death in the Ancient Near East*, edited by S. Campbell and A. Green. Oxbow Monograph 51. Oxford: Oxbow, pp. 155–70.

Gittlen, B. (1981) The cultural and chronological implications of the Cypro-Palestinian trade during the Late Bronze Age. *Bulletin of the American Schools of Oriental Research* 241:49–59.

Given, M. and A.B. Knapp (2003) *The Sydney Cyprus Survey Project: Social Approaches to Regional Archaeological Survey*. Monumenta Archaeologica 21. Los Angeles: The Cotsen Institute of Archaeology, University of California, Los Angeles.

Gjerstad, E. (1926) *Studies on Prehistoric Cyprus*. Uppsala: Uppsala Universitets Arsskrift.

Gjerstad, E. (1980) *Ages and Days in Cyprus*. Studies in Mediterranean Archaeology Pocketbook 12. Göteborg: Paul Åströms Förlag.

Gjerstad, E., J. Lindros, E. Sjöqvist and A. Westholm (1934) *The Swedish Cyprus Expedition. Finds and Results of the Excavations in Cyprus 1927–1931*. Vol. I. Stockholm: Victor Pettersons Bokindustriaktiebolag.

Goldstein, L. G. (1976) Spatial Structure and Social Organization: Regional Manifestations of Mississippian Society. Unpublished PhD dissertation, Department of Anthropology, Northwestern University.

Goldstein, L. G. (1981) One-dimensional archaeology and multi-dimensional people: spatial organisation and mortuary analysis. In *The Archaeology of Death*, edited by R. W. Chapman, I. Kinnes and K. Randsborg. Cambridge: Cambridge University Press, pp. 53–69.

Goldstein, L. G. (1989) The ritual of secondary disposal of the dead. Paper presented at the 1989 Theoretical Archaeology Group Meetings, Newcastle upon Tyne, Great Britain.

Goldstein, L. G. (1995) Landscapes and mortuary practices: a case for regional perspectives. In *Regional Approaches to Mortuary Analysis*, edited by L. A. Beck. New York and London: Plenum Press, pp. 101–21.

Goodenough, W. H. (1965) Rethinking 'status' and 'role': toward a general model of the cultural organization of social relationships. In *The Relevance of Models for Social Anthropology*, edited by M. Banton. ASA Monographs 1. New York: F. A. Praeger, pp. 1–24.

Goody, J. (1962) *Death, Property, and the Ancestors*. London: Tavistock.

Goody, J. (1976) *Production and Reproduction. A Comparative Study of the Domestic Domain*. Cambridge: Cambridge University Press.

Gophna, R. (1992) The intermediate Bronze Age. In *The Archaeology of Ancient Israel*, edited by A. Ben-Tor. New Haven: Yale University Press, pp. 126–58.

Goren, Y., S. Bunimovitz, I. Finkelstein and N. Na'aman (2003) The location of Alashiya: new evidence from petrographic investigation of Alashiyan tablets from El-Amarna and Ugarit. *American Journal of Archaeology* 107:233–55.

Goring, E. (1988) *A Mischievous Pastime: Digging in Cyprus in the Nineteenth Century*. Edinburgh: National Museums of Scotland.

Goring, E. (1989) Death in everyday life: aspects of burial practice in the Late Bronze Age. In *Early Society in Cyprus*, edited by E. J. Peltenburg. Edinburgh: Edinburgh University Press, pp. 95–105.

Goring, E. (1991) Pottery figurines: the development of coroplastic art in Chalcolithic Cyprus. *Bulletin of the American Schools of Oriental Research* 282–3:153–61.

Grace, V. R. (1940) A Cypriote tomb and Minoan evidence for its date. *American Journal of Archaeology* 44:10–52.

Gudeman, S. (1977) Morgan in Africa. *Reviews in Anthropology* 4:575–80.

Hadjisavvas, S. (1985) A Late Cypriote II tomb from Dhenia. *Report of the Department of Antiquities of Cyprus*, pp. 133–6.

Hadjisavvas, S. (1986) Alassa: a new Late Cypriote site. *Report of the Department of Antiquities of Cyprus*, pp. 62–7.

Hadjisavvas, S. (1989) A Late Cypriot community at Alassa. In *Early Society in Cyprus*, edited by E. J. Peltenburg. Edinburgh: Edinburgh University Press, pp. 32–42.

Hadjisavvas, S. (1991) LCIIC to LCIIIA without intruders. In *Cypriot Ceramics: Reading the Prehistoric Record*, edited by J. A. Barlow, D. L. Bolger and B. Kling. University Museum Monograph 74. Philadelphia: University Museum, pp. 173–80.

Hadjisavvas, S. (1994) Alassa Archaeological Project 1991–1993. *Report of the Department of Antiquities of Cyprus*, pp. 107–14.

Hadjisavvas, S. (1996) Alassa: a regional centre of Alasia? In *Late Bronze Age Settlement in Cyprus: Function and Relationship*, edited by P. Åström and E. Herscher. Studies in Mediterranean Archaeology and Literature, Pocketbook 126. Jonsered: Paul Åström's Förlag, pp. 23–38.

Harrison, T. (1962) Borneo death. *Bijdragen Tot de Taal-, Land-en Volkenkunde* 118:1–41.

Held, S.O. (1992) *Pleistocene Fauna and Holocene Humans: A Gazeteer of Paleontological and Early Archaeological Sites on Cyprus*. Studies in Mediterranean Archaeology 92. Jonsered: Paul Åström's Förlag.

Helms, M.W. (1988) *Ulysses' Sail: An Ethnographic Odyssey of Power, Knowledge, and Geographical Distance*. Princeton: Princeton University Press.

Hendon, J. (2000) Having and holding: storage, memory, knowledge, and social relations. *American Anthropologist* 102(1):42–53.

Hennessy, J. B. (1964) *Stephania: A Middle and Late Bronze Age Cemetery in Cyprus*. London: Quaritch.

Hennessy, J. B., K. O. Eriksson and I. C. Kehrberg (1988) *Ayia Paraskevi and Vasilia. Excavations by J.R.B. Stewart*. Studies in Mediterranean Archaeology 82. Göteborg: Paul Åström's Förlag.

Herscher, E. (1975) New light from Lapithos. In *The Archaeology of Cyprus. Recent Developments*, edited by N. Robertson. Park Ridge, New Jersey: Noyes Press, pp. 39–60.

Herscher, E. (1976) South coast ceramic styles at the end of the Middle Cypriote. *Report of the Department of Antiquities of Cyprus*, pp. 11–19.

Herscher, E. (1978) The Bronze Age Cemetery at Lapithos, *Vrysi tou Barba*, Cyprus. Results of the University of Pennsylvania Museum Excavation, 1931. Unpublished PhD dissertation, University of Pennsylvania.

Herscher, E. (1981) Southern Cyprus, the disappearing Early Bronze Age and the evidence from Phaneromeni. In *Studies in Cypriote Archaeology*, edited by J. C. Biers and D. Soren. UCLA Institute of Archaeology, Monograph 18. Los Angeles: UCLA Institute of Archaeology, pp. 79–85.

Herscher, E. (1984) The pottery of Maroni and regionalism in Late Bronze Age Cyprus. In *Cyprus at the Close of the Late Bronze Age*, edited by V. Karageorghis and J. D. Muhly. Nicosia: Leventis Foundation, pp. 23–8.

Herscher, E. (1988) Kition in the Middle Bronze Age: the tombs at Larnaca-*Ayios Prodromos*. *Report of the Department of Antiquities of Cyprus*:141-66.

Herscher, E. (1991) Beyond regionalism: toward an islandwide Middle Cypriot sequence. In *Cypriot Ceramics: Reading the Prehistoric Record*, edited by J. A. Barlow, D. L. Bolger and B. Kling. University Museum Monograph 74. Philadelphia: University Museum, pp. 45–50.

Herscher, E. (1997) Representational relief on Early and Middle Cypriot pottery. In *Proceedings of the Third International Conference of Cypriote Studies, Nicosia, 3–4 May, 1996*, edited by V. Karageorghis, R. Laffineur, and F. Vandenabeele. Nicosia: A.G. Leventis Foundation, pp. 25–36.

Herscher, E. (1998) Archaeology in Cyprus. *American Journal of Archaeology* 102(2):309–54.

Hertz, R. (1960) *Death and the Right Hand*. Translated by R. and C. Needham. New York: Free Press. Originally published in 1907 as Contribution à une étude sur la representation collective de la mort. *Année sociologique* 10:48–137.

Hickerson, H. (1960) The feast of the dead among seventeenth century Algonkians of the upper Great Lakes. *American Anthropologist* 62:81–105.

Hjörtsjö, C.H. (1946–7) To the knowledge of the prehistoric craniology of Cyprus. *Arsberättelse*. Bulletin de la Societé Royale des Lettres de Lund 1946-7. Lund: CWK Gleerup.

Hodder, I. (1982) Sequences of structural change in the Dutch Neolithic. In *Symbolic and Structural Archaeology*, edited by I. Hodder. Cambridge: Cambridge University Press, pp. 162–77.

Hodder, I. (1984) Burials, houses, women and men in the European Neolithic. In *Ideology, Power and Prehistory*, edited by D. Miller and C. Tilley. Cambridge: Cambridge University Press, pp. 51–68.

Hodder, I. (1990) *The Domestication of Europe*. Oxford: Basil Blackwell.

Holmes, Y. L. (1971) The location of Alashiya. *Journal of the American Oriental Society* 91:426–29.

Holmes, Y. L. (1975) The foreign trade of Cyprus during the Late Bronze Age. In *The Archaeology of Cyprus. Recent Developments*, edited by N. Robertson. Park Ridge, NJ: Noyes Press, pp. 90–110.

Hudson, A. B. (1966) Death ceremonies of the Padju Epat Ma'anyan Dayaks. *Sarawak Museum Journal* 13:341–416.

Hult, G. (1992) *Nitovikla Reconsidered*. Medelhavsmuseet Memoir 8. Stockholm: Medelhavsmuseet.

Huntington, W. R. (1973) Death and the social order: Bara funeral customs (Madagascar). *African Studies* 32(2):65–84.

Ilan, D. (1995) Mortuary practices at Tel Dan in the Middle Bronze Age: a reflection of Canaanite society and ideology. In *The Archaeology of Death in the Ancient Near East*, edited by S. Campbell and A. Green. Oxbow Monograph 51. Oxford: Oxbow, pp. 117–39.

Iliffe, J. H. and T. B. Mitford (1953) An ivory masterpiece and treasures of gold and silver from Cyprus of 3000 years ago. *Illustrated London News* (2 May 1953), pp. 710–11.

Johnson, J. (1980) *Maroni de Chypre*. Studies in Mediterranean Archaeology 59. Göteborg: Paul Åström's Förlag.

Johnstone, W. (1971) A Late Bronze Age tholos tomb at Enkomi. In *Alasia* I, edited by C. F. A. Schaeffer. Paris: Klincksieck, pp. 51–122.

Kan, S. (1989) *Symbolic Immortality: The Tlingit Potlatch of the Nineteenth Century*. Washington, DC: Smithsonian Institution Press.

Karageorghis, J. (1977) *La Grande Déesse de Chypre et Son Culte à Travers l'Iconographie de l'Époque Néolithique*. Lyon: Maison de l'Orient.

Karageorghis, V. (1957) The Mycenaean window crater in the British Museum. *Journal of Hellenic Studies* 77:269–71.

Karageorghis, V. (1958) Finds from Early Cypriot cemeteries. *Report of the Department of Antiquities of Cyprus 1940–1948*:115–52.

Karageorghis, V. (1959) Les personnages en robe sur les vases mycéniens. *Bulletin de Correspondance Hellénique* 83:193–205.

Karageorghis, V. (1960a) Chronique des fouilles et découvertes archéologiques à Chypre en 1959. *Bulletin de Correspondance Hellénique* 84:242–99.

Karageorghis, V. (1960b) Fouilles de Kition 1959. *Bulletin de Correspondance Hellenique* 84:504–88.

Karageorghis, V. (1964a) Chronique des fouilles et découvertes archéologiques à Chypre en 1963. *Bulletin de Correspondance Hellénique* 88:289–379.

Karageorghis, V. (1964b) A Late Cypriote tomb at Angastina. *Report of the Department of Antiquities of Cyprus*, pp. 1–28.

Karageorghis, V. (1965a) A Late Cypriote tomb at Tamassos. *Report of the Department of Antiquities of Cyprus*, pp. 11–29.

Karageorghis, V. (1965b) Fouilles des tombes du chypriote récent à Akhera. In *Nouveaux Documents pour l'Étude du Bronze Récent à Chypre*, by V. Karageorghis. Études Chypriotes 3. Paris: Boccard, pp. 71–138.

Karageorghis, V. (1965c) Une nécropole du chypriote récent I à Pendayia. In *Nouveaux Documents pour l'Étude du Bronze Récent à Chypre*, by V. Karageorghis. Études Chypriotes 3. Paris: Boccard, pp. 14–70.

Karageorghis, V. (1965d) Sur quelques formes de vases particulières à la céramiques chypromycénienne. In *Nouveaux Documents pour l'Étude du Bronze Récent à Chypre*, by V. Karageorghis. Études Chypriotes 3. Paris: Boccard, pp. 201–30.

Karageorghis, V. (1967) An early 11th century tomb from Palaepaphos. *Report of the Department of Antiquities of Cyprus*, pp. 1–24.

Karageorghis, V. (1968) Notes on a Late Cypriote settlement and necropolis site near the Larnaca salt lake. *Report of the Department of Antiquities of Cyprus*, pp. 1–11.

Karageorghis, V. (1970) Chronique des fouilles et découvertes archéologiques à Chypre en 1969. *Bulletin de Correspondance Hellénique* 94:191–300.

Karageorghis, V. (1972a) Chronique des fouilles et découvertes archéologiques à Chypre en 1971. *Bulletin de Correspondance Hellénique* 96:1005–88.

Karageorghis, V. (1972b) Two Late Bronze Age tombs from Hala Sultan Tekke. In *Hala Sultan Tekke I*, by P. Åström *et al*. Studies in Mediterranean Archaeology 45(1). Göteborg: Paul Åström's Förlag.

Karageorghis, V. (1974) *Excavations at Kition I: The Tombs*. Nicosia: Department of Antiquities.

Karageorghis, V. (1976) *Kition: Mycenaean and Phoenician Discoveries in Cyprus*. London: Thames and Hudson.

Karageorghis, V. (1977) Chronique des fouilles et découvertes archéologiques à Chypre en 1976. *Bulletin de Correspondance Hellénique* 101:707–79.

Karageorghis, V. (1979) Kypriaka IV. *Report of the Department of Antiquities of Cyprus*, pp. 198–209.

Karageorghis, V. (1980) Chronique des fouilles et découvertes archéologiques à Chypre en 1979. *Bulletin de Correspondance Hellénique* 104:761–803.

Karageorghis, V. (1983) *Palaepaphos-Skales, An Iron Age Cemetery in Cyprus*. Deutsches Archäologisches Institut Ausgrabungen in Alt-Paphos auf Cypern, Band 3. Konstanz: Universitätsverlag Konstanz.

Karageorghis, V. (1984) New light on Late Bronze Age Cyprus. In *Cyprus at the Close of the Late Bronze Age*, edited by V. Karageorghis and J. D. Muhly. Nicosia: Leventis, pp. 19–22.

Karageorghis, V. (1990a) *Tombs at Palaepaphos. 1. Teratsoudhia. 2. Eliomylia*. Nicosia: Leventis Foundation.

Karageorghis, V. (1990b) *The End of the Late Bronze Age in Cyprus*. Nicosia: Pierides Foundation.

Karageorghis, V. (1991) *The Coroplastic Art of Ancient Cyprus*. Vol. I. *Chalcolithic-Late Cypriote I*. Nicosia: Leventis.

Karageorghis, V. (1999) An Anatolian terracotta bull's head from the Late Cypriote necropolis of Agia Paraskevi. *Report of the Department of Antiquities of Cyprus*, pp. 147–50.

Karageorghis, V. and E. T. Vermeule (1982) *Mycenaean Pictorial Vase Painting*. Cambridge, MA: Harvard University Press.

Karageorghis, V. and M. Demas (1984) *Pyla-Kokkinokremos: A Late 13th Century Fortified Settlement in Cyprus*. Nicosia: Department of Antiquities.

Karageorghis, V. and M. Demas (1985) *Excavations at Kition V: The Pre-Phoenician Levels. Areas I and II*. Nicosia: Department of Antiquities.

Karageorghis, V. and M. Demas (1988) *Excavations at Maa-Palaeokastro, 1979–1986*. Nicosia: Department of Antiquities.

Keesing, R. M. (1982) *Kwaio Religion: The Living and the Dead in a Solomon Island Society*. New York: Columbia University Press.

Kehrberg, I. C. (1995) *Northern Cyprus in the Transition from the Early to the Middle Cypriot Period: Typology, Relative and Absolute Chronology of Some Early Cypriot III to Middle Cypriot I Tombs*. Studies in Mediterranean Archaeology and Literature, Pocketbook 108. Jonsered: Paul Åström's Förlag.

Kenna, M. E. (1976) Houses, fields and graves: property and ritual obligation on a Greek island. *Ethnology* 15:21–34.

Kenna, V. E. G. (1968) The seal use of Cyprus in the Bronze Age. *Bulletin de Correspondance Héllenique* 92:142–56.

Kenna, V. E. G. (1971) *Corpus of Cypriote Antiquities*, Vol. 3. *Catalogue of the Cypriote Seals of the Bronze Age in the British Museum*. Studies in Mediterranean Archaeology 20(3). Göteborg: Paul Åström's Förlag.

Kenna, V. E. G. (1972) Glyptic. In *Swedish Cyprus Expedition*. Vol. IV, Part ID. *The Late Cypriote Bronze Age, Other Arts and Crafts*, by L. and P. Åström. Lund: Swedish Cyprus Expedition.

Kennedy, K. A. R. (1989) Skeletal markers of occupational stress. In *Reconstruction of Life from the Skeleton*, edited by M. Y. Iscan and K. A. R. Kennedy. New York: Alan R. Liss, Inc., pp. 129–60.

Keswani, P. S. (1989a) Mortuary Ritual and Social Hierarchy in Bronze Age Cyprus. Unpublished PhD dissertation, University of Michigan.

Keswani, P. S. (1989b) Dimensions of social hierarchy in Late Bronze Age Cyprus: an analysis of the mortuary data from Enkomi. *Journal of Mediterranean Archaeology* 2(1):49–86.

Keswani, P. S. (1993) Models of local exchange in Late Bronze Age Cyprus. *Bulletin of the American Schools of Oriental Research* 289(4):73–83.

Keswani, P. S. (1994) The social context of animal husbandry in early agricultural societies: ethnographic insights and an archaeological example from Cyprus. *Journal of Anthropological Archaeology* 13(3):255–77.

Keswani, P. S. (1996) Hierarchies, heterarchies, and urbanization processes: the view from Bronze Age Cyprus. *Journal of Mediterranean Archaeology* 9(2):211–49.

Keswani, P. S. and A. B. Knapp (2003) Bronze Age boundaries and social exchange in north-west Cyprus. *Oxford Journal of Archaeology* 22(3): 213–23.

Kiliian, K. (1978) Ausgrabungen in Tiryns 1976. *Archäologischer Anzeiger*:449–98.

Kiliian, K. (1988) Ausgrabungen in Tiryns 1982/3. *Archäologischer Anzeiger*:105–51.

Klengel, H. (1984) Near Eastern trade and the emergence of interaction with Crete in the third millennium B.C. *Studi Micenei ed Egeo-Anatolici* 24:7–19.

Kling, B. (1989) *Mycenaean IIIC:1b and Related Pottery in Cyprus*. Studies in Mediterranean Archaeology 87. Göteborg: Paul Åström's Förlag.

Knapp, A. B. (1979) A Re-Examination of the Interpretation of Cypriote Material Culture in the MCIII-LCI Period in the Light of Textual Data. Unpublished PhD dissertation, University of California at Berkeley.

Knapp, A. B. (1980) KBOI 26: Alasiya and Hatti. *Journal of Cuneiform Studies* 32:43–47.

Knapp, A. B. (1985) Alashiya, Caphtor/Keftiu, and eastern Mediterranean trade: recent studies in Cypriote archaeology and history. *Journal of Field Archaeology* 12:231–50.

Knapp, A. B. (1986a) Production, exchange, and socio-political complexity on Bronze Age Cyprus. *Oxford Journal of Archaeology* 5:35–60.

Knapp, A. B. (1986b) *Copper Production and Divine Protection: Archaeology, Ideology and Social Complexity in Bronze Age Cyprus*. Studies in Mediterranean Archaeology and Literature, Pocketbook 42. Göteborg: Paul Åström's Förlag.

Knapp, A. B. (1988a) Ideology, archaeology, and polity. *Man* (N.S.) 23:133–63.

Knapp, A. B. (1988b) Hoards d'oeuvres: of metals and men on Bronze Age Cyprus. *Oxford Journal of Archaeology* 7(2):147–76.

Knapp, A. B. (1990a) Production, location, and integration in Bronze Age Cyprus. *Current Anthropology* 31(2):147–76.

Knapp, A. B. (1990b) Ethnicity, entrepreneurship, and exchange: Mediterranean inter-island relations in the Late Bronze Age. *Annual of the British School at Athens* 85:115–53.

Knapp, A. B. (1994a) The prehistory of Cyprus: problems and prospects. *Journal of World Prehistory* 8(4):377–453.

Knapp, A. B. (1994b) Emergence, development and decline on Bronze Age Cyprus. In *Development and Decline in the Bronze Age Mediterranean*, edited by C. Mathers and S. Stoddart. Sheffield Archaeological Monographs 8. Sheffield: Sheffield Academic Press, pp. 271–304.

Knapp, A. B. (1996a) Settlement and society on Late Bronze Age Cyprus: dynamics and development. In *Late Bronze Age Settlement in Cyprus: Function and Relationship*, edited by P. Åström and E. Herscher. Studies in Mediterranean Archaeology and Literature, Pocketbook 126. Jonsered: Paul Åström's Förlag, pp. 54–80.

Knapp, A. B. (1996b) (ed.) *Near Eastern and Aegean Texts from the Third to the First Millennia BC.* Vol. II in *Sources for the History of Cyprus*, edited by P. W. Wallace and A. G. Orphanides. Albany, NY: Institute of Cypriot Studies, University at Albany, State of New York.

Knapp, A. B. (1997) *The Archaeology of Late Bronze Age Cypriot Society*. University of Glasgow, Dept of Archaeology, Occasional Paper 4. Glasgow: University of Glasgow, Dept. of Archaeology.

Knapp, A. B. (1999) Reading the sites: prehistoric Bronze Age settlements on Cyprus. *Bulletin of the American Schools of Oriental Research* 313: 75–86.

Knapp, A. B. (2001) Archaeology and ethnicity. A dangerous liason. *Archaeologia Cypria* 4:29–46.

Knapp, A. B., J. D. Muhly, and P. M. Muhly (1988) To hoard is human: the metal deposits of LCIIC-LCIII. *Report of the Department of Antiquities of Cyprus*, pp. 233–62.

Knapp, A. B. and J. F. Cherry (1994) *Provenience Studies and Bronze Age Cyprus. Production, Exchange and Politico-Economic Change.* Monographs in World Archaeology 21. Madison, WI: Prehistory Press.

Knapp, A. B., S. O. Held, I. Johnson and P. S. Keswani (1994) The Sydney Cyprus Survey Project (SCSP)—second preliminary season (1993). *Report of the Department of Antiquities of Cyprus*, pp. 329–43.

Knapp, A. B. and L. M. Meskell (1997) Bodies of evidence on prehistoric Cyprus. *Cambridge Archaeological Journal* 7(2):183–204.

Knapp, A. B., M. Donnelly and V. Kassianidou (1998) Excavations at Politiko-*Phorades* 1997. *Report of the Department of Antiquities of Cyprus*, pp. 247–68.

Knapp, A. B., M. Donnelly and V. Kassianidou (1999) Excavations at Politiko-*Phorades* 1998. *Report of the Department of Antiquities of Cyprus*, pp. 125–46.

Knapp, A. B., V. Kassanidou and M. Donnelly (2002) Excavations at Politiko-*Phorades*: a Bronze Age copper smelting site on Cyprus. *Antiquity* 76:319–20.

Knorr-Cetina, K. D. (1981) Introduction: the microsociological challenge of macro-sociology: towards a reconstruction of social theory and methodology. In *Advances in Social Theory and Methodology: Toward an Integration of Micro- and Macro-Sociologies*, edited by K. Knorr-Cetina and A. V. Cicourel. Boston: Routledge and Kegan Paul, pp. 1–47.

Kopytoff, I. (1987) The internal African frontier: the making of African political culture. In *The African Frontier: The Reproduction of Traditional African Societies*, edited by I. Kopytoff. Bloomington and Indianapolis: Indiana University Press, pp. 3–84.

Kristiansen, K. (1984) Ideology and material culture: an archaeological perspective. In *Marxist Perspectives in Archaeology*, edited by M. Spriggs. Cambridge: Cambridge University Press, pp. 72–100.

Kromholz, S. (1982) *The Bronze Age Necropolis at Ayia Paraskevi*. Studies in Mediterranean Archaeology Pocketbooks 17. Göteborg: Paul Åström's Förlag.

Lagarce, J. and E. Lagarce (1985) *Alasia IV: Deux Tombes du Chypriote Recent d'Enkomi (Tombes 1851 et 1907)*. Éditions Recherche sur les Civilisations. Mémoire 51. Paris: A.D.P.F.

Lagarce, J. and E. Lagarce (1986) Les découvertes d'Enkomi et leur place dans la culture internationale du bronze récent. In *Enkomi et le Bronze Récent à Chypre*, by J.-C. Courtois, J. Lagarce and E. Lagarce. Nicosia: Leventis.

de Laguna, F. (1972) *Under Mount Saint Elias: The History and Culture of the Yakutat Tlingit*. Smithsonian Contributions to Anthropology 7. Washington: Smithsonian Institution Press.

Lane, R. and A. Sublett (1972) Osteology of social organization: residence patterns. *American Antiquity* 37:186–200.

Law, R. (1989) 'My head belongs to the king': on the political and ritual significance of decapitation in pre-colonial Dahomey. *Journal of African History* 30:399–415.

Leach, E. R. (1954) *Political Systems of Highland Burma*. Boston: Beacon Press.

Leonard, A. Jr (1981) Considerations of morphological variation in the Mycenaean pottery from the southeastern Mediterranean. *Bulletin of the American Schools of Oriental Research* 241:87–101.

Leonard, A., Jr. (2000) The Larnaka Hinterland Project. A preliminary report on the 1997 and 1998 seasons. *Report of the Department of Antiquities of Cyprus*: 117–46.

Liverani, M. (1975) Communautés de village et palais dans la Syrie du IIème millénaire. *Journal of the Economic and Social History of the Orient* 18:146–64.

Lukacs, J. R. (1989) Dental paleopathology: methods for reconstructing dietary patterns. In M. Y. Iscan and K. A. R. Kennedy (eds), *Reconstruction of Life from the Skeleton*. New York: Alan R. Liss, Inc., pp. 261–86.

Lunt, D. A. (1985) Report on the human dentitions and discussion of the human dentitions. In *Lemba Archaeological Project* I: *Excavations at Lemba Lakkous, 1976–1983*, by E. J. Peltenburg *et al.* Studies in Mediterranean Archaeology 70(1). Göteborg: Paul Åström's Förlag, pp. 54–8, 150–53, 245–9.

Lunt, D. A. (1994) Report on human dentitions from Souskiou-*Vathyrkakas* 1972. In Excavations at Kouklia (Palaipaphos). Seventeenth preliminary report: seasons 1991 and 1992, by F. G. Maier and M. L. von Wartburg, pp. 120–28. *Report of the Department of Antiquities of Cyprus*, pp. 115–28.

Lunt, D. A. (1995) Lemba-*Lakkous* and Kissonerga-*Mosphilia*: evidence from the dentition in Chalcolithic Cyprus. In *The Archaeology of Death in the Ancient Near East*, edited by S. Campbell and A. Green. Oxbow Monograph 51. Oxford: Oxbow, pp. 56–61.

McClellan, M. C., P. J. Russell and I. A. Todd (1988) Kalavasos-*Mangia*: rescue excavations at a Late Bronze Age cemetery. *Report of the Department of Antiquities of Cyprus*, pp. 201–22.

McFadden, G. H. (1954) A Late Cypriote III tomb from Kourion, Kaloriziki no. 40. *American Journal of Archaeology* 58:131–42.

Macintyre, M. (1989) The triumph of the *susu*. Mortuary exchanges on Tubetube. In *Death Rituals and Life in the Societies of the Kula Ring*, edited by F. H. Damon and R. Wagner. DeKalb: Northern Illinois University Press, pp. 133–52.

Maddin, R., J. D. Muhly, and T. S. Wheeler (1983) Metal working. In *Excavations at Athienou, Cyprus, 1971–1972*, by T. K. Dothan and A. Ben-Tor. Jerusalem: Institute of Archaeology, Hebrew University of Jerusalem, pp. 132–8.

Maier, F. G. (1973) Evidence for Mycenaean settlement at Old Paphos. In *Acts of the International Symposium 'The Mycenaeans in the Eastern Mediterranean'*. Nicosia: Department of Antiquities, pp. 68–78.

Maier, F. G. and M.-L. von Wartburg (1985) Reconstructing history from the earth, c. 2800 BC–1600 AD. Excavating at Palaepaphos, 1966–1984. In *Archaeology in Cyprus 1960–1985*, edited by V. Karageorghis. Nicosia: Leventis, pp. 142–72.

Malbran-Labat, F. (1999) Nouvelles données épigraphiques sur Chypre et Ougarit. *Report of the Department of Antiquities of Cyprus*, pp. 121–3.

Malinowski, B. (1925) Magic, science, and religion. In *Magic, Science, and Religion and Other Essays*. New York: Doubleday, pp. 10–87.

Mallowan, M. (1936) Excavations at Chagar Bazar. *Iraq* 3:55–6.

Manning, S. W. (1993) Prestige, distinction, and competition: the anatomy of socioeconomic complexity in fourth to second millennium BCE. *Bulletin of the American Schools of Oriental Research* 292:35–58.

Manning, S. W. (1998a) Changing pasts and sociopolitical cognition in Late Bronze Age Cyprus. *World Archaeology* 30:39–58.

Manning, S. W. (1998b) *Tsaroukkas*, Mycenaeans and Trade Project: preliminary report on the 1996–1997 seasons. *Report of the Department of Antiquities of Cyprus*, pp. 39–54.

Manning, S. W., D. Collon, D. H. Conwell, H.-G. Jansen, D. Sewell, L. Steel, and A. Swinton (1994) *Tsaroukkas*, Mycenaeans and Trade Project: preliminary report on the 1993 season. *Report of the Department of Antiquities of Cyprus*, pp. 83-106.

Manning, S. W. and S. Swiny (1994) Sotira-*Kaminoudhia* and the chronology of the Early Bronze Age in Cyprus. *Oxford Journal of Archaeology* 13(2):149–72.

Manning, S. W. and F. A. De Mita, Jr (1997) Cyprus, the Aegean, and Maroni-*Tsaroukkas*. In *Proceedings of the International Archaeological Conference 'Cyprus and the Aegean in Antiquity.' From the Prehistoric Period to the 7th Century AD. Nicosia 8–10 December 1995*, edited by

D. Christou. Nicosia: Department of Antiquities, pp. 103–41.

Manning, S. W. and S. J. Monks (1998) Late Cypriot tombs at Maroni *Tsaroukkas*, Cyprus. *Annual of the British School at Athens* 93:297–351.

Manning, S. W., B, Weninger, A. K. South, B. Kling, P. I. Kuniholm, J. D. Muhly, S. Hadjisavvas, D. A. Sewell and G. Cadogan (2001) Absolute age range of the Late Cypriot IIC period on Cyprus. *Antiquity* 75:328–40.

Marcus, G. E. (1988) Contemporary problems of ethnography in the modern world system. In *Writing Culture: The Poetics and Politics of Ethnography*, edited by J. Clifford and G. E. Marcus. Berkeley: University of California Press, pp. 165–93.

Marcus, G. E. and M. M. J. Fischer (1986) *Anthropology as Cultural Critique: An Experimental Moment in the Human Sciences*. Chicago: University of Chicago Press.

Masson, E. (1986) Les écritures chyprominoennes: reflet fidèle du brassage des civilisations pendant le bronze récent. In *Acts of the International Archaeological Symposium, Cyprus between the Orient and the Occident*, edited by V. Karageorghis. Nicosia: Department of Antiquities, pp. 180–200.

Maxwell-Hyslop, K. R. (1971) *Western Asiatic Jewellery c. 3000–612 BC*. London: Methuen.

Megaw, A. H. S. (1957) Archaeology in Cyprus, 1956. *Archaeological Reports, Journal of Hellenic Studies 77 Supplement*, pp. 24–31.

Mellink, M. J. (1991) Anatolian contacts with Chalcolithic Cyprus. *Bulletin of the American Schools of Oriental Research* 282/283:167–75.

Mellink, M. J. (1993) The Anatolian south coast in the Early Bronze Age: the Cilician perspective. In *Between the Rivers and Over the Mountains. Archaeologica Anatolica et Mesopotamica Alba Palmieri Dedicata*, edited by M. Frangipane, H. Hauptmann, M. Liverani, P. Matthiae and M. Mellink. Rome: Dipartimento di Scienze Storiche e Antropologiche dell' Antichità Università di Roma 'La Sapienza', pp. 495–508.

Merbs, C. F. (1989) Trauma. In *Reconstruction of Life from the Skeleton*, edited by M. Y. Iscan and K. A. R. Kennedy. New York: Alan R. Liss, Inc., pp. 161–90.

Merrillees, R. S. (1966) Finds from Kalopsidha Tomb 34. In *Excavations at Kalopsidha and Ayios Iakovos*, by P. Åström. Studies in Mediterranean Archaeology 2. Lund: Paul Åström's Förlag, pp. 31–35.

Merrillees, R. S. (1968) *The Cypriote Bronze Age pottery found in Egypt*. Studies in Mediterranean Archaeology 18. Göteborg: Paul Åström's Förlag.

Merrillees, R. S. (1971) The early history of Late Cypriote I. *Levant* 3:56–79.

Merrillees, R. S. (1974) A Middle Cypriote III tomb from Arpera *Mosphilos*. In *Trade and Transcendance in the Bronze Age Levant*, by R. S. Merrillees. Studies in Mediterranean Archaeology 39. Göteborg: Paul Åström's Förlag, pp. 3–79.

Merrillees, R. S. (1984) Ambelikou-*Aletri*: a preliminary report. *Report of the Department of Antiquities of Cyprus*, pp. 1–13.

Merrillees, R. S. (1986) A 16th Century BC tomb group from central Cyprus with links both east and west. In *Acts of the International Archaeological Symposium: Cyprus between the Orient and the Occident*, edited by V. Karageorghis. Nicosia: Department of Antiquities, pp. 114–48.

Merrillees, R. S. (1988) C.F.A. Schaeffer's excavations at Bellapais-*Vounous* in 1933. *Report of the Department of Antiquities of Cyprus*, part I, pp. 63–9.

Merrillees, R. S. (1992) The absolute chronology of the Bronze Age in Cyprus: a revision. *Bulletin of the American Schools of Oriental Research* 288:47–52.

Merrillees, R. S. (1994) Review of G. Hult, *Nitovikla Reconsidered* (Stockholm 1992). *Opuscula Atheniensa* 20:256–8.

Merrillees, R. S. and J. N. Tubb (1979) A Syro-Cilician jug from Middle Bronze Age Cyprus. *Report of the Department of Antiquities of Cyprus*, pp. 223–9.

Meskell, L. M. (2000) Writing the body in archaeology. In *Reading the Body. Representations and Remains in the Archaeological Record*, edited by A. E. Rautman. Philadelphia: University of Pennsylvania Press, pp. 13–21.

Metcalf, P. (1982) *A Borneo Journey into Death: Berawan Eschatology from its Rituals*. Philadelphia: University of Pennsylvania Press.

Metcalf, P and R. Huntington (1991) *Celebrations of Death: The Anthropology of Mortuary Ritual*. Second edition. Cambridge: Cambridge University Press.

Miles, D. (1965) Socio-economic aspects of secondary burial. *Oceania* 35(3):161–74.

Millard, A. R. (1973) Cypriot copper in Babylonia, c. 1745 BC. *Journal of Cuneiform Studies* 25: 211–14.

Moortgart, A. (1969) *The Art of Ancient Mesopotamia*. London: Phaidon Press.

Morris, D. (1985) *The Art of Ancient Cyprus.* Oxford: Phaidon Press.

Morris, I. (1987) *Burial and Ancient Society: The Rise of the Greek City-State.* Cambridge: Cambridge University Press.

Morris, I. (1991) The archaeology of ancestors: the Saxe/Goldstein hypothesis revisited. *Cambridge Archaeological Journal* 1(2):147–69.

Morris, I. (1992) *Death-Ritual and Social Structure in Classical Antiquity.* Cambridge: Cambridge University Press.

Moyer, C. J. (1997) Human remains from Marki-*Alonia*, Cyprus. *Report of the Department of Antiquities of Cyprus*, pp. 111–18.

Muhly, J. D. (1972) The Land of Alashiya: references to Alashiya in the texts of the second millennium BC and the history of Cyprus in the Late Bronze Age. In *Acts of the First International Cyprological Congress*, edited by V. Karageorghis. Nicosia: Department of Antiquities, pp. 201–19.

Muhly, J. D. (1982) The nature of trade in the Late Bronze Age eastern Mediterranean: the organization of the metals' trade and the role of Cyprus. In *Acta of the International Archaeological Symposium: Early Metallurgy in Cyprus, 4000-500 BC*, edited by J. D. Muhly, R. Maddin and V. Karageorghis. Nicosia: Pierides Foundation, pp. 251–66.

Muhly, J. D. (1985) The Late Bronze Age in Cyprus: a 25 year retrospect. In *Archaeology in Cyprus 1960–1985*, edited by V. Karageorghis. Nicosia: Leventis, pp. 20–46.

Muhly, J. D. (1986) The role of Cyprus in the economy of the eastern Mediterranean during the second millennium BC. In *Acts of the International Archaeological Symposium, Cyprus between the Orient and the Occident*, edited by V. Karageorghis. Nicosia: Department of Antiquities, pp. 45–62.

Muhly, J. D. (1989) The organisation of the copper industry in Late Bronze Age Cyprus. In *Early Society in Cyprus*, edited by E. J. Peltenburg. Edinburgh: Edinburgh University Press, pp. 298–314.

Muhly, J. D. (1991) The development of copper metallurgy in Late Bronze Age Cyprus. In *Bronze Age Trade in the Mediterranean*, edited by N. H. Gale. Studies in Mediterranean Archaeology 90. Göteborg: Paul Åström's Förlag, pp. 180–96.

Muller, J.-C. (1976) Of souls and bones: the living and the dead among the Rukuba, Benue-Plateau State, Nigeria. *Africa* 46(3):258–73.

Murray, A. S., A. H. Smith and H. B. Walters (1900) *Excavations in Cyprus.* British Museum: Department of Greek and Roman Antiquities.

Myres, J. L. (1897) Excavations in Cyprus in 1894. *Journal of Hellenic Studies* 17:134–73.

Myres, J. L. (1940–45) Excavations in Cyprus, 1913: a Bronze Age cemetery at Lapithos. *Annual of the British School at Athens* 41:78-85.

Myres, J. L. and M. Ohnefalsch-Richter (1899) *A Catalogue of the Cyprus Museum.* Oxford: Clarendon Press.

Nakou, G. (1995) The cutting edge: a new look at Aegean metallurgy. *Journal of Mediterranean Archaeology* 8(2):1–32.

Nicolaou, I. and K. Nicolaou (1988) The Dhenia 'Kafkalla' and 'Mali' tombs. *Report of the Department of Antiquities of Cyprus*, pp. 71–120.

Nicolaou, I. and K. Nicolaou (1989) *Kazaphani. A Middle/Late Cypriot Tomb at Kazaphani-Ayios Andronikos: T.2A, B.* Nicosia: Department of Antiquities.

Nicolaou, K. (1972) A Late Cypriote necropolis at Ankastina in the Mesaoria. *Report of the Department of Antiquities of Cyprus*, pp. 58–108.

Nicolaou, K. (1983) A Late Cypriote necropolis at Yeroskipou, Paphos. *Report of the Department of Antiquities of Cyprus*, pp. 142–52.

Niklasson, K. (1983) Tomb 23: a shaft grave of the Late Cypriote III period. In *Hala Sultan Tekke 8*, by P. Åström *et al*. Studies in Mediterranean Archaeology 45(8). Göteborg: Paul Åström's Förlag, pp. 169–213.

Niklasson, K. (1985) The graves and Burial customs. In *Lemba Archaeological Project I: Excavations at Lemba Lakkous, 1976–1983*, by E. J. Peltenburg *et al*. Göteborg: Paul Åström's Förlag, pp. 43–53, 134–49, 241–5.

Niklasson, K. (1991) *Early Prehistoric Burials in Cyprus.* Studies in Mediterranean Archaeology 96. Jonsered: Paul Åström's Förlag.

Niklasson-Sönnerby, K. (1987) Late Cypriote III shaft graves: burial customs of the last phase of the Bronze Age. In *Thanatos: les coutumes funéraires en Egée à l'âge du Bronze: actes du colloque de Liège, 21–23 avril 1986*, edited by R. Laffineur. Aegaeum I. Liège: Université de l'Etat, Histoire de l'art archéologie de la Grèce antique, pp. 219–25.

Nyerges, A. E. (1992) The ecology of wealth-in-people: agriculture, settlement, and society on the perpetual frontier. *American Anthropologist* 94(4):860–81.

Ohnefalsch-Richter, M. H. (1893) *Kypros, the Bible, and Homer.* London: Asher and Co.

Oren, E. (1969) Cypriote imports in the Palestinian Late Bronze I context. *Opuscula Atheniensa* 9:127–50.

Orthmann, W. (1980) Burial customs of the 3rd millennium BC in the Euphrates valley. In *Le Moyen Euphrate. Zone de contacts et d'échanges. Actes du Colloque de Strasbourg 10–12 mars 1977*, edited by J.-Cl. Margueron. Strasbourg: Travaux du Centre de Recherche sur le Proche-Orient et la Grèce Antiques.

Orthmann, W. (1981) *Halawa 1977 bis 1979. Vorläufiger Bericht über die 1. bis 3. Grabungskampagne.* Bonn: Rudolf Habelt Verlag GMBH.

O'Shea, J. M. (1981) Social configurations and the archaeological study of mortuary practices: a case study. In *The Archaeology of Death*, edited by R. Chapman, I. Kinnes and K. Randsborg. New York: Cambridge University Press, pp. 39–52.

O'Shea, J. M. (1984) *Mortuary Variability: An Archaeological Investigation.* New York: Academic Press.

O'Shea, J. M. (1988) Social organization and mortuary behavior in the Late Woodland period in Michigan. *Ohio State University Occasional Papers in Anthropology* 3:68–85.

Overbeck, J. C. and S. Swiny (1972) *Two Cypriot Bronze Age Sites at Kafkallia (Dhali).* Studies in Mediterranean Archaeology 33. Göteborg: Paul Åström's Förlag.

Pader, E.-J. (1982) *Symbolism, Social Relations and the Interpretation of Mortuary Remains.* British Arachaeological Reports International Series 130. Oxford: British Arachaeological Reports.

Palgi, P. and H. Abramovitch. (1984) Death: a cross cultural perspective. *Annual Review of Anthropology* 13:385–417.

Parker Pearson, M. (1982) Mortuary practices, society, and ideology: an ethnoarchaeological study. In *Symbolic and Structural Archaeology*, edited by I. Hodder. Cambridge: Cambridge University Press, pp. 99–113.

Parker Pearson, M. (1984) Economic and ideological change: cyclical growth in the pre-state societies of Jutland. In *Ideology, Power and Prehistory*, edited by D. Miller and C. Tilley. Cambridge: Cambridge University Press, pp. 69–92.

Parker Pearson, M. (1993) The powerful dead: archaeological relationships between the living and the dead. *Cambridge Archaeological Journal* 3(2):203–29.

Parker Pearson, M. (1999) *The Archaeology of Death and Burial.* College Station: Texas A&M University Press.

Pearlman, D. (1985) Kalavasos village Tomb 51: Tomb of an unknown soldier. *Report of the Department of Antiquities of Cyprus*, pp. 164–79.

Pecorella, P. E. (1977) *Le Tombe dell'Età del Bronzo Tardo della Necropoli a Mare di Ayia Irini.* Rome: Consiglio Nazionale delle Ricerche Istituto per gli Studi Micenei ed Egeo-Anatolici.

Peebles, C. S. (1971) Moundville and surrounding sites: some structural considerations of mortuary practices II. In *Approaches to the Social Dimensions of Mortuary Practices*, edited by J. A. Brown. Society for American Archaeology Memoir 25, pp. 68–91.

Peebles, C. S. and S. M. Kus (1977) Some archaeological correlates of ranked societies. *American Antiquity* 42:421–48.

Pelon, O. (1973) Les tholoi d'Enkomi. In *Acts of the International Symposium: 'The Mycenaeans in the Eastern Mediterranean'.* Nicosia: Department of Antiquities, pp. 246–53.

Peltenburg, E. J. (1972) On the classification of faience vases from Late Bronze Age Cyprus. In *Acts of the First International Cyprological Congress*, edited by V. Karageorghis. Nicosia: Department of Antiquities, pp. 129–36.

Peltenburg, E. J. (1974) The glazed vases (including a polychrome rhyton). Appendix I in *Kition I*, by V. Karageorghis. Nicosia: Department of Antiquities, pp. 105–43.

Peltenburg, E. J. (1991a) Local exchange in prehistoric Cyprus: an initial assessment of picrolite. *Bulletin of the American Schools of Oriental Research* 282/283:107–26.

Peltenburg, E. J. (1991b) Kissonerga-*Mosphilia*: A major Chalcolithic site in Cyprus. *Bulletin of the American Schools of Oriental Research* 282/283:17–35.

Peltenburg, E. J. (1991c) *Lemba Archaeological Project 2(2): A Ceremonial Area at Kissonerga.* Studies in Mediterranean Archaeology 70. Göteborg: Paul Åström's Förlag.

Peltenburg, E. J. (1992) Birth pendants in life and death: evidence from Kissonerga Grave 563. In *Kypriakai Spoudai, Studies in honour of Vassos Karageorghis*, edited by G. C. Ioannides. Nicosia: Leventis Foundation, pp. 27–36.

Peltenburg, E. J. (1993) Settlement discontinuity and resistance to complexity in Cyprus, ca. 4500–2500 BC. *Bulletin of the American Schools of Oriental Research* 292:9–23.

Peltenburg, E. J. (1994) Constructing authority: the Vounous enclosure model. *Opuscula Atheniensa* 20(10):157-62.

Peltenburg, E. J. (1996) From isolation to state formation in Cyprus, c. 3500–1500 BC. In *The Development of the Cypriot Economy from the Prehistoric Period to the Present Day*, edited by V. Karageorghis and D. Michaelides. Nicosia: Lithographica, pp. 17–44.

Peltenburg, E. J. *et al.* (1985) *Lemba Archaeological Project* I: *Excavations at Lemba Lakkous, 1976–1983*. Studies in Mediterranean Archaeology 70(1). Göteborg: Paul Åström's Förlag.

Peltenburg, E. J. *et al.* (1998) *Lemba Archaeological Project II(1). Excavations at Kissonerga-Mosphilia, 1979–1992*. Studies in Mediterranean Archaeology 70(2). Partille: Paul Åström's Förlag.

Peltenburg, E. J., S. Colledge, P. Croft, A. Jackson, C. McCartney, and M. A. Murray (2000) Agropastoralist colonization of Cyprus in the 10th millennium BP: initial assessments. *Antiquity* 74:844–53.

Petruso, K.M. (1984) Prolegomena to Late Cypriot weight metrology. *American Journal of Archaeology* 88:293–304.

Philip, G. (1991) Cypriot bronzework in the Levantine world: conservatism, innovation, and social change. *Journal of Mediterranean Archaeology* 4:59–107.

Philip, G. (1995) Warrior burials in the ancient Near-Eastern Bronze Age: the evidence from Mesopotamia, western Iran and Syria-Palestine. In *The Archaeology of Death in the Ancient Near East*, edited by S. Campbell and A. Green. Oxbow Monograph 51. Oxford: Oxbow:140–54.

Pickles, S. and E. J. Peltenburg (1998) Metallurgy, society and the Bronze/Iron transition in the eastern Mediterranean and the Near East. *Report of the Department of Antiquities of Cyprus*, pp. 67–100.

Pieridou, A. (1966) A tomb-group from Lapithos 'Ayia Anastasia'. *Report of the Department of Antiquities of Cyprus*, pp. 1–12.

Poldrugo, F. (2002) Some preliminary considerations on the development of Cypriot diadems and mouthpieces during Late Bronze Age. *Report of the Department of Antiquities of Cyprus*: 89–100.

Pollock, S.M. (1983) The Symbolism of Prestige: An Archaeological Example from the Royal Cemetery of Ur. Unpublished PhD dissertation, University of Michigan.

Porada, E. (1948) Cylinder seals of the Late Cypriote Bronze Age. *American Journal of Archaeology* 52, pp. 173–98.

Porada, E. (1986) Late Cypriote cylinder seals between East and West. In *Acts of the International Archaeological Symposium, Cyprus between the Orient and the Occident*, edited by V. Karageorghis. Nicosia: Department of Antiquities, pp. 289–99.

Porter, A. (2002) The dynamics of death: ancestors, pastoralism, and the origins of a third-millennium city in Syria. *Bulletin of the American Schools of Oriental Research* 325:1–36.

Portugali, Y. and A. B. Knapp (1985) Cyprus and the Aegean: a spatial analysis of interaction in the 17th–14th centuries BC. In *Prehistoric Production and Exchange*, edited by A. B. Knapp and T. Stech. UCLA Institute of Archaeology, Monograph 25. Los Angeles: UCLA Institute of Archaeology, pp. 44-69.

Pottier, E. (1907) Documents céramiques du Musée du Louvre. *Bulletin de Correspondance Hellénique* 31:228–55.

Poursat, J.-C. (1977) *Les ivoires mycéniens: essai sur la formation d'un art mycénien*. Bibliothèque des Écoles françaises d'Athènes et de Rome, fascicule 230. Athens: École française d'Athènes.

Powell, M.L. (1988) *Status and Health in Prehistory*. Washington, DC: Smithsonian Institution Press.

Price, T.D. (ed.) (1989) *The Chemistry of Prehistoric Human Bone*. Cambridge: Cambridge University Press.

Pulak, C. (1988) The Bronze Age shipwreck at Ulu Burun, Turkey: 1985 campaign. *American Journal of Archaeology* 92:1–38.

Pulak, C. (1997) The Uluburun shipwreck. In *Res Maritimae: Cyprus and the Eastern Mediterranean from Prehistory to Late Antiquity. Nicosia, Cyprus, October 1994*, edited by S. Swiny, R. L. Hohlfelder and H. W. Swiny. American Schools of Oriental Research Archaeological Reports 4. Atlanta: American Schools of Oriental Research, pp. 232–62.

Pulak, C. (1998) The Uluburun shipwreck: an overview. *The International Journal of Nautical Archaeology* 27(3): 188–224.

Quilici, L. (1985) La mission italienne à Ayia Irini (Kyrenia). In *Archaeology in Cyprus 1960–1985*, edited by V. Karageorghis. Nicosia: Leventis, pp. 182–92.

Quilici, L. (1990) *La Tomba dell'Età del Bronzo Tardo dall'Abitato di Paleokastro presso Ayia*

*Irini*. Roma: Consiglio nazionale delle ricerche, Istituto per gli studi micenei ed egeo-anatolici, Edizione dell'Ateneo.

Radcliffe-Brown, A. R. (1922) *The Andaman Islanders*. London: Cambridge University Press.

Ramsden, P. G. (1990) Death in winter: changing symbolic patterns in southern Ontario prehistory. *Anthropologica* 32:167–81.

Rathje, W. (1973) Models for the mobile Maya: a variety of constraints. In *The Explanation of Culture Change*, edited by C. Renfrew. London: Duckworth, pp. 731–57.

Rattray, R. S. (1927) *Religion and Art in Ashanti*. London: Oxford University Press.

Rawski, E. S. (1988) A historian's approach to Chinese death ritual. In *Death Ritual in Late Imperial and Modern China*, edited by J. L. Watson and E. S. Rawski. Berkeley and Los Angeles: University of California Press, pp. 20–34.

Reese, D. S. (1995) Equid sacrifices/burials in Greece and Cyprus: an addendum. *Journal of Prehistoric Religion* 9:35–42.

Renfrew, C. (1974) Beyond a subsistence economy: the evolution of social organisation in prehistoric Europe. In *Reconstructing Complex Societies: An Archaeological Colloquium*, edited by C. B. Moore. Bulletin of the American Schools of Oriental Research, Supplement 20, pp. 69–95.

Renfrew, C. (1976) Megaliths, territories, and populations. In *Acculturation and Continuity in Atlantic Europe*, edited by S. de Laef. Brugge: De Tempel, pp. 198–220.

Renfrew, C. (1986) Introduction: peer polity interaction and socio-political change. In *Peer Polity Interaction and Socio-Political Change*, edited by C. Renfrew and J. F. Cherry. Cambridge: Cambridge University Press, pp. 1–18.

Ribeiro, E. C. (2002) Altering the body: representations of pre-pubescent gender groups on Early and Middle Cypriot scenic compositions. In *Engendering Aphrodite: Women and Society in Ancient Cyprus*, edited by D. Bolger and N. Serwint. American Schools of Oriental Research Archaeological Reports, Volume 7. Cyprus American Archaeological Research Institute Monographs, Volume 3. Boston: American Schools of Oriental Research, pp. 197–209.

Rosman, A. and P. G. Rubel (1983) The evolution of exchange structures and ranking: some Northwest Coast and Athapascan examples. *Journal of Anthropological Research* 39(1):1–25.

Rosman, A. and P. G. Rubel (1986) The evolution of central Northwest Coast societies. *Journal of Anthropological Research* 42(4):557–72.

Ross, J. F. (1994) The Vounous jars revisited. *Bulletin of the American Schools of Oriental Research* 296:15–30.

Sahlins, M. (1972) *Stone Age Economics*. New York: Aldine.

Sahlins, M. (1981) *Historical Metaphors and Mythical Realities: Structure in the Early History of the Sandwich Islands Kingdom*. Ann Arbor: University of Michigan Press.

Salles, J-F. (1995) Rituel mortuaire et rituel social à Ras Shamra/Ougarit. In *The Archaeology of Death in the Ancient Near East*, edited by S. Campbell and A. Green. Oxbow Monograph 51. Oxford: Oxbow, pp. 171–84.

Sasson, J. M. (1966) A sketch of North Syrian economic relations in the Middle Bronze Age. *Journal of the Economic and Social History of the Orient* 9:161-81.

Saxe, A. A. (1970) Social Dimensions of Mortuary Practices. Unpublished PhD dissertation, University of Michigan.

Schaar, K. W. (1985) House form at Tarsus, Alambra, and Lemba. *Report of the Department of Antiquities of Cyprus*, pp. 37–44.

Schaeffer, C. F. A. (1936) *Missions en Chypre*. Paris: Geuthner.

Schaeffer, C. F. A. (1939) *Ugaritica*, Vol. I. *Mission de Ras Shamra* 3. Paris: Geuthner.

Schaeffer, C. F. A. (1952) *Enkomi-Alasia (1946–1950)*. Paris: Klincksieck.

Schaeffer, C.F.A. (ed.) (1971) *Alasia*, Vol. I. *Mission archéologique d'Alasia* IV. Paris: Klincksieck.

Schoeninger, M. J. and K. Moore (1992) Bone stable isotope studies in archaeology. *Journal of World Prehistory* 6(2):247–96.

Schulte-Campbell, C. (1983) A Late Bronze Age Cypriote from Hala Sultan Tekke and another discussion of artificial cranial deformation. Appendix V in *Hala Sultan Tekke* VIII, by P. Åström *et al*. Studies in Mediterranean Archaeology 45. Göteborg: Paul Åström's Förlag, pp. 249–52.

Schulte-Campbell, C. (1986) Human skeletal remains. In *Vasilikos Valley Project I: The Bronze Age Cemetery in Kalavasos Village*, edited by I. Todd. Studies in Mediterranean Archaeology 71(1). Göteborg: Paul Åström's Förlag, pp. 168–78.

Schulte-Campbell, C. (2003) The human skeletal remains. In *Sotira Kaminoudhia: An Early Bronze*

*Age Site in Cyprus*, edited by S. Swiny *et al.* American Schools of Oriental Research Archaeological Reports, no. 8. CAARI Monograph Series, vol. 4. Boston: American Schools of Oriental Research, pp. 413–38.

Schurr, M. R. (1992) Isotopic and mortuary variability in a Middle Mississippian population. *American Antiquity* 57(2):300–20.

Schwartz, J. (1974) The human remains from Kition and Hala Sultan Tekke: a cultural interpretation. Appendix IV in *Kition* I, by V. Karageorghis. Nicosia: Department of Antiquities, pp. 151–62.

Shanks, M. and C. Tilley (1982) Ideology, symbolic power and ritual communication: a reinterpretation of Neolithic mortuary practices. In *Symbolic and Structural Archaeology*, edited by I. Hodder. Cambridge: Cambridge University Press, pp. 129–54.

Shaw, J. W. (1984) Excavations at Kommos (Crete) during 1982–1983. *Hesperia* 53:251–87.

Shennan, S. (1975) The social organisation at Branc. *Antiquity* 49:279–88.

Shennan, S. J. (1986) Central Europe in the third millennium BC: a evolutionary trajectory for the beginning of the European Bronze Age. *Journal of Anthropological Archaeology* 5:115–46.

Sherratt, A. G. (1981) Plough and pastoralism: aspects of the secondary products revolution. In *Pattern of the Past: Studies in Hounour of David Clarke*, edited I. Hodder, G. Isaac and N. Hammond. Cambridge: Cambridge University Press, pp. 261–305.

Sherratt, A. G. (1983) The secondary exploitation of animals in the Old World. *World Archaeology* 15:90–104.

Sherratt, A. G. (1987) Cups that cheered. In *Bell Beakers of the Western Mediterranean: Definition, Interpretation, Theory and New Site Data. The Oxford International Conference 1986. Part I*, edited by W. H. Waldren and R. C. Kennard. British Archaeological Reports International Series 331(i). Oxford: British Archaeological Reports, pp. 81–114.

Sherratt, A. G. (1990) The genesis of megaliths: monumentality, ethnicity and social complexity in Neolithic north-west Europe. *World Archaeology* 22(2):147–67.

Silverman, H. and D. B. Small (eds) (2002) *The Space and Place of Death*. Archaeological Papers of the American Anthropological Association 11. Arlington, VA: American Anthropological Association.

Sjögren, K.-G. (1986) Kinship, labor, and land in Neolithic Sweden: social aspects of megalithic graves. *Journal of Anthropological Archaeology* 5(3):229–65.

Sjöqvist, E. (1940) *Problems of the Late Cypriote Bronze Age*. Stockholm: The Swedish Cyprus Expedition.

Smith, J. S. (1994) Seals for Sealing in the Late Cypriot Period. Unpublished PhD dissertation, Bryn Mawr College.

Sneddon, A.C. (2002) *The Cemeteries at Marki: Using a Looted Landscape to Investigate Prehistoric Bronze Age Cyprus*. BAR International Series, No. 1028. Oxford: Archaeopress.

South, A. K. (1984) New light on Late Bronze Age Cyprus. In *Cyprus at the Close of the Late Bronze Age*, edited by V. Karageorghis and J.D. Muhly. Nicosia: Leventis Foundation, pp. 11-18.

South, A. K. (1989) From copper to kingship: aspects of Bronze Age society viewed from the Vasilikos valley. In *Early Society in Cyprus*, edted by E.J. Peltenburg. Edinburgh: Edinburgh University Press, pp. 315–24.

South, A. K. (1996) Kalavasos-*Ayios Dhimitrios* and the Organisation of Late Bronze Age Cyprus. In *Late Bronze Age Settlement in Cyprus: Function and Relationship*, edited by P. Åström and E. Herscher. Studies in Mediterranean Archaeology and Literature, Pocketbook 126. Jonsered: Paul Åström's Förlag, pp. 39–45.

South, A. K. (1997) Kalavasos-*Ayios Dhimitrios* 1992–1996. *Report of the Department of Antiquities of Cyprus*, pp. 151–75.

South, A. K. (2000) Late Bronze Age burials at Kalavasos-*Ayios Dhimitrios*. In *Proceedings of the Third International Congress of Cypriot Studies, Nicosia, 16–20 April 1996*, edited by G. C. Ioannides and S. Hadjistyllou. Nicosia: Leventis Foundation, pp. 345–64.

South, A. K., P. Russell and P. Schuster Keswani (1989) *Vasilikos Valley Project* III: *Kalavasos-Ayios Dhimitrios* II: *Ceramics, Objects, Tombs, Specialist Studies*. Studies in Mediterranean Archaeology 71(3). Göteborg: Paul Åström's Förlag.

South, A. K., E. Goring, C. J. Moyer and P. Russell (n.d.) *Vasilikos Valley Project* IV: *Kalavasos-Ayios Dhimitrios* III. *Tombs 8, 9, 11–20*. Studies in Mediterranean Archaeology 71(4). Jonsered: Paul Åström's Förlag.

Stech, T. (1982) Urban metallurgy in Late Bronze Age Cyprus. In *Acta of the International Archaeological Symposium: Early Metallurgy in*

*Cyprus, 4000–500 BC*, edited by J. D. Muhly, R. Maddin and V. Karageorghis. Nicosia: Pierides Foundation, pp. 105–15.

Steel, L. (1993) The establishment of the city kingdoms in Iron Age Cyprus: an archaeological commentary. *Report of the Department of Antiquities of Cyprus*, pp. 147–56.

Steel, L. (1994) Representations of a shrine on a Mycenaean chariot krater from Kalavasos-*Ayios Dhimitrios*, Cyprus. *Annual of the British School at Athens* 89:201–11.

Steel, L. (1995) Differential burial practices in Cyprus at the beginning of the Iron Age. In *The Archaeology of Death in the Ancient Near East*, edited by S. Campbell and A. Green. Oxbow Monograph 51. Oxford: Oxbow, pp. 199–204.

Steel, L. (1997) Four thousand years of images on Cypriote pottery. In *Proceedings of the Third International Conference of Cypriote Studies, Nicosia, 3–4 May, 1996*, edited by V. Karageorghis, R. Laffineur and F. Vandenabeele. Nicosia: A.G. Leventis Foundation, pp. 37–47.

Steel, L. (1998) The social impact of Mycenaean imported pottery in Cyprus. *Annual of the British School at Athens* 93:285–96.

Steponaitas, V. (1978) Location theory and complex chiefdoms: a Mississippian example. In *Mississippian Settlement Patterns*, edited by B. D. Smith. New York: Academic Press, pp. 417–53.

Steponaitas, V. (1981) Settlement hierarchies and political complexity in nonmarket societies: the Formative period of the valley of Oaxaca. *American Anthropologist* 83:320–63.

Stewart, E. and J. R. Stewart (1950) *Vounous 1937–1938. Field report on the excavations sponsored by the British School of Archaeology at Athens*. Svenska Institutet i Rom Skrifter XIV. Lund: Gleerup.

Stewart, J. R. (1939a) Decorated tomb façades, Cyprus. *Antiquity* 13:461–3.

Stewart, J. R. (1939b) An imported pot from Cyprus. *Palestine Exploration Quarterly* 1939: 162–8.

Stewart, J. R. (1957) The Melbourne Cyprus expedition, 1955. *University of Melbourne Gazette* XIII.1:1–3.

Stewart, J. R. (1962a) The Early Cypriote Bronze Age. In *The Swedish Cyprus Expedition*, Vol. IV, Part IA, by P. Dikaios and J.R. Stewart. Lund: The Swedish Cyprus Expedition, pp. 203–401.

Stewart, J. R. (1962b) The tomb of the seafarer at Karmi in Cyprus. *Opuscula Atheniensa* 4:197–204.

Stiebing, W. H. (1971) Hyksos burials in Palestine: a review of the evidence. *Journal of Near Eastern Studies* 30:110–17.

Strathern, A. J. (1981) Death as exchange: two Melanesian cases. In *Mortality and Immortality*, edited by S. C. Humphreys and H. King London: Academic Press, pp. 205–23.

Strathern, A. J. (1984) *A Line of Power*. London: Tavistock.

Stuart-Macadam, P. L. (1989) Nutritional deficiency diseases: a survey of scurvy, rickets, and iron-deficiency anemia. In *Reconstruction of Life from the Skeleton*, edited by M.Y. Iscan and K.A.R. Kennedy. New York: Alan R. Liss, Inc., pp. 201–23.

Swiny, S. (1981) Bronze Age settlement patterns in southwest Cyprus. *Levant* 13:51–87.

Swiny, S. (1982) Correlations between the composition and function of Bronze Age metal types in Cyprus. In *Acta of the International Archaeological Symposium: Early Metallurgy in Cyprus, 4000–500 BC*, edited by J. D. Muhly, R. Maddin and V. Karageorghis. Nicosia: Pierides Foundation, pp. 69–80.

Swiny, S. (1985) Sotira-*Kaminoudhia* and the Chalcolithic/Early Bronze Age transition in Cyprus. In *Archaeology in Cyprus 1960–1985*, edited by V. Karageorghis. Nicosia, pp. 115–24.

Swiny, S. (1986a) The Philia Culture and its foreign relations. In *Acts of the International Archaeological Symposium: Cyprus between the Orient and the Occident*, edited by V. Karageorghis. Nicosia: Department of Antiquities, pp. 29–44.

Swiny, S. (1986b) *The Kent State University Expedition to Episkopi Phaneromeni*. Studies in Mediterranean Archaeology 74(2). Nicosia: Paul Astrom's Forlag.

Swiny, S. (1989) From round house to duplex: a reassessment of prehistoric Bronze Age society. In *Early Society in Cyprus*, edited by E. J. Peltenburg. Edinburgh: Edinburgh University Press, pp. 14–31.

Swiny, S. (1991) Reading the prehistoric record: a view from the south in the late third millennium BC. In *Cypriot Ceramics: Reading the Prehistoric Record*, edited by J. A. Barlow, D. L. Bolger and B. Kling. University Museum Monograph 74. Philadelphia: University Museum, pp. 37–44.

Swiny, S., G. Rapp and E. Herscher (2003) *Sotira Kaminoudhia: An Early Bronze Age Site in Cyprus*. American Schools of Oriental Research Archaeological Reports, no. 8. CAARI Monograph Series, vol. 4. Boston: American Schools of Oriental Research.

Tainter, J.A. (1973) The social correlates of mortuary patterning at Kaloko, North Kona, Hawaii. *Archaeology and Physical Anthropology in Oceania* 8:1–11.

Tainter, J.A. (1975) Social inference and mortuary practices: an experiment in numerical classification. *World Archaeology* 7:1–15.

Talalay, L. E. and T. Cullen (2002) Sexual ambiguity in plank figures from Bronze Age Cyprus. In *Engendering Aphrodite: Women and Society in Ancient Cyprus*, edited by D. Bolger and N. Serwint. American Schools of Oriental Research Archaeological Reports, Volume 7. Cyprus American Archaeological Research Institute Monographs, Volume 3. Boston: American Schools of Oriental Research, pp. 181–95.

Taylor, J. DuPlat (1952) A Late Bronze Age settlement at Apliki, Cyprus. *The Antiquaries Journal* 32:133–67.

Taylor, J. DuPlat (1957) *Myrtou-Pighades. A Late Bronze Age Sanctuary in Cyprus*. London: Headley Bros.

Thomas, J. (1987) Relations of production and social change in the Neolithic of north-west Europe. *Man* 22:405–30.

Thomas, J. (1988) *Rethinking the Neolithic*. Cambridge: Cambridge University Press.

Thwaites, R. G. (ed.) (1896–1901) *The Jesuit Relations and Allied Documents: Travels and Explorations of the Jesuit Missionaries in New France 1610–1791*. Vol. 10. Cleveland: Burrows Bros. Company.

Tilley, C. (1984) Ideology and the legitimation of power in the Middle Neolithic of southern Sweden. In *Ideology, Power and Prehistory*, edited by D. Miller and C. Tilley. Cambridge: Cambridge University Press, pp. 111–46.

Todd, I. A. (1985) A Middle Bronze Age tomb at Psematismenos-*Trelloukas*. *Report of the Department of Antiquities of Cyprus*, pp. 55–77.

Todd, I. A. (ed.) (1986) *Vasilikos Valley Project I: The Bronze Age Cemetery in Kalavasos Village*. Studies in Mediterranean Archaeology 71(1). Göteborg: Paul Åström's Förlag.

Todd, I. A. (1988) The Middle Bronze Age in the Kalavasos area. *Report of the Department of Antiquities of Cyprus*, pp. 133–40.

Todd, I. A. (1993) Kalavasos-*Laroumena*: test excavation of a Middle Bronze Age settlement. *Report of the Department of Antiquities of Cyprus*, pp. 81–96.

Todd, I. A. and A. K. South (1992) The Late Bronze Age in the Vasilikos Valley: recent research. In

*Kypriakai Spoudai, Studies in honour of Vassos Karageorghis*, edited by G.C. Ioannides. Nicosia: Leventis Foundation, pp. 192-204.

Toumazou, M. K. (1987) Aspects of Burial Practices in Early Prehistoric Sites, c. 7000–2500/2300 BC. Unpublished PhD dissertation, Bryn Mawr College.

Traube, E. G. (1980) Affines and the dead: Mambai rituals of alliance. *Bijdragen tot de Taal-Land-en Volkenkunde* 136:90–115.

Trigger, B.G. (1990) Monumental architecture: a thermodynamic explanation of symbolic behaviour. *World Archaeology* 22(2):119–32.

Trinkaus, K. M. (1995) Mortuary behavior, labor organization, and social rank. In *Regional Approaches to Mortuary Analysis*, edited by L. A. Beck. New York and London: Plenum Press, pp. 53–75.

Tylor, E. B. (1979) Animism. In *Reader in Comparative Religion*, edited by W.A. Lessa and E. Z. Vogt. New York: Harper and Row, pp. 9–19. Originally published in 1873 in E. B. Tylor, *Primitive Culture*, 2nd edn. London: John Murray.

Vagnetti, L. and F. LoSchiavo (1989) Late Bronze Age long distance trade in the Mediterranean: the role of Cyprus. In *Early Society in Cyprus*, edited by E. J. Peltenburg. Edinburgh: Edinburgh University Press, pp. 217–43.

Valentine, C.A. (1965) The Lalakai of New Britain. In *Gods, Ghosts and Men in Melanesia*, edited by M. Meggitt. Melbourne: Oxford University Press, pp. 162–97.

Van Gennep, A. (1960) *The Rites of Passage*. Translated by M. B. Vizedom and G. L. Caffee. Chicago: University of Chicago Press. Originally published in 1909.

Vercoutter, J. (1959) The gold of Kush. *Kush* 7: 120–53.

Vermeule, E. T. (1974) *Toumba tou Skourou: The Mound of Darkness. A Bronze Age Town on Morphou Bay in Cyprus*. The Harvard University Cyprus Archaeological Expedition and the Museum of Fine Arts, Boston 1971–1974. Cambridge and Boston, MA: Harvard University and the Museum of Fine Arts.

Vermeule, E. T. (1996) Toumba tou Skourou. In *Late Bronze Age Settlement in Cyprus: Function and Relationship*, edited by P. Åström and E. Herscher. Studies in Mediterranean Archaeology and Literature, Pocketbook 126. Jonsered: Paul Åström's Förlag, pp. 50–53.

Vermeule, E. T. and F. Z. Wolsky (1977) The bone and ivory of Toumba tou Skourou. *Report of the Department of Antiquities of Cyprus*, pp. 80–96.

Vermeule, E. T. and F. Z. Wolsky (1978) New Aegean relations with Cyprus: the Minoan and Mycenaean pottery from Toumba tou Skourou, Morphou. *Proceedings of the American Philosophical Society* 122(5):294–317.

Vermeule, E. T. and F. Z. Wolsky (1990) *Toumb tou Skourou: A Bronze Age Potter's Quarter on Morphou Bay in Cyprus*. Cambridge, MA: Harvard University Press.

Waldron, T. (1987) The relative survival of the human skeleton: implications for palaeopathology. In *Death, Decay and Reconstruction: Approaches to Archaeology and Forensic Science*, edited by A. Boddington, A. N. Garland and R. C. Janeway. Manchester: Manchester University Press, pp. 55–64.

Walter, T. (1994) *The Revival of Death*. London: Routledge.

Walters, H. B. (1912) *A Catalogue of the Greek and Etruscan Vases in the British Museum*. Vol. 1.1–1.2. *Cypriote, Italian, and Etruscan Pottery*. London: The British Museum.

Watkins, T. (1981) The Chalcolithic period in Cyprus: the background to current research. In *Chalcolithic Cyprus and Western Asia*, edited by J. Reade. British Museum Occasional Publication 26, pp. 9–20. London: British Museum.

Watson, R. S. (1988) Remembering the dead: graves and politics in southeastern China. In *Death Ritual in Late Imperial and Modern China*, by J. L. Watson and E. S. Rawski pp. 203–27. Berkeley and Los Angeles: University of California Press.

Webb, J. M. (1992) Funerary ideology in Bronze Age Cyprus: toward the recognition and analysis of Cypriote ritual data. In *Kypriakai Spoudai, Studies in honour of Vassos Karageorghis*, edited by G. C. Ioannides. Nicosia: Leventis Foundation, pp. 87–99.

Webb, J. M. (1995) Abandonment processes and curate/discard strategies at Marki *Alonia*, Cyprus. *The Artefact* 18:64–70.

Webb, J. M. (1999) *Ritual Architecture, Iconography and Practice in the Late Cypriot Bronze Age*. Jonsered: Paul Åström's Förlag.

Webb, J. M. (2002) Device, image, and coercion: the role of glyptic in the political economy of Late Bronze Age Cyprus. In *Script and Seal Use on Cyprus in the Bronze and Iron Ages*, edited by J. S. Smith. Boston: Archaeological Institute of America. Colloquia and Conference Papers 4, pp. 111–54.

Webb, J. M. and D. Frankel (1994) Making an impression: storage and surplus finance in Late Bronze Age Cyprus. *Journal of Mediterranean Archaeology* 7(1):5-26.

Webb, J. M. and D. Frankel (1999) Characterizing the Philia facies: material culture, chronology, and the origin of the Bronze Age in Cyprus. *American Journal of Archaeology* 103:3–43.

Webb, J. M. and D. Frankel (2001) *Eight Middle Bronze Age Tomb Groups from Dhenia in the University of New England Museum of Antiquities*. Studies in Mediterranean Archaeology 20(21). Jonsered: Paul Åström's Förlag.

Weinberg, S. (1956) Exploring the Early Bronze Age in Cyprus. *Archaeology* 9:112–21.

Weinberg, S. (1983) *Bamboula at Kourion: The Architecture*. University Museum Monograph 42 (University of Pennsylvania). Philadelphia: the University Museum.

Weiner, A. B. (1976) *Women of Value, Men of Renown: New Perspectives in Trobriand Exchange*. Austin: University of Texas Press.

Weiner, A. B. (1980) Reproduction: a replacement for reciprocity. *American Ethnologist* 7:71–85.

Weiss, H., M.-A. Courty, W. Wetterstrom, F. Guichard, L. Senior, R. Meadow, and A. Curnow (1993) The genesis and collapse of third millennium north Mesopotamian civilization. *Science* 261:995–1004.

Weiss, K. K. (1972) On the systematic bias in skeletal sexing. *American Journal of Physical Anthropology* 37:239–50.

Westholm, A. (1939a) Some Late Cypriote tombs at Milia. *Quarterly of the Department of Antiquities of Palestine* 8:1–20.

Westholm, A. (1939b) Built tombs in Cyprus. *Opuscula Archaeologica* 2:29–58.

Wheatley, P. (1975) Satyānrta in suvarnadvīpa. From reciprocity to redistribution in ancient southeast Asia. In *Ancient Civilizations and Trade*, edited by J. A. Sabloff and C. C. Lamberg-Karlovsky. Albuquerque: University of New Mexico Press, pp. 227–83.

Wheeler, T. S. (1974) Early Bronze Age burial customs in western Anatolia. *American Journal of Archaeology* 78:415–25.

White, D. (1986 ) 1985 Excavations on Bate's Island, Marsa Matruh. *Journal of the American Research Center in Egypt* 23:51–84.

Whitley, J. (2002) Too many ancestors. *Antiquity* 76:119–26.

Wilkinson, L. (1990) *SYGRAPH: The System for Graphics*. Evanston, Illinois: SYSTAT, Inc.

Williamson, J. W. and P. Christian (n.d.) Notes from Excavations in Cyprus Conducted by Messrs. Williamson and Christian on Behalf of the British Museum. Ms on file, Dept of Greek and Roman Antiquities, British Museum, London.

Witzel, N. (1979) Finds from the area of Dromolaxia. *Report of the Department of Antiquities of Cyprus*, pp. 181–97.

Wobst, H. M. (1974) Boundary conditions for Paleolithic social systems. *American Antiquity* 39:47–78.

Woodburn, J. (1982) Social dimensions of death in four African hunting and gathering societies. In *Death and the Regeneration of Life*, edited by M. Bloch and J. Parry. Cambridge: Cambridge University Press, pp. 187–210.

Wright, G. A. (1978) Social differentiation in the early Natufian. In *Social Archaeology: Beyond Subsistence and Dating*, edited by C. L. Redman *et al*. New York: Academic Press, pp. 201–23.

Wright, G. E. (1940) The Syro-Palestinian jar from Vounous, Cyprus. *Palestine Exploration Quarterly* 1940:154–7.

Wright, H. T. (1984) Pre-state political formations. In *On the Evolution of Complex Societies: Essays in Honor of Harry Hojier 1982*, edited by T. K. Earle. Malibu: Undena, pp. 41–77. Reprinted in 1994 in, *Chiefdoms and Early States in the Near East: The Organizational Dynamics of Complexity*, edited by G. Stein and M. Rothman, pp. 67–84. Madison, WI: Prehistory Press.

Wrong, G. M. (ed.) (1939) *The Long Journey to the Country of the Hurons*, by Father Gabriel Sagard. Translated by H. H. Langton. Toronto: The Champlain Society.

Yon, M. (1999) Chypre et Ougarit à la fin du bronze récent. *Report of the Department of Antiquities of Cyprus*, pp. 113–9.

Zaccagnini, C. (1987) Aspects of ceremonial exchange in the Near East during the late second millennium BC. In *Centre and Periphery in the Ancient World*, edited by M. Rowlands, M. Larsen and K. Kristiansen. Cambridge: Cambridge University Press, pp. 57–65.

# Tables

Table 1.1  Chronology of the Chalcolithic and Bronze periods in Cyprus. Absolute dates and terminology are synthesized from Knapp (1994a: Fig. 1), Webb and Frankel (1999:5), Merrillees (1992: Table 2) and Karageorghis (1990b); see also Manning and Swiny (1994). A recent review of 58 radiocarbon samples suggests that the absolute age range for LCIIC should be revised upward slightly, beginning between 1340–1315 BC and terminating *c.*1200 BC +20/–10 (Manning *et al.* 2001).

| Traditional Terminology | Years BC | Alternative Designation |
|---|---|---|
| Chalcolithic | 3900/3700–2400 | Erimi Culture |
| Philia Phase | 2500–2350 | preBA I |
| Early Cypriot I | ?2300–2150 | preBA I |
| Early Cypriot II | 2150–2100 | preBA I |
| Early Cypriot IIIA | 2100–2025 | preBA I |
| Early Cypriot IIIB | 2025–1950 | preBA II |
| Middle Cypriot I | 1950–1850 | preBA II |
| Middle Cypriot II | 1850–1750 | preBA II |
| Middle Cypriot III | 1750–1650 | proBA I |
| Late Cypriot IA | 1650–1550 | proBA I |
| Late Cypriot IB | 1550–1450 | proBA I |
| Late Cypriot IIA | 1450–1375 | proBA II |
| Late Cypriot IIB | 1375–1300 | proBA II |
| Late Cypriot IIC | 1300–1200/1190 | proBA II |
| Late Cypriot IIIA | 1200/1190–1125/1100 | proBA III |
| Late Cypriot IIIB | 1125/1100–1050 | proBA III |

**Table 3.1** An overview of Philia and Early–Middle Cypriot cemetery investigations. For site locations see Figure 3.1.

| Tomb Locations | Sample Size/Evaluation | Major Publications |
|---|---|---|
| Alambra | 82 tombs of uncertain dates excavated by Cesnola, 16 more by Ohnefalsch-Richter, all minimally recorded. 6 more pit and chamber tombs excavated by Coleman, only 2 intact; 7 burials carefully reported. | Cesnola 1877; Ohnefalsch-Richter 1893; Myres and Ohnefalsch-Richter 1899; Coleman 1996 |
| Arpera *Mosphilos* | Approximately 10 possibly intact but scantily recorded tombs excavated by Markides in 1914; other investigated more recently by Dept of Antiquities. | Gjerstad 1926; Merrillees 1974; Leonard 2000 |
| Ayios Iakovos *Melia* | 14 numbered tomb complexes dating from MC–LC periods investigated by the SCE, 2 tombs with intact, well-reported MC burials (Tombs 6 and 7) and 4 others with informative MC burial deposits (Tombs 1, 4, 12 and 13). | Gjerstad *et al.* 1934; Åström 1962 |
| Bellapais *Vounous* Sites A and B | 164 numbered tomb complexes excavated, some multi-chambered. 107+ chambers appear intact; more than 200 burials alluded to or described. | Schaeffer 1936; Dikaios 1940; Stewart and Stewart 1950; Merrillees 1988; Dunn-Vaturi 2003 |
| Dhali *Kafkallia* | One looted tomb of MCIII–LCIII date excavated, notable for the presence of a bronze shafthole axe and 'warrior' belt seemingly associated with the earliest use of the tomb. | Overbeck and Swiny 1972 |
| Dhenia *Kafkalla* and *Mali* | Extensive cemetery with many looted tombs; 3 large chamber tombs with well-reported finds, presumably used for multiple burials (*Kafkalla* Tombs 6, 48 and G.W.1); several disturbed pit tombs possibly representing exhumed primary burials. | Åström and Wright 1962; Nicolaou and Nicolaou 1988; Webb and Frankel 2001 |
| Episkopi *Phaneromeni* | 12 multi-chambered tomb complexes excavated by Weinberg, some chambers possibly intact but burials and finds incompletely reported. Other tombs excavated by Carpenter remain unpublished. | Weinberg 1956; Carpenter 1981; Herscher 1981 |
| Galinoporni | 3 tombs excavated, 2 by the Dept of Antiquities, briefly reported, and another excavated by the SCE, more extensively described, possibly intact. | Megaw 1957; Åström 1960 |
| Kalavasos | 61 tombs reported as of 1986, 23 more excavated in unpublished rescue excavations, many damaged by construction. Burial data from tombs excavated by Dept of Antiquities and the VVP is sufficiently detailed to suggest secondary treatment. | Karageorghis 1958; Todd 1986; Todd 1988 |
| Kalopsidha | 31 tombs excavated by Myres; little or no reference to skeletal material except for the 'thick layer of bones' in Tomb 28. Also an early ('ECIC') pit grave with a single contracted burial described by Merrillees. | Myres 1897; Åström 1966; Merrillees 1966 |
| Karmi *Palealona* | 13 numbered single and multi-chambered tomb complexes investigated by Stewart. One intact chamber with an intact burial reported in detail. | Stewart 1962b; Kehrberg 1995 |
| Katydhata | Markides excavated approximately 100 tombs including 32 of MC date. Information about the burials and condition of the tombs is scanty. 4 other MC tombs have been excavated since 1965 at nearby Linou *Alonia* and *Ayii Saranta*. | Gjerstad 1926; Åström 1989; Flourentzos 1989 |
| Kition | 3 tombs excavated at *Chrysopolitissa*, 8 at *Kathari*, 6 at *Agios Prodromos*; most badly disturbed with few skeletal remains preserved. | Karageorghis 1974; Herscher 1988 |

| Tomb Locations | Sample Size/Evaluation | Major Publications |
|---|---|---|
| Korovia *Paleoskoutella* | 7 tumulus covered mortuary complexes excavated: 2 elaborate and intact chamber tombs with multiple burials, 2 possibly emptied primary burial chamber tombs, 3 sets of pits and other features presumably associated with mortuary processing and rituals. | Gjerstad *et al.* 1934 |
| Kyra *Kaminia* | One Philia phase pit tomb excavated by Dikaios; burials minimally recorded. | Dikaios 1962; see also Toumazou 1987 |
| Lapithos *Vrysi tou Barba* | 75 scantily reported tombs excavated by Myres and Markides. 61 numbered multi-chambered tomb complexes excavated by the Stewarts and the SCE, with *c.* 96 chambers intact or informative. 253+ reported burials with burial information on *c.* 114 individuals. | Gjerstad 1926; Myres 1940–1945; Gjerstad *et al.* 1934; Grace 1940; Herscher 1975; 1978 |
| Limassol *Ayios Nikolaos* | 4 tombs excavated by the Dept. of Antiquities; condition uncertain; little or no burial information. | Karageorghis 1958 |
| Marki | Over 700 tombs in 5 cemeteries identified, all looted. One intact burial and fragments of several other skeletons from settlement contexts at *Alonia*. | Frankel and Webb 1996; 1997; 1999; 2000a; 2000b; Moyer 1997; Sneddon 2002 |
| Maroni and Psematismenos | One scantily preserved tomb excavated by the Cyprus Survey in Maroni Village in 1966. Four tombs opened at Psematismenos *Trelloukkas* by Ohnefalsch-Richter. One intact tomb now dated to the Early Cypriot period excavated at Trelloukkas in 1982 and reported in detail by Todd 1985. Another partially bulldozed EC pit? tomb excavated at *Trelloukkas* by the Dept of Antiquities in 1999. Two more partially bulldozed pit? tombs, one datable to the EC, excavated by the Dept of Antiquities in 2000 at Maroni *Maraes*. Some MCIII pottery in Maroni Kapsaloudhia Tomb 2 excavated by Cadogan. | Ohnefalsch-Richter 1893; Myres and Ohnefalsch-Richter 1899; Johnson 1980; Herscher 1984; Cadogan 1984; Todd 1985; Georgiou 2000; 2001 |
| Nicosia *Ayia Paraskevi* | Many tombs opened by Cesnola, Ohnefalsch-Richter, Myres, Stewart, and the Dept of Antiquities at two burial locations: Cemetery A near the Hilton and Cemetery B at *Dhasilion Sergides*, mostly looted and/or damaged by recent construction. Burial remains from Stewart's Tomb 13 possibly represent collective secondary treatment. | Cesnola 1877; Ohnefalsch-Richter 1893; Myres and Ohnefalsch-Richter 1899; Myres 1897; Merrillees and Tubb 1979; Kromholz 1982; Hennessy *et al.* 1988; Flourentzos 1988; Georgiou 2002 |
| Philia *Vasiliko/ Laxia tou Kasinou* | Five Philia phase tombs of chamber and possibly pit type excavated by Dikaios; burials minimally recorded; some disturbance. | Dikaios 1962; see also Toumazou 1987 |
| Politiko | *c.* 50 tombs opened by Ohnefalsch-Richter, scantily recorded, but indications of possible collective secondary treatment in *Lambertis* Tomb 18; comparatively detailed enumeration of metal finds from *Chomazoudhia* Tomb 3. | Gjerstad 1926; Buchholz 1973; Buchholz and Untiedt 1996 |
| Pyrgos | Large EC–MC cemetery near settlement at *Mavrorachi*. 19 tombs of EC–MC date investigated by the Dept. of Antiquities 1941–1993; one tomb with a seemingly intact burial recently excavated. | Belgiorno 1997; 2002 |
| Sotira *Kaminoudhia* | 21 rock-cut tombs and cists excavated, some badly eroded, 11 with skeletal remains including 12 adults and one child. | Swiny *et al.* 2003 |
| Vasilia *Kafkallia* | 5 tombs excavated by Stewart at *Kafkallia*, only one not 'totally devastated', no burial data; reports of a cemetery of pit graves nearby. | Stewart 1962a; Hennessy *et al.* 1988 |

**Table 3.2**    An overview of Late Cypriot cemetery investigations. For site locations see Figure 3.2.

| Tomb Locations | Sample Size/Evaluation | Major Publications |
|---|---|---|
| Akaki *Trounalli* | One intact chamber tomb with 4 burials and several other tombs of indeterminate condition excavated by the Dept of Antiquities in rescue operations. | Karageorghis 1960a; 1970 |
| Akhera *Chiflik Paradisi* | 3 intact pit tombs excavated by the Dept of Antiquities. Some osteological analysis. | Karageorghis 1965b |
| Alassa *Pano Mandilaris* | 8 tombs excavated; 3 intact. | Hadjisavvas 1986; 1989; 1991 |
| Angastina *Vounos* | 4 bulldozed but partially preserved chamber tombs excavated by the Dept of Antiquities, plus one pit possibly representing an emptied mortuary feature. | Karageorghis 1964b; Nicolaou 1972 |
| Ayia Irini | 8 looted chamber tombs, some with substantially preserved finds and burials excavated at *Paleokastro;* another partly disturbed tomb with *c.* 37 burials further inland; all investigated by the Istituto per gli Studi Micenei ed Egeo-Anatolici. | Pecorella 1977; Quilici 1990 |
| Ayios *Iakovos Melia* | 3 intact chamber tombs (8, 10, 14) with multiple burials possibly representing collective secondary treatment, plus 2 others with LC deposits (12, 13) excavated by the SCE; other intact MC and looted MC/LC tombs nearby. Some osteological analysis. | Gjerstad *et al.* 1934 |
| Dhali *Kafkallia* | One looted tomb of MCIII–LCIII date excavated; no burial information but preserved finds reported in detail. See also Table 3.1. | Overbeck and Swiny 1972 |
| Dhenia *Kafkalla* and *Mali* | Several tombs with LC finds noted at both localities; very little burial information due to looting or other disturbance. Finds from the looted *Mali* Tomb 8 reported in detail. | Hadjisavvas 1985; Nicolaou and Nicolaou 1988; Webb and Frankel 2001 |
| Dromolaxia *Trypes* | 2 bulldozed and partly looted chamber tombs excavated by the Dept of Antiquities; other scantily reported tombs excavated by the British Museum in 1897–98. | Admiraal 1982; Witzel 1979 |
| Enkomi *Ayios Iakovos*—British Museum excavations | 100 tombs opened, many possibly intact. No burial data recorded, finds incompletely reported. Sample includes 4 ashlar built tombs and 1–2 tholos tombs along with chamber tombs and shaft graves. | Murray *et al.* 1900; Westholm 1939a; Courtois 1986b |
| Enkomi *Ayios Iakovos*—Cypriot excavations | *c.* 30 mortuary features including chamber tombs, shaft and pit graves, and pot burials excavated by the Dept of Antiquities. Of these, 3–4 chamber tombs, 7 shaft and pit graves, and 4 pot burials were intact. | Dikaios 1969 |
| Enkomi *Ayios Iakovos*—French excavations | *c.* 37 chamber tombs, shaft graves, and other mortuary features investigated, including 2 ashlar tombs discovered by the BM and one tholos tomb. *c.* 14 tombs and graves possibly intact, some osteological analysis. | Schaeffer 1936; 1952; Schaeffer 1971; Johnston 1971; Courtois 1981; 1986b; Lagarce and Lagarce 1985 |
| Enkomi *Ayios Iakovos*—SCE excavations | 28 mortuary features investigated, including chamber tombs, shaft graves, and one tholos tomb. *c.* 21 tombs and graves intact, with detailed reporting of burial deposits and some osteological analyses. | Gjerstad *et al.* 1934 |
| Hala Sultan Tekke *Vizaja* | 10 possibly intact, 50–60 looted tombs opened by the British Museum 1897–98; 2 bulldozed tombs excavated by the Dept. of Antiquities; 3 looted/emptied chamber tombs, one intact shaft grave excavated by the Swedish expedition. Some osteological analysis. | Karageorghis 1968; 1972b; Bailey 1972; Niklasson 1983; Åström 1983 |

| Tomb Locations | Sample Size/Evaluation | Major Publications |
|---|---|---|
| Kalavasos | 7 tombs excavated in the village area (1 intact); 8 at *Mangia* (2 intact); 21 mortuary features excavated at *Ayios Dhimitrios* (3 intact chambers, others substantially preserved). Detailed osteological analyses, with 80–90 individuals identified. | South 1997; 2000; South et al. 1989; South *et al.* n.d. |
| Katydhata | Several LC tombs excavated by Ohnefalsch-Richter and Markides, condition indeterminate. Markides' Tomb 42 is described as having been used for 5 seated burials. | Gjerstad 1926; Åström 1989 |
| Kazaphani *Ayios Andronikos* | One partly looted dual-chambered complex excavated by the Cyprus Museum and two other chamber tombs (as yet unpublished) excavated by the Cyprus Survey. | Nicolaou and Nicolaou 1989; Karageorghis 1972a |
| Kition | 2 chamber tombs (one intact) and a lined pit tomb at *Pefkakia;* 1 intact and 2 looted chamber tombs in *Chrysopolitissa* Area I; all excavated by the Dept of Antiquities. Some osteological analysis. | Karageorghis 1960b; 1974 |
| Korovia *Nitovikla* | 2 seemingly intact chamber tombs and one disturbed mortuary feature were excavated by the SCE within a cemetery containing at least 12 other looted tombs. | Gjerstad *et al.* 1934 |
| Kouklia | c. 30 unpublished tombs at *Evreti, Asproyi, Timi,* and *Marcello;* several more at *Teratsoudhia,* and on at *Eliomylia;* 20 unpublished pit graves found at *Kaminia;* other LCIIIB–CGI tombs *Kato Alonia, Hassan-Agha, Xerolimni, Lakkos tou Skarnou* and *Skales.* | Catling 1968; 1979; Karageorghis 1967; 1983; 1990a; Maier and Wartburg 1985 |
| Kourion *Bamboula* | 33 chamber and 2 pit tombs of LC date excavated by Daniel, 9 possibly intact. Detailed but incomplete osteological analyses by Angel. Scantily reported excavations by the BM expedition. | Benson 1972; Murray *et al.* 1900 |
| Lapithos *Ayia Anastasia* | 3 tombs with LC components opened by Markides in 1914–1917; finds from Tomb 2/502 published in detail; burials apparently disturbed by Iron Age reuse and collapse of the roof. | Gjerstad 1926; Gjerstad *et al.* 1934:164; Pieridou 1966 |
| Maroni | Many plundered and 26 'productive' tombs excavated at *Tsaroukkas* by the British Museum in 1897, plus several of indeterminate preservation at *Vournes;* other looted tombs discovered in recent excavations. 2 unpublished tombs excavated at *Kapsaloudhia.* | Johnson 1980; Cadogan 1984; Herscher 1984; Manning *et al.* 1994; Manning 1998a; Manning and Monks 1998 |
| Milia *Vikla Trachonas* | One intact, 2 looted and one looted or emptied chamber tomb excavated by the SCE. | Westholm 1939a |
| Morphou *Toumba tou Skourou* | 6 intact tombs with 12 chambers total and *c.* 97–100 burials excavated by the Harvard Boston Museum; some osteological information. Another looted tomb excavated by the Dept of Antiquities. | Vermeule and Wolsky 1990; Karageorghis 1964a |
| Myrtou *Stephania* | 14 mortuary features including 5 intact chamber tombs and other tombs or cuttings that had been looted or deliberately emptied, investigated by the Sydney University expedition. | Hennessy 1964 |
| Pendayia *Mandres* | One intact and 2 looted chamber tombs excavated by the Dept of Antiquities. | Karageorghis 1965c |
| Politiko | One intact pit tomb excavated by the Dept of Antiquities. | Karageorghis 1965a |
| Yeroskipou *Asproyia* | 3 chamber tombs excavated by the Dept of Antiquites, one looted, 2 others probably disturbed. | Nicolaou 1983 |

**Table 4.1** Examples of secondary treatment and complex ritual practices in Early–Middle Cypriot contexts.

| Site | Evidence for Secondary Treatment | Examples | References |
|------|----------------------------------|----------|------------|
| *Vounous* A | Burials represented primarily or exclusively by skulls | Tombs 82B, 118, 153 | Stewart and Stewart 1950: 55, 164, 186 |
| *Vounous* A | Largely complete/articulated burials with skulls separated from the rest of the skeletons | Tomb 164A | Stewart and Stewart 1950: 226–7, Fig. 165 |
| *Vounous* A | Simultaneous burials | Tombs 84, 90 | Stewart and Stewart 1950: 68, 94 |
| *Vounous* B | Burials represented primarily or exclusively by skulls | Tombs 44, 50A, 125, 141, 143, 146, 156 | Dikaios 1940:88; Dunn-Vaturi 2003:5; Stewart and Stewart 1950:264, 317, 324, 346, 357–58, 328, Fig. 242 |
| *Vounous* B | Grouping of disarticulated body parts | Tomb 31 | Dikaios 1940:65 |
| *Vounous* B | Multiple disarticulated burials, no articulated remains from latest phase of tomb use | Tomb 36, 66, 70B, 71, 72, 75-77 | Dikaios 1940:72–4; Dunn-Vaturi 2003:106, 129, 132, 140, 157, 160, 166 |
| Lapithos *Vrysi tou Barba* | Burials represented by disarticulated bone piles | Tomb 302B | Gjerstad *et al.* 1934:42 |
| Lapithos *Vrysi tou Barba* | Burial represented by head, neck and arms only | Tomb 306A | Gjerstad *et al.* 1934:60 |
| Lapithos *Vrysi tou Barba* | Burial represented by vertebral column only | Tomb 312A | Gjerstad *et al.* 1934:84, 77, Fig. 37:7 |
| Lapithos *Vrysi tou Barba* | Two or more individuals in highly variable states of preservation, some notably incomplete | Tombs 809A, 812A, 813B, 818, 826B, 322A | Herscher 1978:229, 296–7, 322, 413–4, 499; Gjerstad *et al.* 1934:147–56 |
| Lapithos *Vrysi tou Barba* | Burials on top of or mixed with pithos sherds (remnants of broken primary containers?) | Tombs 801, 809A, 817B, 823, 825, 826A, 829C, 831, 833C, 305B, 311B, 318 | Herscher 1978:786–7 and note 27; Gjerstad *et al.* 1934 |
| Lapithos *Vrysi tou Barba* | Emptied pit graves | | Herscher 1978:818–9 |
| Lapithos *Vrysi tou Barba* | Dromos cupboards emptied/plundered | Tombs 804, 829, 835, 307, 315, 316, 318, 322 | Herscher 1978:91–3, 571–2, 661; Gjerstad *et al.* 1934:63, 106, 115, 126, 144 |
| Lapithos *Vrysi tou Barba* | Multiple simultaneous burials | Tombs 803A, 812A, 826A, 832A, 309B, 311, 313A | Herscher 1978:782 and note 2; Gjerstad *et al.* 1934:70, 78, 87–9 |
| Lapithos *Vrysi tou Barba* | Pottery spanning hundreds of years (ECI–MCI) with no evident stratification of burials and architectural features (buttresses) possibly indicating construction late within this timespan (secondary interments of diverse date?) | Tomb 812A | Herscher 1978:296–7, 311–7 |
| Ayios *Iakovos Melia* | Multiple simultaneous burials | Tomb 6 | Gjerstad *et al.* 1934:315–7, Fig. 124:2 |
| Ayios *Iakovos Melia* | Four individuals in highly variable states of preservation, two notably incomplete | Tomb 7 | Gjerstad *et al.* 1934:322–5, Fig. 125:5 |
| Ayios *Iakovos Melia* | Post-interment 'looting' or partial | Tombs 1 and 4 | Gjerstad *et al.* 1934:302– |

| Site | Evidence for Secondary Treatment | Examples | References |
|---|---|---|---|
| | removal of burial remains | | 306, 309–13, Figs 119:1, 124:1 |
| Korovia *Paleoskoutella* | Small emptied primary tombs | Tombs 2 and 5 | Gjerstad *et al.* 1934:421–2, 426–8, Figs 163:8–10, 165, 166:2–3, 169 |
| Korovia *Paleoskoutella* | Multiple simultaneous burials | Tombs 4 and 7 | Gjerstad *et al.* 1934:423–5, 427, 429–31, Fig. 166:1, 7–12; 171 |
| Korovia *Paleoskoutella* | Ritual celebration and/or processing areas | Features associated with Tumuli 1, 3, and 6 | Gjerstad *et al.* 1934:416–9, 423, 428–9, Figs 163:1–7, 11–13; 166:4–6 |
| Kalavasos | Burials represented primarily or exclusively by skulls | Tombs 1, 8, 9, 11 | Karageorghis 1958 |
| Kalavasos | Multiple disarticulated burials, no articulated remains from latest phase of tomb use | Tombs 39, 40 | Todd 1986:28–30 |
| Dhenia *Kafkalla* and *Mali* | Probable small emptied primary tombs | *Kafkalla* Tombs 49, 163, 167–9; *Mali* Tombs 24-26 | Nicolaou and Nicolaou 1988 |
| Nicosia *Ayia Paraskevi* | Multiple simultaneous interments (?) | Tomb 13 | Hennessy *et al.* 1988:20–21 |
| Kalopsidha | Multiple simultaneous interments (?) | Tomb 28 | Myres 1897:146 |
| Politiko *Lambertis* | Multiple simultaneous interments (?) | Tomb 18 | Gjerstad 1926:81 |

**Table 4.2** Burial group statistics for *Vounous* and Lapithos.

| | *Vounous* A | *Vounous* B |
|---|---|---|
| **Mean Burials per Chamber*** | 1.59 | 2.68 |
| **Standard Deviation** | 0.97 | 1.91 |
| **Range** | 1–6 | 1–9 |
| **n** | 39 | 44 |

*The difference between the means for sites A and B is statistically significant (T statistic = 3.217; $p$ = 0.002).

| | Lapithos Overall | Early Group (ECII–IIIA/B) | Middle Group (ECIIIB–MCI) | Late Group (MCI–III) |
|---|---|---|---|---|
| **Mean Burials per Chamber**** | 2.58 | 2.26 | 2.18 | 4.19 |
| **Standard Deviation** | 1.97 | 1.25 | 1.00 | 3.27 |
| **Range** | 1–15 | 1–6 | 1–4 | 1–15 |
| **n** | 98 | 35 | 34 | 21 |

**A one way analysis of variance suggests that the differences between the mean burial group sizes in different periods at Lapithos are statistically significant (F = 9.139, $p$ = 0.000).

**Table 4.3** Tomb size statistics for *Vounous* (dr. = dromos; ch. = chamber; all measurements are in meters). The differences between the mean values at Site A and Site B are statistically significant for several of the variables.

|  |  | *Vounous* A | *Vounous* B | T statistic | *p* value |
|---|---|---|---|---|---|
| **Est. Ch. Floor Area** | Mean | 3.68 | 7.16 | 4.946 | 0.000 |
|  | Std. dev. | 1.73 | 4.30 |  |  |
|  | Range | 1.00–8.50 | 1.10–20.80 |  |  |
|  | n | 44 | 40 |  |  |
| **Ch. Length** | Mean | 2.24 | 2.87 | –4.448 | 0.000 |
|  | Std. dev. | 0.52 | 0.94 |  |  |
|  | Range | 0.80–3.70 | 1.00–6.00 |  |  |
|  | n | 51 | 91 |  |  |
| **Ch. Width** | Mean | 2.32 | 3.12 | –4.939 | 0.000 |
|  | Std. dev. | 0.59 | 1.06 |  |  |
|  | Range | 0.92–3.92 | 0.70–6.10 |  |  |
|  | n | 50 | 92 |  |  |
| **Ch. Height** | Mean | 1.44 | 1.60 | –0.880 | 0.390 |
|  | Std. dev. | 0.33 | 0.44 |  |  |
|  | Range | 0.83–1.74 | 1.00–2.28 |  |  |
|  | n | 7 | 13 |  |  |
| **Dr. Length** | Mean | 2.26 | 2.56 | 2.377 | 0.019 |
|  | Std. dev. | 0.52 | 0.66 |  |  |
|  | Range | 1.42–3.58 | 1.26–3.92 |  |  |
|  | n | 36 | 70 |  |  |
| **Dr. Width** | Mean | 2.11 | 1.68 | 5.876 | 0.000 |
|  | Std. dev. | 0.46 | 0.29 |  |  |
|  | Range | 1.16–3.18 | 1.15–2.75 |  |  |
|  | n | 37 | 69 |  |  |
| **Dr. Height** | Mean | 1.19 | 1.22 | 0.395 | 0.694 |
|  | Std. dev. | 0.37 | 0.48 |  |  |
|  | Range | 0.46–2.18 | 0.20–2.20 |  |  |
|  | n | 36 | 66 |  |  |

**Table 4.4** Tomb size statistics for Lapithos (dr. = dromos; ch. = chamber; all measurements are in meters). The mean values are very similar for the tombs of the early and middle groups, but the means for the late group are consistently higher. A one way analysis of variance suggests that the differences between the means for chamber width and estimated chamber floor area are statistically significant (for chamber width, $F=6.485$, $p=0.002$; for floor area, $F=5.919$, $p=0.007$). Note that the total number of observations for Lapithos overall exceeds the sum of the observations for the three subgroups because a few tombs used over extended periods of time could not be assigned to a single chronological division.

| | | Lapithos overall | Early Group (ECII–IIIA/B) | Middle Group (ECIIIB–MCI) | Late Group (MCI–III) |
|---|---|---|---|---|---|
| **Est. Ch. Floor Area** | Mean | 7.47 | 5.25 | 5.23 | 11.18 |
| | Std. dev. | 5.43 | 3.36 | 1.89 | 7.05 |
| | Range | 1.80–27.00 | 2.60–11.40 | 2.40–8.10 | 1.80–27.00 |
| | n | 32 | 6 | 14 | 12 |
| **Ch. Length** | Mean | 2.57 | 2.58 | 2.39 | 2.69 |
| | Std. dev. | 0.92 | 0.69 | 0.76 | 1.34 |
| | Range | 0.58–5.60 | 1.64–4.45 | 1.30–4.70 | 0.50–5.60 |
| | n | 106 | 32 | 34 | 26 |
| **Ch. Width** | Mean | 2.55 | 2.16 | 2.41 | 3.22 |
| | Std. dev. | 1.19 | 0.75 | 1.01 | 1.68 |
| | Range | 0.81–6.85 | 1.25–5.45 | 0.90–5.05 | 0.81–6.85 |
| | n | 107 | 32 | 34 | 27 |
| **Ch. Height** | Mean | 1.12 | 1.09 | 1.09 | 1.18 |
| | Std. dev. | 0.27 | 0.22 | 0.23 | 0.39 |
| | Range | 0.55–2.50 | 0.78–1.64 | 0.75–1.55 | 0.55–2.50 |
| | n | 101 | 32 | 32 | 23 |
| **Dr. Length** | Mean | 2.87 | 2.81 | 2.72 | 3.13 |
| | Std. dev. | 0.85 | 0.36 | 0.54 | 1.66 |
| | Range | 1.84–7.85 | 2.13–3.55 | 1.90–3.78 | 1.84–7.85 |
| | n | 54 | 18 | 15 | 11 |
| **Dr. Width** | Mean | 1.67 | 1.64 | 1.58 | 1.78 |
| | Std. dev. | 0.28 | 0.22 | 0.33 | 0.35 |
| | Range | 0.92–2.34 | 1.40–2.00 | 0.92–2.34 | 1.24–2.30 |
| | n | 53 | 18 | 15 | 10 |
| **Dr. Height** | Mean | 1.42 | 1.39 | 1.42 | 1.44 |
| | Std. dev. | 0.29 | 0.19 | 0.36 | 0.39 |
| | Range | 0.70–2.20 | 1.13–1.80 | 0.70–2.00 | 0.92–2.20 |
| | n | 47 | 16 | 11 | 11 |

**Table 4.5**   Correlations (Pearson's r) between the variables of burials per chamber, estimated chamber floor area, total pots per chamber and total copper-based artifacts per chamber at *Vounous* A and B.  Correlations were calculated pairwise with *n* values (sample size) of 38–39 for *Vounous* A and *n* values of 30–44, and 91 for *Vounous* B (lower values reflect the smaller number of observations for chamber floor areas and burials respectively).

|  | *Vounous* A | | | |
|---|---|---|---|---|
|  | Burials per Chamber | Est. Chamber Floor Area | Pots per Chamber | Copper Items per Chamber |
| Burials per Chamber | 1 | | | |
| Est. Chamber Floor Area | 0.456 | 1 | | |
| Pots per Chamber | 0.503 | 0.777 | 1 | |
| Copper Items per Chamber | 0.293 | 0.564 | 0.575 | 1 |

|  | *Vounous* B | | | |
|---|---|---|---|---|
|  | Burials per Chamber | Est. Chamber Floor Area | Pots per Chamber | Copper Items per Chamber |
| Burials per Chamber | 1 | | | |
| Est. Chamber Floor Area | 0.427 | 1 | | |
| Pots per Chamber | 0.788 | 0.717 | 1 | |
| Copper Items per Chamber | 0.599 | 0.756 | 0.695 | 1 |

**Table 4.6** Correlations (Pearson's r) between the variables of burials per chamber, estimated chamber floor area, total pots per chamber and total copper-based artifacts per chamber at Lapithos. Correlations were calculated pairwise with *n* values (sample size) of 32–95 for Lapithos overall, 6–36 for the Early Group, 14–35 for the Middle Group, and 12–24 for the Late Group; the lowest *n* values reflect the small number of observations for chamber floor area.

Lapithos Overall

|  | Burials per Chamber | Est. Chamber Floor Area | Pots per Chamber | Copper Items per Chamber |
|---|---|---|---|---|
| Burials per Chamber | 1 | | | |
| Est. Chamber Floor Area | 0.680 | 1 | | |
| Pots per Chamber | 0.451 | 0.325 | 1 | |
| Copper Items per Chamber | 0.780 | 0.769 | 0.430 | 1 |

Early Group (ECII–IIIA/B)

|  | Burials per Chamber | Est. Chamber Floor Area | Pots per Chamber | Copper Items per Chamber |
|---|---|---|---|---|
| Burials per Chamber | 1 | | | |
| Est. Chamber Floor Area | 0.854 | 1 | | |
| Pots per Chamber | 0.720 | 0.888 | 1 | |
| Copper Items per Chamber | 0.295 | 0.745 | 0.680 | 1 |

Middle Group (ECIIIB–MCI)

|  | Burials per Chamber | Est. Chamber Floor Area | Pots per Chamber | Copper Items per Chamber |
|---|---|---|---|---|
| Burials per Chamber | 1 | | | |
| Est. Chamber Floor Area | 0.299 | 1 | | |
| Pots per Chamber | 0.364 | 0.441 | 1 | |
| Copper Items per Chamber | 0.236 | 0.418 | 0.607 | 1 |

Late Group (MCI–III)

|  | Burials per Chamber | Est. Chamber Floor Area | Pots per Chamber | Copper Items per Chamber |
|---|---|---|---|---|
| Burials per Chamber | 1 | | | |
| Est. Chamber Floor Area | 0.565 | 1 | | |
| Pots per Chamber | 0.277 | −0.078 | 1 | |
| Copper Items per Chamber | 0.858 | 0.693 | 0.283 | 1 |

**Table 4.7a** *Vounous* A tomb finds and other variables. Tombs are sorted by copper and pottery totals. Dates are according to Stewart (1962a). HTW stands for hook-tang weapon. * indicates tomb with carved façade. Complex ritual refers to possible cases of secondary treatment. Ceremonial vessels include bowls and chalices or goblets (stemmed bowls) described by Stewart and Stewart (1950) as cult vessels, usually with modelled zoomorphic figures on the rim, and other forms as indicated.

| Tomb No. | Dates of Use | Preservation | Number of Burials | Complex Ritual | Floor Area / sq m | Pots Total | Copper Total | Copper Types | Ceremonial Vessel/ Rare Type | Ceramic Model | Other | Fauna |
|---|---|---|---|---|---|---|---|---|---|---|---|---|
| 115 | ECIC | Intact Eroded | 2 | | 1.3 | 5 | 0 | | bowl | | | |
| 82b | ECI | Intact | 1 | yes | | 6 | 0 | | 2 stemmed bowl | | | |
| 88 | ECI Philia | Intact | 2 | | 2.2 | 7 | 0 | | | | | |
| 87b | ECIB | Intact | 1 | | 3.4 | 8 | 0 | | | | spindle whorl | sheep/goat |
| 81a | ECIB | Intact | 1 | | 2.1 | 10 | 0 | | | | | |
| 93 | ECI | Intact | 1 | | 2.6 | 11 | 0 | | | | spindle whorl | cattle |
| 134 | ECI | Intact | 1 | | 2.7 | 11 | 0 | | | | | |
| 163 | ECIC early | Intact Eroded | 1 | | 3.7 | 11 | 0 | | | | | |
| 92 | ECIC | Intact Eroded | 2 | | 1.0 | 12 | 0 | | conical bowl | spindle | spindle whorl | |
| 81b | ECIB | Intact | 1 | | 2.6 | 13 | 0 | | bowl | | spindle whorl | |
| 103 | ECI | Intact Eroded | 1 | | 3.6 | 13 | 0 | | | | | |
| 104 | ECI | Intact Eroded | 1 | | 2.7 | 13 | 0 | | | | | |
| 114* | ECIB | Looted | 2 | | 4.7 | 13 | 0 | | askos, bowl fragments | dagger, sheath | | |
| 101 | ECIB | Intact Eroded | 1 | | 1.8 | 14 | 0 | | | | | |
| 110a | ECIC early Philia | Intact | 1 | | 1.6 | 14 | 0 | | | | macehead, whetstone | |
| 102 | ECI Philia | Intact Eroded | 1 | | 2.4 | 15 | 0 | | | | | |
| 119 | ECIB | Intact Eroded | 1 | | 1.8 | 16 | 0 | | bowl fragments | bellows–nozzle | | |
| 127 | ECIC Philia ECII | Looted | 6 | | 5.0 | 16 | 0 | | bowl fragments | | | |
| 160b | ECIB or ECIC | Intact | 1 | | 2.4 | 19 | 0 | | bowl | | | |
| 120 | ECIC | Intact | 2 | | 5.1 | 25 | 0 | | | | spindle whorl | |
| 113 | ECI | Intact | 2 | | 2.9 | 26 | 0 | | | | 2 lids | sheep/goat |
| 116* | ECIB | Intact | 2 | | 4.1 | 27 | 0 | | bowl | | 6 spindle whorls | sheep/goat |
| 153 | ECI | Looted | 2 | yes? | 4.1 | 28 | 0 | | 2 bowls | | spindle whorl | sheep/goat |
| 96 | ECI? | Looted | ? | | | 6 | 1 | axe | bowl fragments | | | cattle |
| 100 | ECI | Intact | 1 | | 2.7 | 8 | 1 | knife | | | whetstone | |
| 85a | ECIB ECIC | Looted | ? | | | 9 | 1 | knife | | | whetstone | |

| Tomb No. | Dates of Use | Preservation | Number of Burials | Complex Ritual | Floor Area / sq m | Pots Total | Copper Total | Copper Types | Ceremonial Vessel/ Rare Type | Ceramic Model | Other | Fauna |
|---|---|---|---|---|---|---|---|---|---|---|---|---|
| 107 | ECI | Intact | 2 | | 4.6 | 10 | 1 | knife | | | | cattle |
| 110b | ECIC early | Intact | 1 | | 1.9 | 11 | 1 | knife | | | whetstone, spindle whorl | sheep/goat |
| 109 | ECI | Intact Eroded | 2 | | 2.4 | 12 | 1 | knife | | | | |
| 83 | ECIB ECIC | Intact | 2 | | 2.6 | 17 | 1 | knife | | | | |
| 160a | ECIB | Intact | 1 | | 2.7 | 19 | 1 | knife | 3 bowls (2 stemmed), jug with incised masked dancer | | | |
| 155 | ECIB | Intact | 1 | | 4.8 | 19 | 1 | knife | bowl | | spindle whorl | |
| 82 | ECIB | Intact | 1 | | 3.4 | 30 | 1 | knife | | | spoon | cattle |
| 90 | ECIB | Intact | 2 | yes | 4.4 | 31 | 1 | knife | bowl | | whetstone, spindle whorl | |
| 112 | ECI | Eroded | ? | | | 4 | 2 | knife, HTW | | | | |
| 91 | ECI | Intact | 1 | | 3.1 | 15 | 2 | 2 pins | bowl with incised masked dancer | | | |
| 84 | ECI Philia | Intact | 2 | yes | 2.9 | 31 | 2 | 2 pins | stemmed bowl | | | |
| 154 | ECI | Intact | 1+? | | | 36 | 2 | 2 knives | | | whetstone | |
| 164b | ECIC late | Intact | 2 | | 3.3 | 43 | 2 | points/ needles | | spindle, finger–guard | gold, Levantine vessel, spindle whorl, spoon | sheep/goat |
| 87a | ECIB | Intact | 1 | | 4.0 | 18 | 3 | 3 knives | 2 bowls | | whetstone | sheep/goat |
| 118 | ECIC ECII | Intact | 4 | yes | 8.3 | 44 | 3 | 2 knives 1 HTW | | | | |
| 111 | ECIC early | Intact | 1–2 | | 6.8 | 55 | 3 | 3 knives | 3 bowls (1 stemmed) | hilt | whetstone | cattle |
| 164a | ECIC late | Intact | 2 | yes | 8.5 | 56 | 4 | 4 knives | bowl | horn | | cattle |
| 161 | ECIB | Intact Eroded | 1 | | 6.4 | 44 | 5 | 2 knives, 1 HTW, 2 axes | several bowls, 1 jug | | whetstone | cattle |
| 105 | ECI | Disturbed? | 2 | | 5.0 | 29 | 10 | 4 knives, 1 HTW, 2 axes, 1 chisel, 1 awl, 1 unid. | bowl | | 2 whetstones | cattle, sheep/goat |

**Table 4.7b** *Vounous* B finds and other variables from tombs dating mainly from ECII through ECIIIA. Ceremonial vessels include unusual shapes, jugs and bowls with 'coroplastic' modelled decoration (zoomorphic, anthropomorphic, and pottery motifs); and composite vessels. Dates follow Stewart (1962a) and Dunn-Vaturi (2003).

| Tomb No. | Dates of Use | Preservation | Number of Burials | Complex Ritual | Floor Area sq m | Pots Total | Copper Total | Copper Types | Ceremonial Vessel/ Rare Type | Ceramic Model | Other | Fauna |
|---|---|---|---|---|---|---|---|---|---|---|---|---|
| 146 | ECII | Intact Eroded | 1? | yes | | 5 | 0 | | | | | |
| 145 | ECII | Intact | ? | | | 10 | 0 | | | dagger, sheath | | |
| 123 | ECII | Intact | 1 | | 2.2 | 12 | 0 | | | | | |
| 70 | ECII early | Intact? | | | | 17 | 0 | | | | | |
| 142 | ECII end | Intact | 1 | | 6.0 | 17 | 0 | | | | | |
| 122 | ECIII | Intact Eroded | ? | | | 22 | 0 | | | | | |
| 70a | ECII | Intact | ? | | | 23 | 0 | | | | spindle whorl | |
| 129 | ECIIIA late | Intact | ? | | | 25 | 0 | | | | | |
| 141 | ECII | Intact | 2 | yes | 4.8 | 26 | 0 | | | | spindle whorl | |
| 151 | ECII ECIIIA | Intact | ? | | | 27 | 0 | | 2 stemmed bowls | | | |
| 31a | ECIIIA | Intact | 3 | yes | | 28 | 0 | | | | | |
| 30 | ECIIIA | Intact | 1 | | 5.6 | 30 | 0 | | coroplastic jug | | | |
| 35 | ECII | Intact | 2 | | 3.1 | 33 | 0 | | | | | |
| 51 | ECII | Intact | 2? | | | 34 | 0 | | stemmed bowl | dagger, sheath | | *sp*. not specified |
| 28 | ECIIIA late | Intact | 1? | | | 39 | 0 | | coroplastic bowl, pyxis | | | |
| 125 | ECIIIA | Intact | 1 | yes | 7.8 | 43 | 0 | | | | | *sp*. not specified |
| 132 | ECII ECIIIA | Intact | 2 | | | 54 | 0 | | stemmed bowl | 1 horn, 1 sheath, 2 daggers | | cattle |
| 71 | ECII ECIII | ? | | | | 56 | 0 | | pyxis, bowl | brush model | spindle whorl | *sp*. not specified |
| 41 | ECII ECIIIB | Disturbed | 5 | | | 64 | 0 | | | | | |
| 156 | ECII ECIIIA | Intact | 2 | yes | | 23 | 1 | HTW | | | | cattle |
| 124 | ECII ECIIIA | Intact | 4 | | 5.2 | 46 | 1 | knife | | | | *sp*. not specified |
| 77 | ECII ECIII | Intact? | 4 | ? | | 56 | 1 | pin | conical bowl | | | |
| 121 | ECII ECIIIA | Intact Eroded | 3 | | 4.0 | 47 | 2 | knife, pin | | 2 brush models | | |
| 29 | ECII ECIIIA ECIIIB/ MCI | Intact | Multiple | | | 50 | 2 | knife, pin | 2 composite bowls, 2 coroplastic bowls and 1 jug, 2 ring vases | spindle, dagger, sheath | 2 spindle whorls | |
| 52 | ECII ECIII MCI | Intact? | 6 | | | 52 | 6 | 3 knives, 1 HTW, 1 razor/ scraper, leaf frags | composite bowl, 2 stemmed bowls | | | *sp*. not specified |
| 45 | ECII and later | Intact | 1+ | | | 50 | 7 | 3 knives, 1 HTW, 3 pins | stemmed bowl | dagger, sheath | | |

**Table 4.7c** *Vounous* B finds and other variables from tombs dating mainly from ECIIIB through MCII. Dates follow Stewart (1962a) and Dunn-Vaturi (2003).

| Tomb No. | Dates of Use | Preservation | Number of Burials | Complex Ritual | Floor Area sq m | Pots Total | Copper Total | Copper Types | Ceremonial Vessel/ Rare Type | Ceramic Model | Other | Fauna |
|---|---|---|---|---|---|---|---|---|---|---|---|---|
| 50b | MCII | Intact | 2 | | | 4 | 0 | | | | | |
| 44 | ECIII MCI | Disturbed? | 1 | yes | | 10 | 0 | | | | | |
| 66a | ECIII | Intact | ? | | | 13 | 0 | | | | conch shell | |
| 55 | MCI | Intact | 1? | | | 19 | 0 | | | | spindle whorl | *sp.* not specified |
| 71a | ECIII | Intact | 1 | | | 20 | 0 | | | | | *sp.* not specified |
| 75 | ECIIIB MCI | Intact | 1? | ? | | 20 | 0 | | | | | |
| 65a | ECIII | Intact | ? | | | 24 | 0 | | stemmed bowl | | | |
| 130 | ECIIIB MCI | Intact | 1 | | 2.8 | 28 | 0 | | juglet with plank idol? | | | |
| 157 | ECIIIB MCI | Intact | ? | | | 32 | 0 | | | | | |
| 162 | ECIIIB early | Intact | ? | | | 38 | 0 | | | | | cattle |
| 18 | ECIII MCI | Intact | ? | | | 46 | 0 | | | horn | | |
| 139 | ECIIIB MCI | Intact | 4 | | 5.9 | 59 | 0 | | stemmed bowl | | | cattle |
| 144 | MCI early | Intact Eroded | ? | | | 69 | 0 | | | | | |
| 50 | MCI/II | Intact | 3 | | | 9 | 1 | knife | | | spindle whorl, macehead | |
| 60 | ECIIIB MCI | Intact | 1? | | | 15 | 1 | knife | coroplastic basin | | | |
| 50a | ECIIIB MCI | Intact | 2 | yes | | 24 | 1 | pin | | | spindle whorl, whetstone | sheep/ goat |
| 66 | ECIII | Intact | 2 | ? | | 49 | 1 | knife | stemmed bowl | | | *sp.* not specified |
| 137 | ECIIIB early | Intact | ? | | | 74 | 1 | HTW | composite vessel | | funnel | cattle, sheep/ goat |
| 42 | ECIIIB early | Intact | 2+ | | | 25 | 2 | 2 pins | | | spindle whorl | |
| 69 | MCI | Disturbed? | ? | | | 31 | 2 | 1 knife, 1 pin | | | spindle whorl | |
| 16 | ECIIIB | Intact | ? | | | 47 | 2 | 1 pin, 1 earring | 1 double-necked jug, 1 stemmed bowl, 1 pyxis, coroplastic and composite vessels | | | *sp.* not specified |
| 12 | ECIIIB MCI | Intact | ? | | | 91 | 2 | 1 HTW, 1 tweezer | 2+ coroplastic jugs, 1 composite vessel | | | *sp.* not specified |
| 22 | ECIIIB MCI | Looted | ? | | | 33 | 3 | 2 knives, 1 razor/ scraper | coroplastic and composite vessels | 'sacred enclosure' model | whetstone | |

| Tomb No. | Dates of Use | Preservation | Number of Burials | Complex Ritual | Floor Area sq m | Pots Total | Copper Total | Copper Types | Ceremonial Vessel/ Rare Type | Ceramic Model | Other | Fauna |
|---|---|---|---|---|---|---|---|---|---|---|---|---|
| 57a | ECIIIB MCI MCII | Intact | 2 | | | 35 | 3 | 1 HTW, 1 pin, 1 tweezer | 1 askos | | loom-weight | |
| 78 | ECIIIB | Disturbed? | ? | | | 35 | 3 | 1 knife, 2 pins | 1 stemmed bowl | | spindle whorl | |
| 11 | ECIIIB MCI | Intact | ? | | | 66 | 3 | 1 HTW, 2 pins | 1 coroplastic jug, 1 pyxis | plank idol | | |
| 5 | ECIIIB | Intact | 1+? | | | 26 | 4 | 1 knife, 3 pins | 1 composite vase, 1 coroplastic jug and stemmed bowl | | spindle whorl | |
| 67 | ECIII | Intact | 1 | | | 28 | 4 | 1 knife, 1 HTW, 1 razor/ scraper, 1 tweezer | 1 coroplastic bowl | | whetstone | *sp.* not specified |
| 21 | MCI | Intact | ? | | | 34 | 4 | 1 knife, 2 pins, 1 spiral ornament | | | | |
| 7 | ECIIIB MCI | Intact | ? | | | 74 | 4 | 1 knife, 1 HTW, 2 pins | composite vessel | | spindle whorl | *sp.* not specified |
| 6 | ECII ECIIIB MCI | Intact | ? | | | 110 | 4 | 1 HTW, 2 pins, 1 needle | | | spindle whorl | |
| 36 | ECII– MCI | Intact | 9+? | yes | 7.8 | 121 | 4 | 1 HTW, 1 pin, 2 needles | 3 coroplastic vessels, 1 ring vase, 1 stemmed bowl, 1 double-necked jug | 2 horns | 2 spindle whorls | cattle (substantial portion) |
| 9a/b | ECIIIB MCI | Intact | ? | | | 149 | 4 | 1 HTW, 3 pins | 6 composite vessels, 2 pyxides, 1 stemmed and 1 coroplastic bowl,1 double-necked jug | 2 horns, 1 brush, 1 offering table | 2 spindle whorls | |
| 76 | ECIIIB MCI MCII | Disturbed? | ? | ? | | 50 | 5 | 2 knives, 1 arrow head, 2 pins | 2 coroplastic vessels, 4 kernoi, 2 composite vessels | plank idol | 6 spindle whorls | |
| 8 | ECIIIB MCI | Intact | ? | | | 105 | 5 | 1 HTW, 4 pins | 2 composite vessels, 1 stemmed bowl, 1 double-necked jug | | | cattle |
| 68 | MCI | ? | ? | | | 29 | 6 | 3 knives, ? HTWs, 1 razor/ scraper | askos | | 1 MBIIA pithos | |

| Tomb No. | Dates of Use | Preservation | Number of Burials | Complex Ritual | Floor Area sq m | Pots Total | Copper Total | Copper Types | Ceremonial Vessel/ Rare Type | Ceramic Model | Other | Fauna |
|---|---|---|---|---|---|---|---|---|---|---|---|---|
| 54 | ECIII MCI | Intact | ? | | | 115 | 6 | 3 knives, 2 pins, 1 tweezer | 3 composite bowls, 2 double and 1 triple-necked jugs | | 1-2 spindle whorls | |
| 57 | ECIII MCI | Intact | 4 | ? | | 53 | 7 | 3 knives, 1 HTW, 3 pins | 1 askos, 1 pyxis, 1 coroplastic jug | | spindle whorl | |
| 56 | ECIII MCI MCII | Intact | ? | | | 79 | 7 | 1 knife, 3 pins, 1 tweezer, 1 needle, 1 bracelet | 2 askoi, 1 kernos, 1 tulip bowl | | 6-7 spindle whorls | |
| 13 | ECIIIB MCI | Intact | ? | | | 108 | 7 | 2 knives, 2 HTWs, 1 tweezer, 2 pins | 1 coroplastic vessel, 2 composite vases, 1 double-necked jug | | 1 ladle | cattle (substantial portion) |
| 17 | ECIIIB MCI | Intact | 5+ | | 8.0 | 63 | 8 | 2 knives, 1 HTW, 1 razor/scraper, 4 pins | 1 ring vase, 2 pyxides, 1 composite vase | | 4 spindle whorls | *sp.* not specified |
| 19 | ECIIIB MCI | Intact | 2 | | 10.2 | 93 | 8 | 3 knives, 1 HTW, 1 Minoan dagger, 1 tweezer, 2 pins | 1 coroplastic vessel, 4 composite vessels, 4 pyxides | | faience necklace, 3 spindle whorls | *sp.* not specified |
| 26 | ECIIIB MCI | Intact | 4 | | | 95 | 9 | 3 knives, 5 HTWs, 1 razor/scraper | 1 stemmed bowl | | whetstone | *sp.* not specified |
| 65 | ECIII | Intact | 3 | | | 103 | 9 | 3 knives, 2 HTWs, 2 tweezers, 1 awl, 1 pin | 2 coroplastic vessels, 1 kernos, 1 stemmed bowl | | 3 spindle whorls | *sp.* not specified |
| 15 | ECIIIB | Intact | ? | | | 109 | 11 | 3 knives, 1 HTW, 7 pins | 5 composite vessels, 1 pyxis, 1 double-necked jug, 1 jug with incised stags | | 2 faience necklaces | |
| 143 | ECIII MCI | Intact | 8 | yes | 8.1 | 136 | 11 | 3 knives, 1 HTW, 1 razor/scraper, 1 Minoan dagger, 4 pins, 1 needle | 2 kernoi, 1 stemmed bowl | | stone tablet?, stone axe-amulet | sheep/goat |
| 47 | ECIIIB MCI | Intact | 4+? | | | 70 | 13 | 3 knives, 2 HTWs, 2 tweezers, 5 pins, 1 ring | 1 triple-necked jug, 1 pyxis | | spindle whorl | |

| Tomb No. | Dates of Use | Preservation | Number of Burials | Complex Ritual | Floor Area sq m | Pots Total | Copper Total | Copper Types | Ceremonial Vessel/ Rare Type | Ceramic Model | Other | Fauna |
|---|---|---|---|---|---|---|---|---|---|---|---|---|
| 2 | ECIIIB MCI | Intact | Mul-tiple | | | 143 | 14 | 3 knives 3 HTWs, 8 pins | 4 coroplastic vessels, 1 pyxis, 1 composite vase | 2 plank idols | 4 faience beads, 3 spindle whorls | *sp*. not specified |
| 64 | ECIII MCI MCII | Intact | Multiple | | | 165 | 14 | 7 knives, 2 razor/ scrapers, 1 tweezer, 1 axe, 2 pins, 1 chisel/awl | 2 composite vessels, 1 coroplastic vessel, 1 boat pyxis, 1 pyxis/askos, 1 kernos, 1 triple and 1 double-necked jug, 1 stemmed bowl | plank idol | 1 MBIIA pithos, 5 faience beads, 10 spindle whorls, 1 bone tube, 1 mace-head, 1 rubber, 1 ladle | |
| 27 | ECIIIB MCI | Intact | 6 | | | 105 | 15 | 7 knives, 4 HTWs, 2 razor/ scrapers, 2 tweezers | 1 coroplastic vessel, 3 composite vases, 1 stemmed bowl | | 1 mace-head, 1 spindle whorl | |
| 72 | ECII ECIII MCI | ? | Multiple | | | 149 | 24 | 9 knives, 3 HTWs, 1 razor/ scraper, 1 tweezer 7 pins, 1 bracelet, 1 earring, 1 needle | 2 composite vases, 1 coroplastic jug | horn | lamp, 4 faience beads, 1 serpentine bead, 4 spindle whorls | |
| 59 | MCI | Looted | ? | | | ? | ? | 2 pins, 1 tweezer, 1 razor/ scraper | several coro-plastic and composite vessels | horn | 1 gold earring, 3 faience beads, spindle whorl, whetstone | |

**Table 4.8** Comparison of wealth indices at *Vounous* A and B. The differences between the mean values for the two sites are statistically significant except in the case of copper items per burial; however, the figures for *Vounous* B are greatly affected by the general scarcity of copper in tombs of ECII–ECIIIA date (see Table 4.9).

|  |  | *Vounous* A | *Vounous* B | T statistic | *p* value |
|---|---|---|---|---|---|
| Pots per Chamber | Mean | 20.85 | 50.81 | 4.836 | 0.000 |
|  | Std. Dev. | 13.27 | 37.64 |  |  |
|  | Range | 6–57 | 2–165 |  |  |
|  | n | 39 | 91 |  |  |
| Pots per Burial | Mean | 14.38 | 18.67 | 2.217 | 0.029 |
|  | Std. Dev. | 7.90 | 9.39 |  |  |
|  | Range | 3.50–44.00 | 2.00–46.50 |  |  |
|  | n | 38 | 44 |  |  |
| Copper Items per Chamber | Mean | 1.15 | 3.24 | 2.726 | 0.007 |
|  | Std. Dev. | 1.93 | 4.61 |  |  |
|  | Range | 0–10 | 0–24 |  |  |
|  | n | 39 | 91 |  |  |
| Copper Items per Burial | Mean | 0.74 | 0.82 | 0.286 | 0.775 |
|  | Std. Dev. | 1.24 | 1.13 |  |  |
|  | Range | 0–5 | 0–4 |  |  |
|  | n | 38 | 44 |  |  |

**Table 4.9** Comparison of wealth indices for earlier and later tombs at *Vounous* B. The differences between the means are statistically significant except in the case of pots per burial; the per capita expenditure on pottery was high in the earlier period even though copper appears to have been scarce.

|  |  | *Vounous* B early(ECII–ECIII) | *Vounous* B late(ECIIIB–MCII) | T statistic | *p* value |
|---|---|---|---|---|---|
| Pots per Chamber | Mean | 33.66 | 58.84 | 3.114 | 0.002 |
|  | Std. Dev. | 17.23 | 41.82 |  |  |
|  | Range | 5–70 | 2–165 |  |  |
|  | n | 29 | 62 |  |  |
| Pots per Burial | Mean | 17.90 | 19.20 | 0.445 | 0.658 |
|  | Std. Dev. | 8.58 | 10.04 |  |  |
|  | Range | 8.67–43 | 2–46.5 |  |  |
|  | n | 18 | 26 |  |  |
| Copper Items per Chamber | Mean | 0.59 | 4.48 | 4.075 | 0.000 |
|  | Std. Dev. | 1.38 | 5.05 |  |  |
|  | Range | 0–6 | 0–24 |  |  |
|  | n | 29 | 62 |  |  |
| Copper Items per Burial | Mean | 0.22 | 1.23 | 3.223 | 0.002 |
|  | Std. Dev. | 0.40 | 1.28 |  |  |
|  | Range | 0–1.33 | 0–4 |  |  |
|  | n | 18 | 26 |  |  |

**Table 4.10** Intensity of association between broad categories of artifact types at *Vounous* as measured by the *phi* coefficient, which is interpreted on the same scale as Pearson's r (1.0 = perfect correlation, 0 = no correlation, −1.0 = perfect negative correlation). Note that while there is some evident association between the occurrence of copper items and ceremonial vessels among the later tombs of *Vounous* B, both of these types were present in the majority of tombs in the sample (see Table 4.7c), suggesting widespread availability. For an explanation of *phi*, see Doran and Hodson (1975:147).

*Vounous* A

|  | Copper | Large Fauna | Ceremonial Vessel | Ceramic Model |
|---|---|---|---|---|
| Copper Item(s) | 1 | | | |
| Large Fauna | 0.282 | 1 | | |
| Ceremonial Vessel | 0.120 | 0.081 | 1 | |
| Ceramic Model | −0.009 | 0.150 | 0.317 | 1 |
| n=43 | | | | |

*Vounous* B (ECIIIB–MCII Tombs)

|  | Copper | Large Fauna | Ceremonial Vessel | Ceramic Model |
|---|---|---|---|---|
| Copper Item(s) | 1 | | | |
| Large Fauna | 0.161 | 1 | | |
| Ceremonial Vessel | 0.613 | 0.219 | 1 | |
| Ceramic Model | 0.147 | −0.136 | 0.179 | 1 |
| n=52 | | | | |

Table 4.11a Lapithos finds and other variables from tombs dating mainly from ECII–ECIIIA/B. Dates are according to Stewart (1962a), Åström (1972a), and Herscher (1978). Note that tombs are sorted by total copper-based objects and total pots per chamber.

| Tomb No. | Dates of Use | Preservation | Number of Burials | Complex Ritual | Floor Area sq m | Pots Total | Copper Total | Copper Types | Ceremonial Vessel/Rare Type | Imported Goods | Other | Fauna |
|---|---|---|---|---|---|---|---|---|---|---|---|---|
| 312b | ECIIIA | Intact | 1 | | 4.7 | 2 | 0 | | | | | sheep/ goat |
| 829b | ECII | Intact | 1 | | | 4 | 0 | | | | | |
| 809a | ECII ECIIIA | Intact | 3 | yes, pithos sherds | | 6 | 0 | | | | | |
| 822a | ECII early | Disturbed? | 1? | | | 6 | 0 | | | | | |
| 835c | ECIIIA | Intact | 1? | | | 6 | 0 | | | | | |
| 812b | ECIIIA | Intact | 1 | | | 7 | 0 | | | | 1 spindle whorl | *sp.* not specified |
| 312a | ECIIIA | Intact | 2 | yes | 3.4 | 7 | 0 | | | | | |
| 821b | ECII | Intact | 1 | | | 8 | 0 | | | | | |
| 830b | ECIII | Intact | 1 | | | 8 | 0 | | | | | |
| 822b | ECII early | Disturbed? | | | | 12 | 0 | | | | | |
| 831 | ECII ECIIIB | Intact | 2 | yes, pithos sherds | | 13 | 0 | | | | | sheep/ goat |
| 305b | ECIII? | Intact | 1? | yes, pithos sherds | | 14 | 0 | | | | | |
| 825 | ECII ECIIIB | Intact | 3 | yes, pithos sherds | | 16 | 0 | | | | 3 stone axes, 1 stone grinder? | |
| 821a | ECII | Intact | 3 | | | 18 | 0 | | | | | |
| 838 | ECIIIA | Intact | 2 | | | 19 | 0 | | | | 1 bone needle | sheep/ goat |
| 824a | ECI late ECIIIB | Intact | 4 | | | 21 | 0 | | | | | |
| 834b | ECII ECIIIA | Disturbed? | 2? | | | 21 | 0 | | | | | *sp.* not specified |
| 311b | ECIIIA | Intact | 3 children | yes, pithos sherds | 6.6 | 21 | 0 | | | | | |
| 834a | ECIII? | Intact? | 4 | | | 22 | 0 | | | | | |
| 303a | ECIC ECII | Intact | 2 | | 5.6 | 22 | 0 | | | | 1 whetstone | cattle (substantial portion) dog |
| 826b | ECII ECIIIB | Intact | 2 | | | 25 | 0 | | | | 1 stone tool | |
| 304 | ECIII | Intact | 1 | | 2.6 | 3 | 1 | knife | | | | |
| 823 | ECII early | Intact | 4 | yes, pithos sherds | | 18 | 1 | HTW | | | | |
| 801 | ECIIIA MCI | Intact | 2 | yes, pithos sherds | | 22 | 1 | HTW | | | | sheep/ goat |

| Tomb No. | Dates of Use | Preservation | Number of Burials | Complex Ritual | Floor Area sq m | Pots Total | Copper Total | Copper Types | Ceremonial Vessel/Rare Type | Imported Goods | Other | Fauna |
|---|---|---|---|---|---|---|---|---|---|---|---|---|
| 829a | ECII ECIIIB | Intact | 3 | | | | 27 | 1 | 1 razor/scraper | | | 1 whetstone, 1 bone tool | |
| 826a | ECII ECIIIB | Intact | 4 | yes, pithos sherds | | | 32 | 1 | 1 razor/scraper | | | spindle whorl, [4 stone axes in dromos] | small fauna in bowl |
| 811a | ECII ECIII | Disturbed? | 2 | | | | 38 | 1 | awl | | | 3 whetstones | |
| 817b | ECII | Intact? | 1 | yes, pithos sherds | | | 9 | 2 | 1 knife, 1 HTW | | | | |
| 305a | ECIII? | Intact | 3 | | 2.8 | | 24 | 2 | 1 knife, 1 HTW | | | ceramic figurine | |
| 836a | ECIIIA | Intact | 1 | | | | 10 | 3 | 2 knives, 1 HTW | | | | sheep/goat |
| 817a | ECII ECIIIB MCI | Disturbed | 2+ | | | | 27 | 4 | 2 knives, 1 HTW, 1pin | | | | |
| 815b | ECIIIA ECIIIB | Intact | 2? | | | | 20 | 5 | 1 HTW, 1 razor/scraper, 1 tweezer, 2 pins | | | | |
| 813a | ECII ECIIIB | Disturbed? | ? | | | | 54 | 5 | 3 knives, 1 razor/scraper, 1 needle | 2 double-necked jugs | | 1 spindle whorl, 2 stone pounders, 1 stone quern | |
| 829c | ECII | Intact | 2 | yes, pithos sherds | | | 38 | 6 | 1 knife, 1 tweezer, 4 pins | 2 coro-plastic jugs, 1 askos | | 2 spindle whorls, 1 stone disk | |
| 314b | ECIIIA ECIIIB | Intact | 6 | | 11.4 | | 81 | 6 | 1 knife, 1 HTW, 1 tweezer, 2 pins | 1–2 coro-plastic jugs, 1 composite jug, 1 askos | | 2 spindle whorls, 1 stone amulet, 1 stone bead | |

Table 4.11b Lapithos finds and other variables from tombs dating mainly from ECIIIB–MCI. Dates are according to Stewart (1962a), Åström (1972a), and Herscher (1978). Note that tombs are sorted by total copper-based objects and total pots per chamber.

| Tomb No. | Dates of Use | Preservation | Number of Burials | Complex Ritual | Floor Area sq m | Pots Total | Copper Total | Copper Types | Ceremonial Vessel /Rare Type | Imported Goods | Other | Fauna |
|---|---|---|---|---|---|---|---|---|---|---|---|---|
| 832b | ECIIIB | Intact | 1 child | | | 3 | 0 | | | | | |
| 306c | ECIIIB MCI | Intact | 3 | | 3.5 | 3 | 0 | | | | 1 spindle whorl | |
| 809b | ECIIIB | Intact | 2 | | | 5 | 0 | | | | | |
| 833c | ECIIIB MCI | Intact | 1? | yes, pithos sherds | | 7 | 0 | | | | | |
| 301b | ECIIIB? | Intact | 3? | | 2.4 | 8 | 0 | | | | | |
| 819b | ECIIIA? ECIIIB MCI | Intact | 1 | | | 9 | 0 | | | | | |
| 809c | ECIIIB MCI | Intact | 1 | | | 11 | 0 | | 1 askos? | | | |
| 301a | ECIIIB or later | Intact | 2 | | 2.4 | 12 | 0 | | | | 1 spindle whorl | |
| 818 | ECIIIB | Intact | 3 | yes | | 23 | 0 | | | | | |
| 308 | ECIIIB | Intact | 4? | | 6.1 | 28 | 0 | | 1 coroplastic jug, 1 double-necked jug | | | |
| 810c | ECIIIB MCI | Intact | 1 | | | 12 | 1 | HTW | | | | |
| 828b | ECIIIB MCI | Intact | 1 | | | 13 | 1 | pin | | | | |
| 827a | ECIIIB MCI | Intact | 2 | | | 15 | 1 | HTW | | | | |
| 314a | ECIIIB MCI | Intact | 3 | | 5.2 | 25 | 1 | knife | 1 composite jug | | | *sp.* not specified |
| 810a | ECIC ECII, ECIIIB MCI | Intact | 2 | | | 28 | 1 | HTW | | | 2 stone spindle whorls | *sp.* not specified |
| 832a | ECIIIB | Intact | 2 | yes | | 9 | 2 | 1 razor/ scraper, 1 pin | | | | |
| 317 | ECIIIB | Intact | 2? | | 5.9 | 13 | 2 | 2 pins | | | | |
| 816 | ECIIIB MCI | Intact | 2 | | | 16 | 2 | 1 knife, 1 razor/ scraper | 1 double-necked jug | | 1 stone tool | |
| 306a | ECIIIB MCI early | Intact | 1 | yes | 8.0 | 19 | 2 | 2 knives | 1 double-necked jug | faience necklace | 1 ceramic figurine | |
| 813b | ECIIIB MCI | Intact | 2 | yes | | 27 | 2 | 2 pins | | | 1 spindle whorl, 1 stone disk | |
| 835a | ECIIIA ECIIIB MCI | Intact | 4 | | | 28 | 2 | 2 knives | | | | |
| 302c | ECIIIB late | Intact | 2 | | 3.5 | 38 | 2 | 1 HTW, 1 ring | | | 1 spindle whorl | |

| Tomb No. | Dates of Use | Preservation | Number of Burials | Complex Ritual | Floor Area sq m | Pots Total | Copper Total | Copper Types | Ceremonial Vessel /Rare Type | Imported Goods | Other | Fauna |
|---|---|---|---|---|---|---|---|---|---|---|---|---|
| 803b | ECIIIA MCI MCII early | Intact | 2–3 | | | 21 | 3 | 1 pin, 2 needles | 1 double-necked jug | | 4 spindle whorls, 1 stone disk | |
| 828a | ECIIIA ECIIIB MCI early | Intact | 4 | | | 33 | 3 | 1 knife, 1 razor/scraper, 1 tweezer | 1 double-necked jug | 1 silver ring | | |
| 806b | ECIIIB MCI | Intact | 2+ | | | 39 | 3 | 3 pins | 2 double-necked jugs | | 1 spindle whorl | cattle? |
| 302a | ECIIIB | Intact | 2–3 | | 5.9 | 36 | 4 | 1 knife, 1 HTW, 1 razor/scraper, 1 tweezer | 1 coroplastic jug | | 2 spindle whorls | |
| 309b | ECIIIB MCI | Intact | 3 | yes | 5.9 | 41 | 4 | 4 pins | 1 coroplastic jug, 1 pyxis | | 3 spindle whorls | |
| 827b | ECIIIB MCI | Intact | 2 | | | 72 | 4 | 1 knife, 1 HTW, 1 razor/scraper, 1 tweezer | 1 double-necked jug | | 1 spindle whorl, 2 stone axes, 1 stone pounder | cattle |
| 837 | ECIIIB MCI | Intact | 1 | | | 12 | 5 | 1 knife, 1 HTW, 1 razor/scraper, 1 tweezer, 1 awl | | | 1 whet-stone | |
| 809d | ECIIIB MCI | Intact | 2 | | | 42 | 5 | 1 HTW, 1 razor/scraper, 1 tweezer, 2 pins | | | 1 whet-stone, 1 stone pounder | |
| 302b | ECIIIA? ECIIIB MCI | Intact | 4 | yes | 6.8 | 44 | 5 | 1 knife, 1 HTW, 1 axe, 1 razor/scraper, 1 tweezer | 1 composite vessel | | 3 whet-stones | |
| 301c | ECIIIB or later | Intact | 1 | | 3.4 | 7 | 8 | 3 knives, 2 HTWs, 1 razor/scraper, 2 tweezers | | | | |
| 806c | ECIIIB MCI | Intact | 1+ | | | 35 | 8 | 1 knife, 2 HTWs, 1 razor/scraper, 1 tweezer, 3 pins | | 5 faience beads | 1 spindle whorl | |
| 322b | ECIII? MCI | Looted? | ? | | | 20 | 9 | 2 knives, 2 HTWs, 1 razor/scraper, 3 tweezers, 1 needle | 2 composite vessels, 1 pyxis | 1 silver pin, 1 silver ring, 1 faience necklace | 1 whet-stone, 15 spindle whorls | horse? |

| Tomb No. | Dates of Use | Preservation | Number of Burials | Complex Ritual | Floor Area sq m | Pots Total | Copper Total | Copper Types | Ceremonial Vessel /Rare Type | Imported Goods | Other | Fauna |
|---|---|---|---|---|---|---|---|---|---|---|---|---|
| 318 | ECIIIB MCI | Intact | 4? | yes, pithos sherds | 8.1 | 19 | 10 | 1 knife, 1 HTW, 1 razor/ scraper, 1 tweezer, 6 pins | | | 2 spindle whorls | |
| 309a | ECIIIB MCI | Intact | 3+? | | 6.1 | 46 | 13 | 12 pins, sheet fragment | | | 1 ceramic figurine, 1 spindle whorl | |
| 322e | ECIIIA ECIIIB MCI | Intact? | 4 | | 10.6 | 97 | 13 | 1 knife, 8 pins, 2 rings, 2 needles | 1 composite vessel, 1 pyxis | 2 faience necklaces | 9 spindle whorls, 1 bone pin | cattle |
| 806a | ECIIIB MCI | Intact | 3 | | | 110 | 14 | 2 knives, 1 HTW, 1 tweezer, 10 pins | 6 composite vessels (varied forms) | 3 sheet gold ornaments, 1 silver band, 2 silver pins, 1 silver ?clasp, 4 groups faience beads, 1 Minoan jar | 5 spindle whorls | cattle |

**Table 4.11c** Lapithos finds and other variables from tombs dating mainly from MCI–MCIII. Dates are according to Stewart (1962a), Åström (1972a), and Herscher (1978). Note that tombs are sorted by total copper-based objects and total pots per chamber. Gale *et al.* (1996:397) have indicated that a number of items described in publication as silver-lead alloys are actually pure lead.

| Tomb No. | Dates of Use | Preservation | Number of Burials | Complex Ritual | Floor Area sq m | Pots Total | Copper Total | Copper Types | Ceremonial Vessel /Rare Type | Imported Goods | Other | Fauna |
|---|---|---|---|---|---|---|---|---|---|---|---|---|
| 303b | MCII early | Intact | 1? | | | 11 | 0 | | | | | |
| 319b | MCI late MCII early | Intact | 2 | | 2.8 | 62 | 1 | ring | 3 composite vessels, 1 zoomorphic askos | 1 faience bead | 1 spindle whorl | |
| 802b | MCI early | Intact | 1 | | | 32 | 3 | 1 knife, 2 HTWs | | | | |
| 323d | MCII early | Disturbed? | ? | | | 7 | 4 | 1 HTW, 1 razor/ scraper, 1 tweezer, 1 pin | | | 1 spindle whorl | |
| 307b | MCI | Intact | 1 | | 1.8 | 12 | 4 | 1 knife, 1 pin, 2 rings | | | ceramic figurine | |
| 323c | MCI | Disturbed? | 1+? | | | 20 | 4 | 1 HTW, 3 pins | 1 double-necked jug | | | |
| 311a | MCI MCII | Intact | 4 | | 10.3 | 44 | 5 | 2 knives, 1 tweezer, 2 pins | | 1 silver/ lead ring | 1 spindle whorl | |
| 815c | MCI | Intact | 1 | | | 19 | 6 | 6 pins | | | 2 spindle whorls | |
| 307a | MCI | Intact | 4 | | 4.8 | 15 | 7 | 2 knives, 2 HTWs, 2 pins, 1 ring | 1 composite vessel (kernos) | 1 silver lead ring | ceramic figurine | |
| 322d | ECIIIB? MCI MCII | Intact | 1? | | | 85 | 7 | 1 knife, 4 pins, 2 rings | 2 ring vases, 2 composite vessels | 12 faience beads | stone 'menhir', marble plank idol, 3 spindle whorls | nearly complete dog skeleton |
| 802a | ECIIIB MCI MCII | Intact | 2 | | | 49 | 8 | 3 knives, 1 HTW, 1 razor/ scraper, 3 pins | 2 double-necked jugs, 1 askos | 1 silver ring | 2 ceramic figurines, 1 whetstone, 3 spindle whorls | |
| 803a | MCI | Intact | 2 | yes | | 25 | 9 | 2 knives, 3 HTWs, 2 razor/ scrapers, 2 tweezers | | | 1 spindle whorl | |
| 804d | MCII MCIII late | Disturbed | ? | | | 23 | 11 | 4 knives, 1 tweezer, 1 pin, 5 rings | | | 1 whetstone, 2 spindle whorls, 1 bone pin | |

| Tomb No. | Dates of Use | Preservation | Number of Burials | Complex Ritual | Floor Area sq m | Pots Total | Copper Total | Copper Types | Ceremonial Vessel /Rare Type | Imported Goods | Other | Fauna |
|---|---|---|---|---|---|---|---|---|---|---|---|---|
| 313cd | MCI | Intact | 3–5 | | 4.0 (d) | 29 | 15 | 3 HTWs, 1 axe, 2 razor/ scrapers, 3 tweezers, 2 pins, 1 ring, 1 needle, 1 chisel | | 1 Minoan dagger | 2 spindle whorls, 1 iron lump | |
| 315a | MCII MCIII | Intact | 6 | | 11.2 | 73 | 20 | 5 knives, 2 HTWs, 1 axe, 3 tweezers, 5 pins, 3 needles, 1 spatula | 1 double-necked bottle | | 1 whetstone, 8 spindle whorls, 1 stone loomweight | |
| 315bc | MCII | Intact | 3+? | | 13.4 | 26 | 22 | 5 knives, 3 HTWs, 3 axes, 4 tweezers, 2 pins, 1 needle, 2 awls, 1 chisel, 1 spatula | 2 zoo-morphic askoi | | 1 ceramic figurine, 1 spindle whorl | |
| 313b | MCI MCII early | Intact | 6 | | 9.4 | 47 | 40+ | 5 knives, 4 HTWs, 2 axes, 4 razor/ scrapers, 4 tweezers, 13 pins, 1 ring, 4 needles, 2 chisels, 1 spatula, leaves | 1 composite vessel | 1 silver ring, 2 faience necklaces, 4 faience beads | 4 ceramic figurines, 1 whetstone, 9 spindle whorls | |
| 805 | ECIIIB MCI MCII early | Intact? | 6–7 | | | 34 | 44 | 5 knives, 7 HTWs, 3 razor/ scrapers, 5 tweezers, 21 pins, 1 ring, 1 needle, 1 awl | | 1 silver ring, 21 faience beads, 1 Syro-Cilician? jug | 1 spindle whorl, 2 stone disks, 1 bone rod | |
| 322a | MCI | Intact | 4–5? | yes | 27.0 | 14 | 53 | 9 knives, 8 axes, 3 razor/ scrapers, 3 tweezers, 19 pins, 2 rings, 1 needle, 3 awls, 3 chisels | | 3 gold spirals, 1 silver spiral, 1 faience necklace, 1 Minoan dagger, 1 Minoan razor | 1 ceramic figurine, 2 whetstones, iron lump, 3 spindle whorls | |

| Tomb No. | Dates of Use | Preservation | Number of Burials | Complex Ritual | Floor Area sq m | Pots Total | Copper Total | Copper Types | Ceremonial Vessel /Rare Type | Imported Goods | Other | Fauna |
|---|---|---|---|---|---|---|---|---|---|---|---|---|
| 316 | MCII late | Intact | 6+ | | 12.6 | 135 | 53 | 11 knives, 5 HTWs, 2 axes, 1 razor/ scraper, 4 tweezers, 9 pins, 14 rings, 6 needles, 1 awl | | 7 silver/ lead rings in dromos cupboard | 1 ceramic figurine, 2 whetstones 1 macehead, 3 spindle whorls | |
| 804a | MCI late MCII MCIII early | Disturbed | 6+ | | | 36 | 61 | 9 knives, 3 HTWs, 4 axes, 3 razor/ scrapers, 1 tweezer, 23 pins, 5 rings, 6 needles, 6 awls, 1 hook? | | 3 faience beads | 1 ceramic figurine, 1 whetstone, 11 spindle whorls, 2 flaked stone tools, 1 stone core | |
| 320 | MCII late MCIII | Intact | 7? | | 9.9 | 85 | 64 | 11 knives, 12 HTWs, 3 axes, 2 razor/ scrapers, 7 tweezers, 18 pins, 4 rings, 5 needles, 2 awls | 1 ring vase | 1 faience necklace, [8 silver/ lead rings in 2 dromos cupboards] | 1 ceramic figurine, 1 macehead, 8 spindle whorls | |
| 313a | MCI | Intact | 15 | yes | 20.4 | 34 | 80 | 10 knives, 10 HTWs, 1 dagger, 3 axes, 7 razor/ scrapers, 7 tweezers 27 pins, 2 rings, 9 needles, 3 awls, 1 chisel | 1 askos | 2 silver bracelets, 1 silver ring, faience necklace fragments | 2 ceramic figurines, 6 whetstones, 6 spindle whorls, 1 flint chip | |

**Table 4.12** Lapithos wealth indices overall and by period. The means increase consistently and the differences among them are statistically significant.

| | | Lapithos overall | Early Group (ECII–IIIA/B) | Middle Group (ECIIIB–MCI) | Late Group (MCI–III) | F Statistic | p value |
|---|---|---|---|---|---|---|---|
| Pots per | Mean | 25.97 | 19.67 | 25.14 | 42.25 | 7.358 | 0.001 |
| Chamber | Std. Dev. | 23.92 | 15.45 | 21.38 | 32.14 | | |
| | Range | 2–135 | 2–81 | 3–110 | 7–135 | | |
| | n | 101 | 36 | 35 | 24 | | |
| Pots per | Mean | 10.49 | 8.35 | 12.19 | 13.69 | 3.968 | 0.022 |
| Burial | Std. Dev. | 7.10 | 4.14 | 9.03 | 9.00 | | |
| | Range | 1.00–36.67 | 2–19 | 1.00–36.67 | 2.27–32.00 | | |
| | n | 96 | 35 | 34 | 21 | | |
| Copper Items | Mean | 6.56 | 1.22 | 3.30 | 22.25 | 24.175 | 0.000 |
| per Chamber | Std. Dev. | 14.51 | 1.85 | 3.78 | 24.00 | | |
| | Range | 0–78 | 0–6 | 0–14 | 0–80 | | |
| | n | 101 | 36 | 35 | 24 | | |
| Copper Items | Mean | 1.70 | 0.53 | 1.57 | 4.92 | 31.932 | 0.000 |
| per Burial | Std. Dev. | 2.52 | 0.89 | 2.10 | 3.03 | | |
| | Range | 0–10.60 | 0–3 | 0–8 | 0–10.60 | | |
| | n | 96 | 35 | 34 | 21 | | |

**Table 4.13** Intensity of association between broad categories of artifact types at Lapithos as measured by the *phi* coefficient.

Lapithos ECII–ECIIIA/B Tombs

| | Copper | Large Fauna | Ceremonial Vessel | Figurine/ Idol | Imported Goods |
|---|---|---|---|---|---|
| Copper | 1 | | | | |
| Large Fauna | −0.062 | 1 | | | |
| Ceremonial Vessel | 0.375 | −0.139 | 1 | | |
| Figurine/ Idol | 0.210 | −0.078 | −0.053 | 1 | |
| Imported Goods | — | — | — | — | 1 |
| n=35 | | | | | |

Lapithos ECIIIB–MCI Tombs

| | Copper | Large Fauna | Ceremonial Vessel | Figurine/ Idol | Imported Goods |
|---|---|---|---|---|---|
| Copper | 1 | | | | |
| Large Fauna | 0.233 | 1 | | | |
| Ceremonial Vessel | 0.238 | 0.482 | 1 | | |
| Figurine/ Idol | 0.141 | −0.092 | 0.051 | 1 | |
| Imported Goods | 0.259 | 0.472 | 0.389 | 0.221 | 1 |
| n=38 | | | | | |

| Lapithos MCI–MCIII Tombs | | | | | |
| --- | --- | --- | --- | --- | --- |
| | Copper | Large Fauna | Ceremonial Vessel | Figurine/ Idol | Imported Goods |
| Copper | 1 | | | | |
| Large Fauna | — | 1 | | | |
| Ceremonial Vessel | 0.187 | — | 1 | | |
| Figurine/ Idol | 0.204 | — | 0.389 | 1 | |
| Imported Goods | 0.223 | — | 0.137 | 0.394 | 1 |
| n=23 | | | | | |

**Table 4.14**   Wealth indices for EC–MC tombs in Kalavasos Village.

| | | |
| --- | --- | --- |
| Pots per Chamber | Mean | 37.20 |
| | Std. Dev. | 30.45 |
| | Range | 6–100 |
| | n | 10 |
| Pots per Burial | Mean | 21.97 |
| | Std. Dev. | 17.86 |
| | Range | 7–64 |
| | n | 9 |
| Copper Items per Chamber | Mean | 8.71 |
| | Std. Dev. | 6.75 |
| | Range | 2–20 |
| | n | 7 |
| Copper Items per Burial | Mean | 3.92 |
| | Std. Dev. | 2.58 |
| | Range | 1.60–8.00 |
| | n | 7 |
| Estimated Chamber Floor Area (sq m) | Mean | 3.56 |
| | Std. Dev. | 1.22 |
| | Range | 1.60–5.17 |
| | n | 9 |

**Table 4.15** Percentages of tombs at *Vounous* and Lapithos in which particular artifact types occur. Numbers in parentheses represent the percentage of tombs with any copper at all in which the specified type is present. The latter values merit consideration inasmuch as the presence or absence of copper, especially in the earlier tomb groups, may have been affected by the genders and/or ages of the individuals buried.

|  | *Vounous* A ECI Group [N=45] | *Vounous* B ECII– ECIIIA Group [N=26] | *Vounous* B ECIIIB–MCII Group [N=52] |
|---|---|---|---|
| Copper present | 49 | 27 | 75 |
| Knife | 40(82) | 19(71) | 52(69) |
| HTW | 7 (14) | 12(43) | 44(59) |
| Axe | 4 (9) | 0 | 2(3) |
| Razor/scraper | 0 | 4(14) | 19(26) |
| Tweezer | 0 | 0 | 25(33) |
| Pin | 4(9) | 15(57) | 56(74) |
| Other ornament | 0 | 4(14) | 10(13) |
| Awl | 2(5) | 0 | 2(3) |
| Needle | 2(5) | 0 | 10(13) |
| Chisel | 2(5) | 0 | 0 |
| Spatula | 0 | 0 | 0 |
| Ceremonial vessel | 47 | 35 | 71 |
| Model or figurine | 13 | 27 | 19 |
| Large fauna | 33 | 27 | 38 |

|  | Lapithos ECII– ECIIIA/B Group [N=35] | Lapithos ECIIIB– MCI Group [N=38] | Lapithos MCI– MCIII Group [N=23] |
|---|---|---|---|
| Copper present | 40 | 74 | 96 |
| Knife | 20(50) | 39(54) | 74(77) |
| HTW | 23(57) | 37(50) | 65(68) |
| Axe | 0 | 0 | 39(41) |
| Razor/scraper | 11(29) | 32(43) | 48(50) |
| Tweezer | 9(21) | 29(39) | 61(63) |
| Pin | 9(21) | 34(46) | 83(86) |
| Other ornament | 10(14) | 5(7) | 57(59) |
| Awl | 0 | 3(4) | 30(32) |
| Needle | 3(5) | 5(7) | 43(45) |
| Chisel | 0 | 0 | 22(23) |
| Spatula | 0 | 0 | 13(14) |
| Ceremonial vessel | 9 | 39 | 43 |
| Model or figurine | 3 | 3 | 48 |
| Large fauna | 23 | 18 | 0 |

**Table 5.1** Examples of secondary treatment and complex ritual practices in Late Cypriot contexts.

| Site | Evidence for Secondary Treatment | Examples | References |
|------|----------------------------------|----------|-----------|
| Pendayia *Mandres* | Possibly two episodes of collective secondary treatment or 'mass burials' | Tomb 1 | Karageorghis 1965c |
| Pendayia *Mandres* | Two tombs either looted or perhaps deliberately emptied in antiquity, but with numerous metal objects left behind (cached?) | Tombs 2 and 3 | Karageorghis 1965c |
| Katydhata | Possible episode of collective secondary burial represented by 5 apparently squatting burials arranged in a circle | Tomb 42 | Gjerstad 1926:42 |
| Akhera *Chiflik Paradisi* | 3 intact tombs with no articulated skeletal remains, possibly indicating sequential or collective secondary treatment for one or more individuals including the latest burial in each tomb | Tombs 1–3 | Karageorghis 1965b |
| Politiko | Intact tomb with no articulated burials; last burial in chamber missing skull; probable sequential or collective secondary treatment for one or more individuals including the last burial | Tomb 6 | Karageorghis 1965a |
| Morphou *Toumba tou Skourou* | Defleshing of skeletons on dromos floor followed by seemingly haphazard redeposition of bones and goods in adjacent chambers | Tomb 1 | Vermeule and Wolsky 1990: 164 |
| Morphou *Toumba tou Skourou* | Individuals preponderantly represented by skulls, suggesting probable secondary treatment | Tomb 6 | Vermeule and Wolsky 1990: 309 |
| Morphou *Toumba tou Skourou* | Possible pre–interment ritual processing in area of 'funeral pyre' | Above Tombs 1 and 2 | Vermeule and Wolsky 1990: 169, 245–6 |
| Myrtou *Stephania* | Collective secondary treatment possibly represented in 'mass burial' | Tomb 12 | Hennessy 1964:31–3 |
| Myrtou *Stephania* | Possible secondary treatment involving relocation of goods and burials from 'niche' and side-chambers to main tomb chambers; attested by pottery joins | Tombs 3 and 5, side-chamber to main chamber of Tomb 14 | Hennessy 1964:3, 44 |
| Myrtou *Stephania* | Possible temporary burial features, emptied | Tombs 1, 11, 6, 8, (10 and 13?) | Hennessy 1964 |
| Ayia Irini *Palaeokastro* | Possible secondary treatment and collective burial | Tombs 20 and 21 | Pecorella 1977:103–7, 133–40 |
| Ayios Iakovos *Melia* | Probable episodes of collective secondary burial | Tombs 8, 10, 14 | Gjerstad *et al.* 1934 |
| Ayios Iakovos *Melia* | Possible secondary treatment and caching of metal goods | Tomb 10 side-chamber | Gjerstad *et al.* 1934:338 |
| Korovia *Nitovikla* | Possible secondary treatment suggested by poorly preserved, disarticulated burials | Tombs 1 and 2 | Gjerstad *et al.* 1934:407–14 |

| Site | Evidence for Secondary Treatment | Examples | References |
|------|----------------------------------|----------|-----------|
| Korovia *Nitovikla* | Possible emptied temporary grave yielding finds but no skeletal remains | Tomb 3 | Gjerstad *et al.* 1934:414–5 |
| Enkomi *Ayios Iakovos* | Deliberately emptied mortuary features | Cypriot Tombs 15, 16, 17, 21 | Dikaios 1969:422–3 |
| Enkomi *Ayios Iakovos* | Incomplete skeletal remains with possibly redeposited grave goods | Swedish Tomb 8 | Gjerstad *et al.* 1934:500–504 |
| Enkomi *Ayios Iakovos* | Collective secondary treatment possibly represented by squatting burials arranged along the chamber walls | Swedish Tombs 2, 6, and 17 | Gjerstad *et al.* 1934 |
| Enkomi *Ayios Iakovos* | Some burials represented mainly by skulls, suggesting secondary treatment | Swedish Tomb 11 | Gjerstad *et al.* 1934:510–25 |
| Enkomi *Ayios Iakovos* | Disarticulated remains of a single burial | Cypriot Tomb 24 (shaft grave) | Dikaios 1969:433 |
| Enkomi *Ayios Iakovos* | Lower left or right arm bones seemingly removed | Swedish Tombs 12 and 15 (15 a shaft grave), Cypriot Tomb 4A (shaft grave), French Tomb 5? | Gjerstad *et al.* 1934:522–6, 538, Figs 201, 204; Dikaios 1969:432; Schaeffer 1952: Pl. 36 |
| Milia *Vikla Trachonas* | Possible emptied burial feature | Tomb 12 | Westholm 1939a:14 |
| Milia *Vikla Trachonas* | 3 skeletons apparently missing skulls and other upper body parts, suggesting secondary treatment | Tomb 11 | Westholm 1939a:10 |
| Angastina *Vounos* | Possible emptied burial feature | Tomb 4 | Nicolaou 1972:58, note 2 |
| Kalavasos *Mangia* | Two seemingly simultaneous burials (at least one secondary?) | Tomb 6 | McClellan *et al.* 1988:208 |
| Kalavasos *Ayios Dhimitrios* | Emptied tombs (burial remains deliberately removed and reinterred?) | Tombs 2, 3, 7, 10, and 16 | South *et al.* 1989; South 1997; 2000 |
| Kalavasos *Ayios Dhimitrios* | Incomplete remains of Skeleton II suggest probable secondary burial | Tomb 11 | South 1997:161; South 2000:349–53; Goring 1989: Fig. 13:1 |
| Kalavasos *Ayios Dhimitrios* | Disarticulated remains of 3 infants and one child; secondary burials | Tomb 11 | as above |
| Kalavasos *Ayios Dhimitrios* | Possible partial removal of skeletal remains for reburial elsewhere | Tomb 14 | South 1997:165, South 2000:353–4 |
| Kalavasos *Ayios Dhimitrios* | Selective placement of skeletal remains in dromos niche | Tombs 11 and 20 | South 1997:162, 167–70; South 2000:352, 356–61 |
| Kourion *Bamboula* | Disarticulated and incomplete skeletal remains | Tomb 6 | Benson 1972:13–4; Angel 1972:160 |
| Kourion *Bamboula* | Carefully arranged piles of disarticulated bone arranged along chamber walls | Tomb 21 | Benson 1972:24 |
| Kouklia *Eliomylia* | Selective placement of skeletal remains in dromos niche | | Karageorghis 1990a:77–8 |

**Table 5.2**  Burial group statistics for *Toumba tou Skourou*, Enkomi, and Kourion *Bamboula*.

|  | Morphou *Toumba tou Skourou* | |
| --- | --- | --- |
| Mean Burials per Chamber | 7.90 | |
| Standard Deviation | 7.40 | |
| Range | 3–29 | |
| n | 10 | |
|  | Enkomi Chamber Tombs Overall | Enkomi Shaft Graves |
| Mean Burials per Chamber | 10.73 | 1.73 |
| Standard Deviation | 11.46 | 0.80 |
| Range | 1–55 | 1–3 |
| n | 26 | 15 |
|  | Kourion *Bamboula* Overall | |
| Mean Burials per Chamber | 10.75 | |
| Standard Deviation | 13.99 | |
| Range | 3–52 | |
| n | 12 | |

**Table 5.3** Age-sex distributions of burials from Ayios Iakovos, Enkomi, and Kourion. Each letter represents one individual. Individuals of borderline age are assigned to the older group (e.g. an individual aged 18 is included in the 18–24 rather than the 12–18 group). Identifications are based on Fischer (1986) and Angel (1972).

a. Ayios Iakovos *Melia*

| Age Group | Count by Sex |
|---|---|
| 0 | |
| 1–6 | |
| CHILD | ? |
| 6–12 | M?-?-? |
| 12–18 | M-F? |
| 18–24 | M?-M?-F-F-F |
| 24–30 | F |
| 30–36 | M-M-M-M-M-M? |
| 36–42 | M-M-M-M |
| 42–48 | M-M-M |
| 48–54 | M-M-M-F |
| 54–60 | F |
| 60+ | M-M-M |
| UNKNOWN | M-M-M-M-M-M-F-? |

b. Enkomi Ayios *Iakovos*

| Age Group | Count by Sex |
|---|---|
| 0 | |
| 1–6 | |
| CHILD | ? |
| 6–12 | ?-?-?-?-?-?-? |
| 12–18 | |
| YOUNG | F |
| 18–24 | F-F |
| 24–30 | M-M-M-M?-F-F-F-F-F |
| 30–36 | M-M-M-M-M-M?-M?-F-F-?-? |
| 36–42 | M-M-M-M-M?-M?-M?-M?-F?-F? |
| 42–48 | M-M-M-M-M?F-F? |
| 48–54 | M-M-M-M-M-F |
| 54–60 | M-M |
| 60+ | M |
| OLD | M-M-M |
| UNKNOWN | M-M-M-M-M?-M?-M?-M?-F-F-F-F?-F?-? |

c. Kourion *Bamboula*

| Age Group | Count by Sex |
|---|---|
| 0–1 | F?*-?-?-?-? |
| 2–6 | M*-M?*-F?*-F?*-F?*-?-?-?-? |
| 6–12 | M*-M?* |
| 12–18 | M?-F-F?-? |
| 18–24 | F? |
| 24–30 | M-M-M-M-M-M-M-F-F-F?-F?-F? |
| 30–36 | M-M-M-M-M-M-M-M-M-M?-M?-M?-F-F-F-F?-F?-F? |
| 36–42 | M-M-M-M-M-M-M-M-M-M?-M?-F-F-F |
| 42–48 | M-M-M-M-F |
| 48–54 | |
| 54–60 | |
| 60+ | |
| UNKNOWN | M-M-M-M-M-M?-F-F-F-F?-F? |

*Note that the attribution of sex to juveniles is highly speculative (see Bass 1981:21).

**Table 5.4**   Tomb size statistics for Morphou *Toumba tou Skourou*, Ayia Irini *Palaeokastro*, and Myrtou *Stephania* (dr. = dromos; ch. = chamber; all measurements are in meters).

| | | Morphou *Toumba tou Skourou* | Ayia Irini *Palaeokastro* | Myrtou *Stephania* |
|---|---|---|---|---|
| **Est. Ch. Floor Area** | Mean | 4.38 | 7.89 | 3.97 |
| | Std. dev. | 2.35 | 3.06 | 1.57 |
| | Range | 1.20–9.70 | 2.70–11.70 | 2.30–7.00 |
| | n | 12 | 8 | 10 |
| **Ch. Length** | Mean | 1.99 | 2.86 | 2.06 |
| | Std. dev. | 0.59 | 0.50 | 0.36 |
| | Range | 1.12–2.80 | 1.80–3.40 | 1.55–2.75 |
| | n | 11 | 8 | 10 |
| **Ch. Width** | Mean | 2.59 | 3.19 | 2.48 |
| | Std. dev. | 0.89 | 0.70 | 0.68 |
| | Range | 1.30–4.10 | 2.20–4.25 | 1.65–3.86 |
| | n | 11 | 8 | 10 |
| **Ch. Height** | Mean | 1.47 | 1.54 | — |
| | Std. dev. | 0.27 | 0.42 | — |
| | Range | 1.12–1.70 | 0.90–2.20 | — |
| | n | 4 | 7 | — |
| **Dr. Length** | Mean | — | 2.12 | 1.69 |
| | Std. dev. | — | 0.77 | 0.55 |
| | Range | 1.00–1.85 | 1.42–3.30 | 0.90–2.70 |
| | n | 2 | 7 | 7 |
| **Dr. Width** | Mean | — | 1.16 | 1.27 |
| | Std. dev. | — | 0.22 | 0.60 |
| | Range | 0.70–1.30 | 0.90–1.50 | 0.45–2.56 |
| | n | 2 | 7 | 8 |
| **Dr. Height** | Mean | — | 1.23 | 0.63 |
| | Std. dev. | — | 0.48 | 0.08 |
| | Range | 1.00 | 0.90–2.10 | 0.50–0.70 |
| | n | 1 | 7 | 7 |

**Table 5.5** Tomb size statistics for Enkomi (dr. = dromos; ch. = chamber; all measurements are in meters). The mean dimensions of tombs cut in LCII are consistently larger than those cut in earlier periods, but only the difference in the means for estimated chamber floor area appear to be statistically significant (T = −2.354, p = 0.034). The dimensions of the LCIII shaft graves would seem to represent a major decrease in energy expended in tomb construction.

| | | Enkomi Chamber Tombs Overall | Chamber Tombs Cut MCIII–LCI | Chamber Tombs Cut LCII | LCIII Shaft Graves |
|---|---|---|---|---|---|
| Est. Ch. Floor Area | Mean | 4.94 | 4.57 | 6.31 | 1.90 |
| | Std. dev. | 2.13 | 2.08 | 1.93 | 1.01 |
| | Range | 0.90–10.20 | 0.90–10.20 | 3.40–9.20 | 0.85–4.47 |
| | n | 42 | 32 | 9 | 12 |
| Ch. Length | Mean | 2.44 | 2.37 | 2.70 | 1.87 |
| | Std. dev. | 0.62 | 0.65 | 0.46 | 0.23 |
| | Range | 1.10–4.05 | 1.10–4.05 | 1.90–3.28 | 1.60–2.35 |
| | n | 41 | 31 | 9 | 12 |
| Ch. Width | Mean | 2.42 | 2.39 | 2.56 | 0.99 |
| | Std. dev. | 0.64 | 0.68 | 0.54 | 0.42 |
| | Range | 0.94–4.44 | 0.94–4.44 | 1.65–3.19 | 0.60–1.90 |
| | n | 41 | 31 | 9 | 12 |
| Ch. Height | Mean | 1.31 | 1.26 | 1.44 | 0.43 |
| | Std. dev. | 0.33 | 0.29 | 0.44 | 0.31 |
| | Range | 0.75–2.00 | 0.95–1.85 | 0.75–2.00 | 0.20–1.05 |
| | n | 27 | 20 | 7 | 11 |
| Dr. Length | Mean | 1.53 | 1.53 | 1.58 | |
| | Std. dev. | 0.45 | 0.45 | 0.46 | |
| | Range | 0.95–2.60 | 1.00–2.60 | 0.95–2.40 | |
| | n | 34 | 25 | 8 | |
| Dr. Width | Mean | 0.97 | 0.98 | 1.01 | |
| | Std. dev. | 0.15 | 0.16 | 0.10 | |
| | Range | 0.71–1.40 | 0.71–1.40 | 0.85–1.14 | |
| | n | 34 | 25 | 8 | |
| Dr. Height | Mean | 1.03 | 1.03 | 1.03 | |
| | Std. dev. | 0.30 | 0.33 | 0.21 | |
| | Range | 0.53–1.85 | 0.53–1.85 | 0.75–1.35 | |
| | n | 31 | 24 | 7 | |

**Table 5.6** Correlations (Pearson's r) between the variables of burials per chamber, estimated chamber floor area, total pots per chamber, total copper-based artifacts per chamber, and total gold weight per chamber at Enkomi. Correlations were run pairwise with the following ranges of n values: Enkomi overall: 23–39; MCIII–LCIB/LCIIA group: 8–13; LCIA–LCIIB group: 6–9; LCIA–LCIIC group: 9–17; LCIIC–LCIII group: 13–25.

| | Enkomi Chamber Tombs Overall | | | | |
| --- | --- | --- | --- | --- | --- |
| | Burials per Chamber | Est. Chamber Floor Area | Pots per Chamber | Copper Items per Chamber | Gold Weight per Chamber |
| Burials per Chamber | 1 | | | | |
| Est. Chamber Floor Area | 0.297 | 1 | | | |
| Pots per Chamber | 0.734 | 0.250 | 1 | | |
| Copper Items per Chamber | 0.684 | 0.470 | 0.408 | 1 | |
| Gold Weight per Chamber | 0.173 | 0.406 | 0.294 | 0.394 | 1 |

| | Enkomi Chamber Tombs in Use from MCIII–LCIB/LCIIA | | | | |
| --- | --- | --- | --- | --- | --- |
| | Burials per Chamber | Est. Chamber Floor Area | Pots per Chamber | Copper Items per Chamber | Gold Weight per Chamber |
| Burials per Chamber | 1 | | | | |
| Est. Chamber Floor Area | 0.110 | 1 | | | |
| Pots per Chamber | 0.289 | 0.472 | 1 | | |
| Copper Items per Chamber | 0.530 | 0.370 | 0.158 | 1 | |
| Gold Weight per Chamber | 0.479 | 0.343 | 0.164 | 0.075 | 1 |

| | Enkomi Chamber Tombs in Use from LCIA–LCIIB | | | | |
| --- | --- | --- | --- | --- | --- |
| | Burials per Chamber | Est. Chamber Floor Area | Pots per Chamber | Copper Items per Chamber | Gold Weight per Chamber |
| Burials per Chamber | 1 | | | | |
| Est. Chamber Floor Area | –0.246 | 1 | | | |
| Pots per Chamber | –0.124 | –0.045 | 1 | | |
| Copper Items per Chamber | –0.278 | 0.022 | 0.978 | 1 | |
| Gold Weight per Chamber | 0.423 | 0.283 | –0.070 | 0.096 | 1 |

| | Enkomi Chamber Tombs in Use from LCIA–LCIIC | | | | |
|---|---|---|---|---|---|
| | Burials per Chamber | Est. Chamber Floor Area | Pots per Chamber | Copper Items per Chamber | Gold Weight per Chamber |
| Burials per Chamber | 1 | | | | |
| Est. Chamber Floor Area | −0.228 | 1 | | | |
| Pots per Chamber | 0.634 | −0.448 | 1 | | |
| Copper Items per Chamber | 0.563 | 0.234 | 0.097 | 1 | |
| Gold Weight per Chamber | −0.161 | 0.268 | 0.164 | 0.471 | 1 |

| | Enkomi Shaft Graves LCIIC–LCIII | | | | |
|---|---|---|---|---|---|
| | Burials per Chamber | Est. Chamber Floor Area | Pots per Chamber | Copper Items per Chamber | Gold Weight per Chamber |
| Burials per Chamber | 1 | | | | |
| Est. Chamber Floor Area | 0.754 | 1 | | | |
| Pots per Chamber | 0.854 | 0.642 | 1 | | |
| Copper Items per Chamber | 0.809 | 0.666 | 0.913 | 1 | |
| Gold Weight per Chamber | 0.938 | 0.673 | 0.931 | 0.310 | 1 |

**Table 5.7** Tomb size statistics for Kourion *Bamboula* (dr. = dromos; ch. = chamber; all measurements are in meters).

| | | Kourion *Bamboula* |
|---|---|---|
| **Est. Ch. Floor Area** | Mean | 5.37 |
| | Std. dev. | 2.00 |
| | Range | 3.00–10.20 |
| | n | 16 |
| **Ch. Length** | Mean | 2.40 |
| | Std. dev. | 0.57 |
| | Range | 1.40–4.20 |
| | n | 22 |
| **Ch. Width** | Mean | 2.83 |
| | Std. dev. | 0.75 |
| | Range | 1.80–5.14 |
| | n | 22 |
| **Ch. Height** | Mean | 1.62 |
| | Std. dev. | 0.35 |
| | Range | 0.92–2.20 |
| | n | 10 |
| **Dr. Length** | Mean | 1.29 |
| | Std. dev. | 0.14 |
| | Range | 1.00–1.40 |
| | n | 7 |
| **Dr. Width** | Mean | 1.28 |
| | Std. dev. | 0.13 |
| | Range | 1.00–1.40 |
| | n | 7 |
| **Dr. Height** | Mean | 0.67 |
| | Std. dev. | 0.29 |
| | Range | 0.45–1.10 |
| | n | 4 |

**Table 5.8** Finds and other variables from well-reported tombs in northern Cyprus dating mainly from MCIII/LCIA–LCIIA (except Kazaphani). Note that finds from the multi-chambered tomb complexes at *Toumba tou Skourou* have been pooled. Abbreviations: Kaz = Kazaphani, TtS = *Toumba tou Skourou*, AIr = Ayia Irini (AIr& is the tomb reported by Quilici 1990), MS = Myrtou *Stephania*, Pen = Pendayia, AIak = Ayios Iakovos, Nit = Nitovikla, Mil = Milia; TeY = Tell el-Yahudiyeh ware, Myc = Mycenaean; other pottery abbreviations are standard.

| Tomb No. | Dates of Use | Preservation | Number of Burials | Complex Ritual | Floor Area sq m | Pots Total | Gold and Silver Types | Copper/ Bronze Types | Other Prestige Emblems | Other Exotic and Local Valuables |
|---|---|---|---|---|---|---|---|---|---|---|
| Kaz 2a | MCIII/ LCIA– LCIIC | Partly looted | ? | | 9.5 | 456 | silver: 2 rings, 1 earring | 2 daggers, 5? knives, spatula fragment, ring fragments | 1 Pastoral crater, 2 WPV rhyta, 1 BR rhyton | 1 LMIIIB jar, faience beads, ostrich egg, 2 alabaster jugs, worked bone |
| Kaz 2b | LCIA– LCIIC | Partly looted | ? | | 10.5 | 547 | silver: 1 ring, 5 earring fragments | 2 daggers, 2? knives, 1 spatula, 5+ rings, hook, rod fragment | 1 cylinder seal, 1 stamp seal, 1 boat model, 1 sandstone weight, 5 WPV rhyta, 15 BR rhyta | carnelian, andesite, faience beads, glass vase, ivory bead and pin fragment |
| TtS1 (3 chambers) | MCIII/ LCIA– LCIB | Intact | 24–29 (ch1) | yes | 9.7, 5.4, 6.6 | 516+ | gold: 2 studs, silver: 14 rings, rivets and other small objects | 90+ objects including 1 Aegean style razor, 1 Near Eastern style dagger, 5 other knives/ daggers, 2 spikes?, 6 tweezers, many spiral rings, beads, bracelets, pins, needles | 3 cylinder seals, 1 scarab, 2 hematite weights, bone gaming pieces and box fragments (1 w/ figure in stippled robe, 1 w/ Hathor head), numerous coroplastic vessels and askoi/ rhyta | LMIB pottery, carnelian, rock crystal, amethyst, faience, glass beads, small bone and ivory objects, ?horse bones, dog/ fox skull |
| TtS 2 (4 chambers) | LCIB LCIIB | Intact | 9, 3, 3, 3 | yes | 2.47, 1.70, 3.24, 4.64 | 31+? | gold: 2 cylinder caps, 2 beads | 1 stud, 1 knob | lapis lazuli cylinder seal | Minoan, Myc pottery, faience and glass beads, 2 ostrich eggs, 2 glass bottles, small ivories, worked bone |
| TtS 3 | LCIA | Intact | 5 | | 4.1 | 17+ | | 10 including 1 knife, 2 pins, 2 spirals | mace-head, 1 WPV coroplastic jug | lead bead/ whorl |

| Tomb No. | Dates of Use | Preservation | Number of Burials | Complex Ritual | Floor Area sq m | Pots Total | Gold and Silver Types | Copper/ Bronze Types | Other Prestige Emblems | Other Exotic and Local Valuables |
|---|---|---|---|---|---|---|---|---|---|---|
| TtS 4 | LCIA LCIB | Intact | 14 | | 6.1 | 125 | silver/ lead coil | 7 spiral beads | 2 BS coroplastic askoi | |
| TtS 5 (2 chambers) | MCIII/ LCIA | Intact | 5, 11 | | 1.15, 1.98 | 45 + ? | | several, including 2 axes, 1 knife, 2 spear- heads, 1 spike, 1 razor, 1 fish hook, tweezers, pins, rings, beads, needles | maceheads | TeY pottery |
| TtS 6 | LCIB | Intact | 5 | yes | 4.6 | 22 + ? | | several, including 1 axe, 1 dagger, tweezers, pins, spirals, beads, needle | 1 askos, 1 composite vase | |
| AIr 3 | LCIA– LCIB | Partly looted | 3 + ? | | 7.0 | 73 | 3 silver rings | 2 bowls, 6 pins, 7 spirals, 1 spiral bead, 2 bracelets | 11 hematite weights, 2 cylinder seals, 1 WPV–VI askos/ rhyton | 2 LHII cups, carnelian, faience beads, worked bone |
| AIr 10 | LCIA– LCIB | Partly looted | 3 + ? | | 7.1 | 22 | | 1 axe, 1 dagger, 2 pins, 2 pins/ awls?, 1 spiral, 2 bracelets | 2 cylinder seals, 1 scarab | |
| AIr 11 | LCIA– LCIB | Partly looted | ? | | 10.9 | 46 + | 1 silver ring, 1 gold bead | 1 blade fragment, 2 pins, 2 rings, 10 spirals | 1 hematite weight, 5 other weights, 3 cylinder seals | faience beads |
| AIr 12 | LCIA– LCIB | Partly looted | 2 + ? | | 5.7 | 11 + | silver: 1 spiral ring, 2 tubular neck fragments | 1 tweezer, 1 spiral ring, 1 bracelet?, 1 pin | WPVI coroplastic vessel(s) | faience beads, 4 lead rings, worked bone |
| AIr 20 | LCIA– LCIB | Partly looted | 15 | yes? | 10.8 | 70 | gold necklace fragment; silver: 2 earrings, 2 spirals | 2 blade frag- ments, 3 pins, 1 ring, 7 spirals, balance pan fragments | 1 cylinder seal, 1 stone weight | 1 LHII cup, faience, glass and lead beads, worked bone |

| Tomb No. | Dates of Use | Preservation | Number of Burials | Complex Ritual | Floor Area sq m | Pots Total | Gold and Silver Types | Copper/ Bronze Types | Other Prestige Emblems | Other Exotic and Local Valuables |
|---|---|---|---|---|---|---|---|---|---|---|
| AIr 21 | LCIA–LCIB | Partly looted | 14 | | 9.8–11.7 | 140 | | dagger fragment, tongs, ingot fragment, balance pan fragments, 2 ferrules, 3 pins, 24 spiral beads, 3 spiral rings, 1 ring/bracelet, other items | 7 hematite weights, 1 cylinder seal, 1 WPVI coroplastic jug | carnelian, faience beads, ostrich egg, worked bone and ivory, iron? |
| AIr & | LCIA–LCIB | Partly looted | 37 | | 4.3 | 300+ | small silver objects | 1 Aegean razor, 3 HTWs 4 knives/daggers, juglet, bowl and mirror fragments, rings, spirals, pins, beads | several hematite and other stone weights, 1 cylinder seal, mace-heads, WPWM burner, WPV askos | 2 LHII cups, glass, faience beads, 2 faience vases, small lead objects, worked bone, iron slag? |
| MS 4 | LCIA–LCIIA/LCIIB | Intact | 5 (+3 in side-chamber) | yes | 7.0 | 45 (+7 in side-chamber) | silver/lead spiral ring | 1 ring (+4 spirals in side-chamber) | | Myc IIIB closed containers, lead spiral in side-chamber |
| MS 5 | LCIA | Intact? | 18? | yes | 4.3 | 34 | | tweezer and pin fragments | | lead ring fragments |
| MS 7 | LCIA–LCIIB | Intact | 3+? | | 4.8 | 41 | gold ring | tweezer fragment, pin | 2 WPV coroplastic vessels | Myc IIIB closed container |
| MS 10 | MCIII | Intact? | 1? | | 2.5 | 21 | | 1 axe, 1 HTW, 1 knife, 1 razor, 1 tweezer, 1 pin | 2 WPV coroplastic vessels | |
| MS 12 | LCIA | Intact | 14+ | yes | 2.6 | 13 | | 1 axe, 2 spirals, 1 tweezer or spatula fragment | | Myc IIIB closed container, lead ring |
| MS 13 | MCIII | Intact? | 1? | | 3.5 | 7 | | | | |
| MS 14 | LCIA–LCIB | Intact | 2+? | yes | 3.7 | 46 | | 3 spirals, 2 beads, 2 awls | 1 copper cylinder seal | 2 lead spirals |

| Tomb No. | Dates of Use | Preservation | Number of Burials | Complex Ritual | Floor Area sq m | Pots Total | Gold and Silver Types | Copper/ Bronze Types | Other Prestige Emblems | Other Exotic and Local Valuables |
|---|---|---|---|---|---|---|---|---|---|---|
| Pen 1 | MCIII/LCIA | Intact | 34 | yes | 5.4 | 77 | silver/lead ring | 9 HTWs, 14 knives, 3 razors, 1 tweezer, bowl fragment, 10 pins, 3 needles, 6 spirals/ rings | 4 mace-heads, 1 BSIII rhyton | paste/ faience beads, worked bone |
| Pen 2 | MCIII/ LCIA | Looted | ? | | 3.3 | — | silver/ lead ring frags | 1 knife, 5 pins, 1 pin/ needle, 1 awl, 4 spirals | | |
| Pen 3 | MCIII/ LCIA | Looted | ? | | 3.8 | — | | 5 HTWs, 5 knives, 3 pins, 1 bracelet | | |
| AIak 8 | LCIA– LCIIA | Intact | 62 | yes | 15.6 | 66 | | 1 axe, 4 knives, 1 scraper, 3 pins, 5 needles, 3 bracelets, 1 ring, 2 fragments leaf | 1 cylinder seal, 3 maceheads | Myc closed containers, ivory pin |
| AIak 10 | LCIB | Looted? | 10 (+3 in side-chamber) | | 24.8 | 12 | | 2 knives, 1 dagger, 2 tweezers, (1 axe, 1 knife, 1 tweezer, 1 awl, 2 spirals, 4 bracelets in side-chamber) | | |
| AIak 12 | MCIII– LCIA | Looted? | ? | | 12.6 | 27 | | 2 knives, 1 tweezer, 7 pins | macehead | faience beads |
| AIak 13 | MCIII, LCIIA | Intact? | 8+? | | 13.7 | 37 | 3 gold beads | 1 pin, 2 spirals, other fragments | macehead, BR rhyton | Myc closed containers, faience bead and plain cylinder |
| AIak 14 | LCIB– LCIIA | Intact | 40 | yes | 20.7 | 41 | | 2 knives, 1 HTW, 1 socketed spearhead, 1 tweezer, 5 needles, 1 bracelet, 3 rings, 2 earrings | 1 mace-head, 1 ceramic figurine, 3 BR rhyta | Myc closed containers, faience beads |

| Tomb No. | Dates of Use | Preservation | Number of Burials | Complex Ritual | Floor Area sq m | Pots Total | Gold and Silver Types | Copper/ Bronze Types | Other Prestige Emblems | Other Exotic and Local Valuables |
|---|---|---|---|---|---|---|---|---|---|---|
| Nit 1 | MCIII/ LCIA | Intact? | 3 | ? | 4.8 | 25 | | 4 knives, 1 scraper, 8 pins, 5 pins/ needles, 1 ring | 2 mace-heads | |
| Nit 2 | MCIII– LCIA | Intact? | 7 | ? | 5.8 | 37 | | 5 knives, 1 scraper, 14+ pins, several rings, 2 bracelets, chain of 3 rings | 4 mace-heads | faience beads |
| Mil 10 | LCIA– LCIB/ LCIIA | Looted | ? | | ? | 94 | | 1 axe, 2 knives, 1 sword/ HTW?, 1 tweezer, 5 needles, 8 rings, 1 bracelet, 1 bead, rod fragments | cylinder seal | TeY juglets, Myc closed containers and kylikes, faience statuette and beads, lead ring |
| Mil 11 | LCIA– LCIB/ LCIIA | Intact | 4 | ? | 3.0 | 22 | gold: 1 earring, 1 hair ring? | 2 knives, 1 dagger?, 3 pins, 6+ rings, 1 bracelet | stone axe | TeY juglet, faience beads |
| Mil 13 | LCIA– LCIB/ LCIIA | Looted? | ? | | 4.3 | 61 | | 2 knives, bowl frag-ments, 3 rings, 1 bracelet | | Myc bowls, faience bottle and beads, worked bone |

**Table 5.9a** Enkomi finds and other variables from well-preserved tombs with principal components of use dating mainly to MCIII, LCI, and early LCIIA. Tombs are listed in order of increasing overall wealth. Abbreviations: Myc= Mycenaean, TeY = Tell el-Yahudiyeh ware.

| Tomb No. | Dates of Use | Preservation | Location | Number of Burials | Complex Ritual | Floor Area sq m | Pots Total | Total Gold Wt g | Gold and Silver Types | Bronze Types | Other Prestige Emblems | Other Exotic and Local Valuables |
|---|---|---|---|---|---|---|---|---|---|---|---|---|
| S12 | MCIII LCIA | Intact | 7W? | 1 | yes | 2.7 | 2 | 0 | | | | |
| C14 | LCIA LCIB | Intact | 4W | 1 | | 4.7 | 4 | 0 | | | | |
| S4 | LCIA LCIB | Intact | 3W? | 4 | | 0.9 | 7 | 0 | | | | |
| B52 | MCIII LCIA | Intact | West Area? | ? | | | 13–14 | 0 | | | | |
| S20 | MCIII LCIA | Intact | 6W? | 1 | | 2.7 | 4 | 0 | | | | 1 TeY Juglet |
| F12 (1946) | MCIII LCIA | Intact | 8E | 2? | | 2.5 | 9 | 0 | | | | 1 TeY? Juglet |
| F11 (1934) | MCIII LCIA | ? | 8W/ 9W? | 4+? | | | 8 | 0 | | 1 ring | | 4 TeY juglets, 15 faience beads |
| C3 | MCIII LCIA | Intact | 4W | 4 | | 3.5 | 10 | 0 | | 6 spiral rings | | 1 lead spiral ring |
| F15 (1949) | LCIB | Intact | ? | 2+? | | 3.0 | 50 | 0 | 2 pr. silver earrings, 1silver pin/ nail | 1 pin | | |
| F1851 | LCIB | Intact | 2W | 7 | | 3.8 | 24 | | | pr. balance pans | 1 rock crystal weight | ostrich egg |
| F32 (1957) | MCIII LCIA LCIB LCIIA? | ? | 5W | ? | | | 59–66 | 0 | | 24 rings/ spirals, 1 axe | 2 stone maceheads, 4 hematite weights, 1 faience scarab | 11 TeY juglets, 7 Red Burnished juglets, 3 faience beads |
| F126 | LCIA LCIB LCIIA | ? | 6W | ? | | ? | 57 | <10? | gold: 2 earrings, 1 pin casing, 1 band; 1 electrum earring | | 1 hematite weight | MycIIIA: 1 pyxis |
| S8 | LCIA LCIB LCIIA? | Intact | 3W? | 5–6? | yes | 4.4 | 22 | 70+ | gold: 6 earrings, 2 beads, 2 pins, 19 bands; 1 silver/ lead ring, 1 silver earring, 1 silver/gold pin | 1 strip, 1 knife, 1 HTW, 4 silver/ bronze pins | | carnelian, amethyst, faience beads, 1 faience vase |

**Table 5.9b** Enkomi finds and other variables from well-preserved tombs with principal components of use dating mainly to LCIB, LCIIA, and LCIIB. Tombs are listed in order of increasing wealth in gold. In tombs with numerous jewelry items I have for the sake of brevity subsumed commonly occurring types such as earrings, hair spirals, beads and bands under the designation 'various' and itemized only the more remarkable, higher order objects (types not found in 'poorer' tombs). Note that the gold bands listed in French Tomb 11 were fairly common, occurring in all the other tombs with gold listed in this table. Abbreviations: Myc= Mycenaean; RLWM= Red Lustrous Wheelmade.

| Tomb No. | Dates of Use | Preservation | Location | Number of Burials | Complex Ritual | Floor Area sq m | Pots Total | Total Gold Wt g | Gold and Silver Types | Bronze Types | Other Prestige Emblems | Other Exotic and Local Valuables |
|---|---|---|---|---|---|---|---|---|---|---|---|---|
| F3(34) | LCIB LCIIA | Intact | 8W/ 9W? | 6+? | | 2.8 | 53 | 0 | | | | |
| S2 | LCIB LCIIA LCIIB? | Intact | 3W? | 11 | yes | 4.6 | 66 | 0 | | | 2 cylinder seals | |
| C19 | LCIIA | Intact | 4W | 5 | | 3.6 | 38 | 0 | | | | ceramic figurine |
| F11 (1949) | LCIIA | Intact | ? | 4+? | | 4.4 | 175 | <30? | 9 gold bands | 3 pr. earrings, 1 unid. | 1 hematite weight | amber, carnelian, faience beads, ivory box, alabaster vases |
| B92 | LCIB LCIIA LCIIB? | ? | 5W | ? | | | ? | 67 | gold: 5 elaborate pins, other items; silver: 1 cup, 2 bowls, other items | 3 knives (1 w/ silver handle) | 5 hematite weights | |
| F2(49) | LCIIA | Intact | 5E | 3 | ? | 6.4 | 36 | 100+ | gold: 2 large bands w/ sphinxes, scepter (?), other items; silver: 2 signet rings, 2 bowls (1 w/ gold inlay), other items | | 2 Myc craters, RLWM arm vessel and other RLWM vases | alabaster vases |
| B67 | LCIB LCIIA LCIIB LCIIC? | ? | 1E/ 2E | ? | | | 21+? | 154 | gold: 2 elaborate pins, 2 signet rings, 2 scarab rings, other items; silver: 1 signet ring, other items | 1 knife, 1 signet ring | 3 Myc craters, 1 Myc bucranial rhyton, 1 cylinder seal, 6 hematite weights | agate, carnelian, faience beads, ivory, ceramic figurines |

| Tomb No. | Dates of Use | Preservation | Location | Number of Burials | Complex Ritual | Floor Area sq m | Pots Total | Total Gold Wt g | Gold and Silver Types | Bronze Types | Other Prestige Emblems | Other Exotic and Local Valuables |
|---|---|---|---|---|---|---|---|---|---|---|---|---|
| S17 | LCIB LCIIA LCIIB | Intact | 7W? | 12 | yes | 4.1 | 68 | 235+ | gold: 1 bowl, 1 elaborate pin, other items | 1 knife | 1 Myc crater ('Zeus'), 1 cylinder seal | 1 faience bead, small ivories |
| B19 | LCIA? LCIB LCIIA LCIIB | ? | 4E | ? | | ? | ? | 385 | gold: several elaborate necklaces, pins and rings, 1 signet ring, other items; various silver items | | Myc crater fragments, faience female head cup, 1 cylinder seal, 7 hematite weights? | 3 faience beads, faience disc, faience dish, small ivories |

**Table 5.9c**  Enkomi finds and other variables from well-preserved tombs with principal components of use dating from LCI–LCIIC. French Tomb 5, with a major LCIII component, is also included as many of the valuables may derive from the LCIIC period. Tombs are listed in order of increasing wealth in gold. As in Table 5.9b, commonly occurring gold jewelry types such as earrings, hair spirals, beads and bands are subsumed in richer tombs under the designations 'various' or 'other'. Abbreviations: see Table 5.9b caption.

| Tomb No. | Dates of Use | Preservation | Location | Number of Burials | Complex Ritual | Floor Area sq m | Pots Total | Total Gold Wt g | Gold and Silver Types | Bronze Types | Other Prestige Emblems | Other Exotic and Local Valuables |
|---|---|---|---|---|---|---|---|---|---|---|---|---|
| S22 | LCIIB LCIIC | Intact? | 5W | 2 | | 3.4 | 10 | 0 | | | | |
| S13 | LCIA? LCIB–LCIIC | Intact? | 6W? | ? | | 5.9 | 215 | <4 | 3 gold beads, 1 gold earring | | Pastoral crater, cylinder seal | faience dish, bead |
| F110 | LCIA–LCIIC | ? | 5W | ? | | ? | 313 | 4? | 2 gold beads, 2 gold earrings | 4 rings, 3 spirals, 3 bracelets | Myc piriform rhyton, carnelian scarab | faience cup and beads, alabaster vase, 2 Minoan jars |
| S6 | LCIIB? LCIIC LCIIIA | Intact? | 3W? | 13–15 | yes | 7.8 | 82 | 5+ | 3 gold bands, 1 silver ring | 5 bracelets | | faience beads, glass bottle, small ivories, stone mortar/pestle |
| S10 | LCIIB LCIIC LCIIIA | Intact? | 6W? | 13 | | 9.2 | 48 | 9 | 1 gold earring, 2 gold bands | 6 bowls | cylinder seal | small ivories |

| Tomb No. | Dates of Use | Preservation | Location | Number of Burials | Complex Ritual | Floor Area sq m | Pots Total | Total Gold Wt g | Gold and Silver Types | Bronze Types | Other Prestige Emblems | Other Exotic and Local Valuables |
|---|---|---|---|---|---|---|---|---|---|---|---|---|
| F5(49) | LCIB–LCIIIB | Intact? | 5E | 55+? | | 5.2 | 283 | 20 | gold: 1 ring, 4 earrings, 3 bands, other small items | 10 rings, 4 earrings, 1 bracelet, 1 spatula, 1 mirror, 1 dagger, 11 bowls | | carnelian, quartz, faience beads, 4 faience vases, 7–8 glass bottles, small ivories, worked bone, alabaster vase, stone mortar/pestle |
| B91 | LCII | ? | 4W | ? | | ? | 15+? | 20 | gold: 2 bands w/sphinxes, other items; unid. silver | unid. | 1 Myc crater | ceramic figurine, unid. faience and glass, alabaster vase? |
| B45 | LCI–LCII | ? | 5W | ? | | ? | 59? | 20 | gold: 3 spirals, 7 bands, other small items | ring fragments | 2 Myc craters, 4+? Pastoral craters, 1–2 BR bull rhyta, cylinder seal | ostrich egg, faience vase, stone mortar/pestle |
| S19 | LCIA LCIB LCIIC | Intact | 5W? | 17 | | 6.3 | 132 | 45 | 1 gold scarab ring and other items, 1 silver bracelet | 6 rings, 6 earrings, 1 bronze/ivory chisel, 1 mirror, 5 bowls | 3 Pastoral craters, macehead | faience bowl and bead, ivory mirror handle w/female fig., 2 ceramic figurines |
| B86 | LCIA–LCIIC | ? | 6W | ? | | ? | ? | 57 | gold: 2 beads, 6 spirals, 13 earrings, 1 band | 2 earrings? | 1 Pastoral crater, 2 faience female head cups, 1 faience ram's head rhyton, 1 cylinder seal | 1 TeY juglet, 1 amber bead, 1 faience pendant, 1 glass bottle, small ivories, ceramic figurine, stone mortar/pestle |

| Tomb No. | Dates of Use | Preservation | Location | Number of Burials | Complex Ritual | Floor Area sq m | Pots Total | Total Gold Wt g | Gold and Silver Types | Bronze Types | Other Prestige Emblems | Other Exotic and Local Valuables |
|---|---|---|---|---|---|---|---|---|---|---|---|---|
| B84/84 a | LCIA? LCIB– LCIIC | ? | 6W | ? | | ? | 26? | 59 | gold: 1 bead, 2 spirals, 3 earrings, 7–9 bands | bracelets, mirror, unid. | Myc crater, 3 cylinder seals (1 w/ cuneiform inscrip., 1 Mitannian?), 1 faience scarab | faience cup and gourd-shaped vase, ivory comb and unid., ceramic figurine |
| S11 | LCIIA LCIIB LCIIC | Intact | 7W? | 21+? | yes? | 5.1 | 249 | 65 | gold: 2 beads, 2 spirals, 8 earrings, 4 bands, 1 bracelet | 2 bowls | 2 Myc craters, 1 faience deer's head rhyton, 3 cylinder seals, 1 stamp seal | ostrich egg, 2 faience dishes, small ivories |
| B79 | LCIIA LCIIB LCIIC | ? | 6W | ? | | ? | 43? | 66 | gold: 6 earrings, 11 bands, 1 necklace | spearhead | Myc crater, 2 scarabs | 2 faience vessels, small ivories, BR bull ?figurine, stone mortar/ pestle |
| B69 | LCIB– LCIIC | ? | 2W | ? | | ? | 47? | 97 | gold: 1 necklace, 1 elaborate ring, other items; silver: various | 1 signet ring | Myc conical rhyton, RLWM arm vessel | 2 faience veases, small ivories, 2 ceramic figurines, stone mortar/ pestle |
| C10 | LCIA– LCIIC | Partly looted? | 4W | 21+? | | 4.1 | 407 | 145 + | gold: 1 bead, 2 spirals, 10 earrings, 15 bands, 1 bracelet, 1 necklace, other small items | 1 ring, 3 bracelets, 1 bowl, 2 helmets | 2 Myc craters, 1 Myc piriform rhyton, BR and WPV bull rhyta, snake-house? | ostrich egg, 4 faience vases and ring, 2 glass bottles, small ivories |
| S3 | LCIB– LCIIC | Partly looted? | 3W? | 15+ | | 7.3 | 172 | 150+ ? | gold: 1 signet ring, 1 other elaborate ring, necklace, other items; silver: 1 cup/bowl and other items | 3 daggers | 19 Myc and 2 Pastoral craters, 1 Minoan crater, 1 faience female head cup, 1 BR bull rhyton | carnelian, faience beads, 2 faience vases, 2 glass bottles, small ivories, ceramic figurine |

| Tomb No. | Dates of Use | Preservation | Location | Number of Burials | Complex Ritual | Floor Area sq m | Pots Total | Total Gold Wt g | Gold and Silver Types | Bronze Types | Other Prestige Emblems | Other Exotic and Local Valuables |
|---|---|---|---|---|---|---|---|---|---|---|---|---|
| S18 | LCIIC | Intact | 5W | 10–12 | ? | 8.5 | 102 | 300+ | gold: 3 signet rings and 2 other elaborate rings, 1 elaborate necklace and other items; silver: 1 bowl, 2 bracelets | 4 rings, 2 anklets, 1 chisel, 2 mirrors, 2 spear-heads, 1 Naue II sword, 1 greave, 13 bowls, 1 jug | 10 Myc and 1 Pastoral crater, 1 Minoan crater | ostrich egg, faience dish, 6 glass bottles, small ivories, 2 ivory combs (1 w/ carved roebuck), stone mortar/ pestle |
| B66 | LCIIB LCIIC | Intact? | 4E | ? | | (Ashlar) | 39? (30 Myc) | 384 | gold: 2 signet rings, several other elaborate rings, pins, and 1 necklace, 1 bowl or mask, other items; 1 silver bowl | bracelet?, 5 spikes, 1 lamp, 5 daggers, 2 spear-heads, 1 mirror and bowl of Aegean types, 2 other bowls | 1 Myc crater, 1 lapis lazuli cylinder seal (Aegean), 1 Mitan-nian cylinder seal | carnelian, amber, faience beads, faience: mortar/ pestle, pail, 2 bottles, 4 bowls, 1 pictorial dish; 5 glass bottles, 2 alabaster vases |
| B93 | LCI? LCIIA LCIIB LCIIC | Disturbed | 9E/ 10E | ? | | ? | 5?? | 800+ | gold: 1 cup and various lumps, 1 signet ring, 1 scarab ring, several other elaborate rings (1 w/ soldered lions' heads), 2 elaborate necklaces, other items; silver: 2 signet rings and other items | | 2 Myc and 1 Pastoral crater, 1 ceramic chariot group, 2 Mitan-nian? cylinder seals, 1 faience scarab, 1 Cretan lapis lazuli engraved prism | carnelian, amber, faience beads, 1 glass bottle, 2 ceramic figurines |

**Table 5.9d** Enkomi finds and other variables from well-preserved tombs or shaft graves with principal components of use dating to LCIIC (late) and LCIII. Tombs are listed in order of increasing wealth in gold. All features with floor areas were shaft graves except French Tomb 6.

| Tomb No. | Dates of Use | Preservation | Location | Number of Burials | Complex Ritual | Floor Area sq m | Pots Total | Total Gold Wt g | Gold and Silver Types | Bronze and Iron Types | Other Prestige Emblems | Other Exotic and Local Valuables |
|---|---|---|---|---|---|---|---|---|---|---|---|---|
| S13C | LCIII | Intact | 6W? | 1 | | 0.9 | 1 | 0 | | | | |
| C4A | LCIII | Intact | 4W | 1 | yes | 2.0 | 0 | 0 | | | | |
| C4 | LCIII | Intact | 4W | 2 | | 0.9 | 0 | 0 | | | | faience beads |
| C24 | LCIII | Intact | 1W | 1 | yes | 1.3 | 3 | 0 | | pin | | |
| C23 | LCIII | Intact | 1W | 1 | | 1.4 | 3 | 0 | | 1 bowl | | |
| S11A | LCIII | Intact | 7W? | 2 | ? | 1.6 | 0 | 0 | | 1 bowl | | |
| S19A | LCIII | Intact | 5W? | 3 | | 2.0 | 1 | 0 | | 1 bowl | | stone mortar/ pestle |
| S7A | LCIIC? LCIII | Intact | 7W? | 2 | | 2.2 | 4 | 0 | | 1 bowl | cylinder seal | |
| F1 (1947) | LCIII | Intact | 11E/12E | 1? | | ? | 9 | 0 | | 6 earrings, 1 cup, 3 bowls | | stone mortar/ pestle, stone vase |
| S15 | LCIIC? LCIII | Intact | 5W? | 1 | yes | 1.2 | 3 | 0 | | | | ivory tube, glass bottle |
| S16 | LCIII | Intact | 5W? | 2 | | 1.9 | | 0 | | 2 bowls | | ivory comb and button, stone mortar/ pestle |
| F108 | LCIII | ? | 5W | mult | | ? | 14 | 0 | | 2 knives/ daggers, 8 bowls, 2 iron knives | | small bone or ivory |
| S14 | LCIII | Intact | 3W? | 3 | | 4.5 | 4 | <5 | 8 gold beads | 1 bowl | | carnelian, faience beads, stone mortar/ pestle |
| F15 (1934) | LCIII | Intact | 5W? | 3 | | 3.0 | 4 | <10 | gold earring | knife, spearhead, 3 bowls | | whetstone? |
| F16 (1934) | LCIII | Intact | 5W? | 2 | | ? | 9 | <10 | 1 band | 4 bracelets, 1 knife, 3 bowls | | |
| F13 (1934) | LCIII | ? | 5W? | 1 | | ? | 15 | <10 | gold ring, band | 2 bracelets, chain fragment | | small ivories, stone mortar/ pestle |

| Tomb No. | Dates of Use | Preservation | Location | Number of Burials | Complex Ritual | Floor Area sq m | Pots Total | Total Gold Wt g | Gold and Silver Types | Bronze and Iron Types | Other Prestige Emblems | Other Exotic and Local Valuables |
|---|---|---|---|---|---|---|---|---|---|---|---|---|
| B47 | LCIIC? LCIII | ? | 5W | mult | | ? | ? | 15 | 3 gold bands | 1 Naue II sword, 1 bowl, 1 jug | Pastoral crater? | 6 faience vases, faience? disc, 1 glass bottle, small ivories, stone mortar/ pestle, ceramic figurine |
| B15 | LCIII | ? | 4E/ 5E | ? | | ? | ? | 17 | 2 gold bands | 2 knives/ daggers, spearhead, 2 greave fragments, 3+ bowls, tripod fragment, jug | | stone mortar/ pestle |
| B16 | LCIII | ? | east scarp | ? | | ? | ? | 17 | 3 gold earrings, 1 band | 5 knives/ daggers, 1 spearhead, bowls? | | 2 carved ivory mirrors (1 w/ goat and stag, 1 w/ warrior), stone vases, stone mortar/ pestle |
| B58 | LCIII | ? | 3W/ 4W | ? | | ? | ? | 18 | 4 gold earrings, 3 bands | 6 knives, 1 mirror, 1 bowl, 1 tripod stand, 1–2 iron knives | | carved ivory draught box (hunt scene), carved ivory mirror fragment, 2 alabaster vases |
| B22 | LCIIC? LCIII | ? | east scarp | ? | | ? | 5? | 29 | 1 gold earring, 6 bands | 4 knives/ daggers | cylinder seal, stamp seal | small ivories, stone mortar/ pestle |
| B24 | LCIIC? LCIII | ? | east scarp | ? | | ? | ? | 54 | 8+gold earrings, 3–7 bands, 1 elaborate ring and 1 other ring | 2 knives, 1 wheel (from offering stand?) | 2 stone stamp seals, 1 faience scarab | 1 carnelian bead, small ivories, carved ivory pyxis and mirror (both w/ man vs. griffin themes), stone vase |

| Tomb No. | Dates of Use | Preservation | Location | Number of Burials | Complex Ritual | Floor Area sq m | Pots Total | Total Gold Wt g | Gold and Silver Types | Bronze and Iron Types | Other Prestige Emblems | Other Exotic and Local Valuables |
|---|---|---|---|---|---|---|---|---|---|---|---|---|
| F6 (1934) | LCIII | Intact? | 7E/ 8E | 10 | | 4.5 | 47 | 75+ | 2 gold beads, placket, 3 rings, 11 bands | 1 pin, 4 spatulae, 3 mirrors, 14 bowls, 1 iron knife | 3 cylinder seals, 1 scarab | carnelian, faience beads, small ivories, ivory comb, small bone items, alabaster and other stone vases, stone mortar/ pestle |
| B61 | LCIII | ? | 3E | mult | | ? | ? | 107 | 16 gold earrings, 14 bands | 1 bracelet, several knives, 1 spearhead, 3 mirrors, 1 bowl | | carnelian, faience beads, faience pictorial vases, ivory fragments, alabaster and other stone vases, stone mortar/ pestle |
| B75 | LCIII | ? | 6W | mult | | ? | ? | 137 | 15 gold earrings, 6 bands, 2 signet rings and 4 other rings | 1 knife, 1 bowl | 2 cylinder seals | faience dish, small ivories, carved ivory pyxis w/ man and sphinx, stone mortar/ pestle |

Table 5.10  Enkomi wealth indices overall and by period.

| | | Enkomi Chamber Tombs Overall | Chamber Tombs MCIII–LCIB/LCIIA | Chamber Tombs LCIA–LCIIB | Chamber Tombs LCIA–LCIIC | Shaft Graves LCIIC–LCIII | F Statistic | p value |
|---|---|---|---|---|---|---|---|---|
| Pots per Chamber | Mean | 94.13 | 15.38 | 72.67 | 170.00 | 7.29 | 15.327 | 0.000 |
| | Std. Dev. | 105.66 | 16.27 | 51.91 | 120.88 | 11.22 | | |
| | Range | 2–407 | 2–50 | 36–175 | 10–407 | 0–47 | | |
| | n | 24 | 8 | 6 | 10 | 17 | | |
| Pots per Burial | Mean | 9.19 | 5.79 | 13.98 | 9.01 | 3.23 | 3.423 | 0.028 |
| | Std. Dev. | 9.35 | 7.81 | 14.77 | 4.71 | 3.91 | | |
| | Range | 1.75–43.75 | 1.75–25.00 | 5.67–43.75 | 5.00–19.38 | 0–15 | | |
| | n | 23 | 8 | 6 | 9 | 16 | | |
| Copper Items per Chamber | Mean | 4.72 | 2.92 | 1.56 | 7.77 | 4.15 | 2.234 | 0.093 |
| | Std. Dev. | 7.93 | 6.64 | 2.40 | 9.77 | 4.91 | | |
| | Range | 0–29 | 0–24 | 0–7 | 0–29 | 0–22 | | |
| | n | 39 | 13 | 9 | 17 | 27 | | |
| Copper Items per Burial | Mean | 0.45 | 0.33 | 0.31 | 0.66 | 1.60 | 1.515 | 0.227 |
| | Std. Dev. | 0.67 | 0.53 | 0.71 | 0.79 | 2.52 | | |
| | Range | 0–2.55 | 0–1.50 | 0–1.75 | 0–2.55 | 0–10 | | |
| | n | 24 | 9 | 6 | 9 | 16 | | |
| Gold Weight per Chamber | Mean | 79.07 | 6.15 | 100.10 | 117.90 | 20.16 | 3.808 | 0.014 |
| | Std. Dev. | 150.98 | 19.38 | 126.16 | 195.03 | 35.88 | | |
| | Range | 0–800 | 0–70 | 0–385 | 0–800 | 0–137 | | |
| | n | 42 | 13 | 10 | 19 | 25 | | |
| Gold Weight per Burial | Mean | 5.15 | 1.30 | 10.07 | 5.72 | 1.72 | 2.593 | 0.068 |
| | Std. Dev. | 9.23 | 3.89 | 13.73 | 8.77 | 3.14 | | |
| | Range | 0–33.33 | 0–11.67 | 0–33.33 | 0–27.27 | 0–10 | | |
| | n | 24 | 9 | 6 | 9 | 16 | | |

**Table 5.11** Finds and other variables from Late Cypriot tombs in southern Cyprus. Abbreviations: LK = Livadhia *Kokotes,* DT = Dromolaxia *Trypes,* HST = Hala Sultan Tekke, Kit = Kition, KV = Kalavasos Village, KMan = Kalavasos *Mangia,* KAD = Kalavasos *Ayios Dhimitrios,* KB = Kourion *Bamboula,* KT = Kouklia *Teratsoudhia,* KElio = Kouklia *Eliomylia,* KEv = Kouklia *Evreti,* YA= Yeroskipou *Asproyia.* Objects in ninth column are bronze unless specifically noted as iron.

| Tomb No. | Dates of Use | Preservation | Number of Burials | Complex Ritual | Floor Area sq m | Pots Total | Gold and Silver Types | Copper/ Bronze and Iron Types | Other Prestige Emblems | Other Exotic and Local Valuables |
|---|---|---|---|---|---|---|---|---|---|---|
| LK 1 | MCIII/ LCIA | Looted | ? | | | 26+ | silver ring | | WPV rhyton | Red Burnished juglet, faience bead |
| DT 1 | MCIII– LCII | Bulldozed, looted | 3+? | | | 85+ | 1 gold earring, 2 silver earrings | 3 tweezers, 1 ring, 1 pin, 4 earrings, other fragments, iron lumps | Myc and Pastoral craters, BCWM pictorial tankard, cylinder seal, ceramic figurine | Minoan and Myc pottery, carnelian and faience beads, small ivories, worked bone |
| DT 2 | MCIII– LCI | Bulldozed, looted? | ? | | | 27+ | | 1 dagger, 1 awl, 1 pin, 4 rings, 2 beads | Syrian cylinder seal | Red Burnished and TeY juglets |
| HST 1 | LCIIB– LCIIC | Bulldozed, looted? | 12+ | | 9.9 | 107 | gold: 1 band, 3 beads; 1 silver ring | | Myc and Pastoral craters, cylinder seal | LMIIIA jar, Myc pottery, faience bead and vases, small ivories, ostrich egg |
| HST 2 | LCIB/ LCIIA– LCIIC | Bulldozed, looted? | 5+ | | 8.5 | 225 | 1 gold bead, cylinder seal cap | 1 ring, 2 earrings, copper slag | Minoan and Myc craters, 3 cylinder seals, Egyptian amulet, ceramic figurine | LMIII and Myc pottery, carnelian bead, faience vases, beads and bezel, small ivories, lead fragments, ostrich egg |
| HST 23 | LCIIIA | Intact | 1 | | | 2 | gold: 3 earrings, 2 beads, 2 pendants, 1 ring, 1 bead, mountings; silver: signet ring, other fragment | 1 trident, 1 dagger, 4 arrowheads, 1 jug, 1 bowl, 1 platter, ? 2 pins, other fragments | faience scarab, hematite weight | agate, carnelian, turquoise, lapis lazuli, faience beads, faience gaming pieces and bowl fragment, 2 ivory gaming boxes, small ivories |
| Kit 1 | LCIIB, LCIIIA | Looted? | 6+? | | 6.5 | 33 | gold: 1 band, 1 earring, 3 beads; silver: 2 spiral bracelets | 2 bowls, 1 ring, 1 earring, 4 studs | Myc crater, stamp seal | Myc pottery, faience vases, glass bottle, ivory lid?, worked bone |

| Tomb No. | Dates of Use | Preservation | Number of Burials | Complex Ritual | Floor Area sq m | Pots Total | Gold and Silver Types | Copper/ Bronze and Iron Types | Other Prestige Emblems | Other Exotic and Local Valuables |
|---|---|---|---|---|---|---|---|---|---|---|
| Kit 3 | LCIIIA | Intact | 3+? | | 3.6 | 14 | | 1 dagger, 1 bowl | | 5 alabaster vases, worked bone, stone mortar/ pestle |
| Kit 4/5 | LCIIC | Looted | ? | | 5.0 +6.9 | 191+ | | | Myc and Pastoral craters, Myc bull figurine | LMIII and Myc pottery, faience beads, glass bottle, small ivories, alabaster vase, stone pestle |
| Kit 9 | LCIIC | Intact? | 48 | | 12.0 | 289+ | gold: 2 signet rings, 5 other elaborate rings, 3+ neck-laces, 14 bands, cylinder cap, other items; silver bowl fragment | 14 daggers, 2 knives, 3 spearheads, 1 spike, 40 bowls, 3 mirrors, 1 pin, 1 bracelet | Myc and Pastoral craters, 2 cylinder seals, 3 scarabs, Myc psi figurine | Anatolian, Minoan, Myc pottery, carnelian and faience beads, numerous faience and alabaster vases, glass and glazed bottles, 3 ivory mirror handles (1 pictorial), small ivories, worked bone, stone mortar/ pestle |
| KV 51 | LCIA | Intact | 1 | | 3.7 | 12 | | sword | | Canaanite jar |
| KMan 5 | LCIIB–LCIIC | Bulldozed | 7 | | 4.5 | 32 | silver: bowl and wire frag-ments | 1 bowl, 1 bracelet | | Myc closed containers and bowl, faience beads, small ivories, worked bone, gypsum cup |
| KMan 6 | LCIIC | Bulldozed | 5 | | 2.0 | 12 | | spearhead | cylinder seal | Myc closed containers, Egyptian blue bead, gypsum alabastron, stone mortar/ pestle |
| KAD 1 | LCIIB–LCIIC | Looted | 9+? | | 6.6 | 39 | | 1 dagger, 1 ring | 2 BR rhyta | Myc closed container, faience bead, small ivories, stone pestle |
| KAD 4 | LCIB–LCIIA | Looted | 2+? | | 6.3 | 72 | | | cylinder seal | glass frag |
| KAD 5 | LCIIA–LCIIC | Looted | 11 | | 5.7 | 19 | 13 gold beads, 1 silver ring | bracelet, other fragments | | glass bead, small ivories |
| KAD 6 | LCIIC | Intact | 2 | | 6.6 | 7 | | ring | | Myc closed containers, small ivories |

| Tomb No. | Dates of Use | Preservation | Number of Burials | Complex Ritual | Floor Area sq m | Pots Total | Gold and Silver Types | Copper/ Bronze and Iron Types | Other Prestige Emblems | Other Exotic and Local Valuables |
|---|---|---|---|---|---|---|---|---|---|---|
| KAD11 | LCIIA/ LCIIB | Intact | 7 | | 10.1 | 44 | gold: 3 signet rings, 3-4 necklaces, 3 bracelets, 12 earrings, 8 spirals, 4 funnels, other ornaments; silver: 4 toerings | 1 dagger, slag | 2 Myc craters, 2 BR rhyta, stamp seal | Myc pottery, amber, faience, glass beads, glass bottles and elaborate box, small ivories, worked bone, alabaster vase |
| KAD 12 | LCIIC | Intact | 4 (3 infants and child) | | | 7 | gold earring, silver Hittite figurine | | | knucklebones |
| KAD 13 | LCIIA– LCIIC | Looted | 1+? | | | 90 | | 3 blades | Myc craters | Myc pottery, faience and alabaster vases, ivory pictorial disc and other small ivories, worked bone |
| KAD 14 | LCIIB | Disturbed? | 2+? | | | | gold bands, foil, rings, beads | 3 daggers, other fragments | Myc crater, 3 RLWM arm vessels | Myc pottery, faience gaming pieces, glass plaques/ gaming box?, small ivories |
| KAD 18 | LCIIB– LCIIC, Iron Age | Disturbed? | Several | | | | various gold and silver items | | Myc and Pastoral crater fragments | Minoan and Myc pottery, glass, ivory |
| KAD 19 | LCIIB, LCIIC | Looted | 6+? | | 8.6 | | 13 gold beads, 1 earring | silver or bronze pin | | Myc containers, cup and bowl, ivory pyxis |
| KAD 21 | LCII | Looted/ emptied | | | | | 5 gold diadems or mouthpieces | | Myc craters | |
| KB 3 | LCIIA, LCIIIA | Intact | 5 | | 3.0 | 23 | | bowl | | small ivory and bone? objects |
| KB 6 | LCIIB/ LCIIC | Intact? | 5+? | yes? | 4.4 | 5 | silver ring | ring | Myc craters | Myc closed container, faience bead, ivory disc, stone pestle |
| KB 12 | LCIA– LCIIB | Intact | 15? | | 5.3 | 67 | gold earring | 1 dagger, 1 tweezer, 1 chisel, 1 awl, 1 ring | 2 cylinder seals | Myc closed containers, other imported pottery?, small ivories, worked bone |
| KB 13 | LCI– LCIIA | Looted | ? | | 10.2 | 39 | gold: 1 earring, 1 bead | bracelet | | Myc closed containers, other imported pottery?, glass bead |
| KB 16 | LCIB– LCIIIA? | Disturbed | 4 | | 6.3 | 28 | 2 gold beads | bowl, iron knife | Myc crater, crystal seal, 2 scarabs | Myc pottery, small ivories, glass vase fragments, stone mortar/ pestle |

| Tomb No. | Dates of Use | Preservation | Number of Burials | Complex Ritual | Floor Area sq m | Pots Total | Gold and Silver Types | Copper/ Bronze and Iron Types | Other Prestige Emblems | Other Exotic and Local Valuables |
|---|---|---|---|---|---|---|---|---|---|---|
| KB 17/ 17A | LCIIA– LCIIIA | Looted | ? | | 7.5+ 8.2 | 12 | gold: 6 earrings, 21 beads, leaf fragments; silver: 1 ring, 1 earring | dagger, fibula | Myc craters, RLWM arm vessel?, cylinder seal, stamp seal | Myc pottery, glass vases, stone mortar/ pestle |
| KB 18 | LCIIB, LCIIIA | Intact | 7-8 | | 3.3 | 11 | silver ring | 1 dagger, 1 spearhead, 1 bowl | | Myc closed container, carnelian and glass beads, ivory disc, stone pestle |
| KB 19 | LCIIA– LCIIIB | Intact? | 52? | ? | 3.7 | 21 | gold: 1 bead, cylinder caps; silver bead | 1 bowl, 1 ring, 1 bracelet, 1 needle, 2 rods | 2 cylinder seals, 3 stamp seals, ceramic figurine | Myc closed container, faience and glass beads, small ivories |
| KB 21 | LCIIA | Intact | 5 | yes | ? | 8 | | ring | | Myc closed containers |
| KB 26 | LCII | Looted | ? | | ? | -- | 3 gold ornaments | | ceramic figurine | Minoan and Myc pottery, glass alabastron, small ivory and bone objects |
| KB 33 | LCIB, LCIII | Intact? | ? | | 4.8 | 37 | gold bead, wire, silver ring | 2 rings, 1 rod | | Myc closed containers, alabaster vase, stone mortar/ pestle |
| KB 36 | LCIIA/ LCIIB– LCIIC | Intact | 4? | | 4.4 | 24 | 3 gold beads, plated cyl seal | 4 earrings | | Myc pottery, faience and glass beads, small ivories |
| KB 40 | LCIIA, LCIIIA | Intact | 20+ | ? | 3.3 | 69 | | 1 bowl, 1 ring, 1 wire fragment | | glass bead, ivory button |
| KT 104B | LCIIC/ LCIIIA | Intact | ? | | 4.7 | 19 | silver ring | needle | | Myc juglet, small ivories, worked bone |
| KT 104N | LCIIC/ LCIIIA | Intact | ? | | 4.5 | 12 | gold earring, other small ornaments | 1 spearhead, 14 arrow- heads, 1 crater, 1 laver, 2 jugs, 1 mirror, other fragments, 1 iron knife | carved (pictorial) ivory disc | Myc bowl, faience gaming pieces, glass bead, small ivories, worked bone, ostrich egg, alabaster vases, stone mortar/ pestle |
| KElio 119 | LCIIC/ LCIIIA | Intact | ? | | 5.9 | 35 | 3 gold bands, silver ring fragment | 1 spearhead, 11 bowls, 1 mattock, 1 chisel, other tool, 1 mirror; 1 iron knife | 1 Pastoral crater | Myc pottery, Canaanite jar, small ivories, alabaster vase, stone mortar/ pestle |

| Tomb No. | Dates of Use | Preservation | Number of Burials | Complex Ritual | Floor Area sq m | Pots Total | Gold and Silver Types | Copper/ Bronze and Iron Types | Other Prestige Emblems | Other Exotic and Local Valuables |
|---|---|---|---|---|---|---|---|---|---|---|
| KEv 8 | LCIIIA | Intact | ? | | | ? | gold: signet ring, 6 other elaborate rings, 2 heavy bracelets, 14 earrings, 1 band, 4 toerings; silver: 2 bowls | 2 daggers, 4 bowls, 1 mirror, 2 iron knives, 2 iron spatulae | carved (pictorial) ivory pyxis and mirror handle | small ivories, stone mortar/ pestle |
| YA 1 | LCIIC | Looted | ? | | 3.3 | 4+ | gold: 1 band, 3 rings | 2 spearheads, 20 bowls, 1 mirror | scarab | Myc bowls, stone mortar/ pestle |
| YA 2A | LCIIB | ? | ? | | 2.9 | 1 | 3 gold earrings | | | small ivories |
| YA 2B | LCIIB | ? | ? | | 4.8 | 6 | gold: 1 ring, 2 earrings | | cylinder seal | |

**Table 5.12** Kourion *Bamboula* wealth indices.

| | | Kourion *Bamboula* Overall |
|---|---|---|
| Pots per Chamber | Mean | 24.06 |
| | Std. Dev. | 20.04 |
| | Range | 4-69 |
| | n | 16 |
| Pots per Burial | Mean | 3.33 |
| | Std. Dev. | 2.08 |
| | Range | 0.40–7.25 |
| | n | 12 |
| Copper Items per Chamber | Mean | 2.92 |
| | Std. Dev. | 2.02 |
| | Range | 1–8 |
| | n | 13 |
| Copper Items per Burial | Mean | 0.33 |
| | Std. Dev. | 0.27 |
| | Range | 0.15–1.00 |
| | n | 9 |

**Table 5.13** Finds and other variables from Late Cypriot tombs in inland areas (northern Troodos and Mesaoria). Abbreviations: Ang = Angastina, APar = Nicosia *Ayia Paraskevi* (APar& is Ohnefalsch-Richter's tomb [Kargeorghis 1999]), DKG = Dhali *Kafkallia* Tomb G, Pol = Politiko, Akh = Akhera, Kat = Katydhata, DM = Dhenia *Mali;* OB = Old Babylonian.

| Tomb No. | Dates of Use | Preservation | Number of Burials | Complex Ritual | Floor Area sq m | Pots Total | Gold and Silver Types | Copper/ Bronze Types | Other Prestige Emblems | Other Exotic and Local Valuables |
|---|---|---|---|---|---|---|---|---|---|---|
| Ang 1 | LCIIA–LCIIC | Bulldozed | 9 | | 9.2 | 217 | silver: 1 earring, 2 bracelets | bowl fragments, 3 slugs?, 2 rings, 1 earring, 3 bracelets | 3 cylinder seals, 7 Pastoral craters, 3 BR rhyta | Myc closed containers and bowls, faience beads, small bone and ivory objects |
| Ang 2 | MCIII/ LCIA–LCIIA? | Bulldozed | 2+? | | 9.9 | 42 | | 1 HTW, 1 knife/ dagger, 2 tweezers, bowl fragments, 1 pin, 14+ rings | cylinder seal | faience beads |
| Ang 3 | LCIIB–LCIIC | Bulldozed | 2+? | | 12.9 | 36 | 2 gold beads | 4 rings, 3 earrings, 2 bracelets | 3 cylinder seals, 1 ceramic figurine | Myc closed containers, cups and bowls, faience beads and button, ivory disc |

| Tomb No. | Dates of Use | Preservation | Number of Burials | Complex Ritual | Floor Area sq m | Pots Total | Gold and Silver Types | Copper/ Bronze Types | Other Prestige Emblems | Other Exotic and Local Valuables |
|---|---|---|---|---|---|---|---|---|---|---|
| Ang 5 | LCIIB–LCIIC | Bulldozed | 5 | | 10.2? | 80 | | 3 earrings, 2 bracelets | 2 cylinder seals, BR rhyton, ceramic figurine, stag horn | Myc closed containers |
| APar 1 | LCIA | ? | ? | | | 9? | 2 cyl caps | 1 axe, 2 daggers, 2 tweezers, 2 awls | 1 hematite OB cylinder seal, 1 other cylinder seal | paste/ faience bead |
| APar 6 | MCIII–LCIII | Intact? | ? | | | 362 | 1 gold bead | 1 dagger, 1 awl, 2 pins, 3 rings, 1 earring, 1 knuckle-bone | 3 cylinder seals, 3 ceramic figurines, WPIII ring vase, BR rhyton | Myc closed containers, carnelian, faience and glass beads, small ivories, alabaster vase |
| APar& | | ? | ? | | | ? | | | Hittite bull's head, ceramic figurine, BR bull rhyta | Myc pictorial crater |
| DKG | MCIII/ LCIA–LCIII | Looted | ? | | 14.1 | 75 + | | 1 shafthole axe, 1 belt, 1 projectile point, 3 pointed implements | 1 cylinder seal, 1 mace-head, 2 Pastoral craters | Myc closed containers and cups, alabaster chalice, frit and glass beads, small ivories |
| Pol 6 | LCIIB | Intact | 22 | yes? | 5.0 | 55 | 2 HTWs, 1 dagger, 1 spearhead, 2 chisels, 5 rings, 2 earrings, 4 bracelets | 2 cylinder seals, 1 macehead, 1 BR rhyton | Myc closed containers and cup, paste/ faience bead and lid, small ivories | |
| Akh1 | LCIA | Intact | 2+? | ? | 4.9 | 80 | silver spiral | 6 HTWs, 3 knives, 1 razor, 2 tweezers, 2 awls, 5 pins, 1 needle, 2 rings, 10 spiral beads/coils | 1 scarab, 1 BCWM rhyton, 2 WPV coroplastic jugs, 1 WPV composite vessel | faience beads |
| Akh 2 | LCIIC | Intact | 14 | ? | 4.0 | 32 | silver: 1 ring, 1 earring, 1 bracelet | 1 ring, 6 earrings | 2 cylinder seals | Myc closed containers and bowl, faience beads, ivory button, alabaster vase |
| Akh 3 | LCIIC | Intact | 6+? | ? | 5.2 | 40 | 3 gold beads | 2 pins, 3 earrings, 1 chisel, 1 bowl | cylinder seal | Myc closed containers and bowls, carnelian bead, faience pendant, alabaster vase |

| Tomb No. | Dates of Use | Preservation | Number of Burials | Complex Ritual | Floor Area sq m | Pots Total | Gold and Silver Types | Copper/ Bronze Types | Other Prestige Emblems | Other Exotic and Local Valuables |
|---|---|---|---|---|---|---|---|---|---|---|
| Kat 11 | LCI–LCII | ? | ? | | | 73 | gold cyl cap, silver ring | 1 dagger, 1 tweezer, 2 rings, 11 spirals | 2 cylinder seals | Myc closed containers and bowls, faience beads |
| Kat 26 | LCIIA | ? | ? | | | 9 | 1 gold earring, 2 silver earrings | 2 pins, 1 ring or bracelet | cylinder seal | Myc closed container |
| Kat 42 | LCI | ? | ? | | | 16 | | knife, spear-head, bracelet | | |
| Kat 50 | LCIIB | ? | ? | | | 27 | | 1 axe, 4 rings, 5 bracelets | | Myc closed containers |
| Kat 80 | MCII–LCI | ? | 2 | | | 28 | | 3 knives, 1 HTW, 2 spearheads, 1 needle, 2 rings | | faience beads |
| Kat 81 | LCIIB/ LCIIC | ? | ? | | | 10 | gold earring | knife, needle, ring/ bracelet | 2 cylinder seals, ceramic figurines | Myc closed containers, faience bottle, worked bone |
| Kat 83 | LCIIB/ LCIIC | ? | ? | | | 4 | gold: 3 small circular bands | | | Cyp-Myc juglet |
| Kat 85 | MCIII/ LCIA LCIIB | ? | ? | | | 34 | | needle | ceramic figurine | Myc closed containers |
| Kat 88 | LCIIC | ? | ? | | | 2 | | | | Cyp-Myc bowl |
| Kat 89 | LCIII | ? | 2 | | | -- | gold ring | 2 fibulae, iron knife | ceramic figurine | faience bead |
| Kat 90 | LCIIA | ? | ? | | | 6 | | ring, bull figurine | ceramic figurine | Myc closed container |
| DM 8 | LCIIB | Looted | 1? | | 3.1 | 7 | gold diadem, silver bracelet | 1 dagger | Myc bull figurine | Myc closed containers, ivory disc |

# Index